Nonlinear Systems and Complexity

Volume 20

Series editor

Albert C.J. Luo
Southern Illinois University
Edwardsville, IL, USA

Nonlinear Systems and Complexity provides a place to systematically summarize recent developments, applications, and overall advance in all aspects of nonlinearity, chaos, and complexity as part of the established research literature, beyond the novel and recent findings published in primary journals. The aims of the book series are to publish theories and techniques in nonlinear systems and complexity; stimulate more research interest on nonlinearity, synchronization, and complexity in nonlinear science; and fast-scatter the new knowledge to scientists, engineers, and students in the corresponding fields. Books in this series will focus on the recent developments, findings and progress on theories, principles, methodology, computational techniques in nonlinear systems and mathematics with engineering applications. The Series establishes highly relevant monographs on wide ranging topics covering fundamental advances and new applications in the field. Topical areas include, but are not limited to: Nonlinear dynamics Complexity, nonlinearity, and chaos Computational methods for nonlinear systems Stability, bifurcation, chaos and fractals in engineering Nonlinear chemical and biological phenomena Fractional dynamics and applications Discontinuity, synchronization and control.

More information about this series at http://www.springer.com/series/11433

Igor S. Aranson • Arkady Pikovsky
Nikolai F. Rulkov • Lev S. Tsimring
Editors

Advances in Dynamics, Patterns, Cognition

Challenges in Complexity

Springer

Editors
Igor S. Aranson
Pennsylvania State University
University Park, PA, USA

Nikolai F. Rulkov
BioCircuits Institute
University of California San Diego
La Jolla, CA, USA

Arkady Pikovsky
Institute for Physics and Astronomy
University of Potsdam
Potsdam, Germany

Lev S. Tsimring
BioCircuits Institute
University of California San Diego
La Jolla, CA, USA

ISSN 2195-9994 ISSN 2196-0003 (electronic)
Nonlinear Systems and Complexity
ISBN 978-3-319-53672-9 ISBN 978-3-319-53673-6 (eBook)
DOI 10.1007/978-3-319-53673-6

Library of Congress Control Number: 2017936388

Printed on acid-free paper

This Springer imprint is published by Springer Nature
The registered company is Springer International Publishing AG
The registered company address is: Gewerbestrasse 11, 6330 Cham, Switzerland

This book is dedicated to the 75th birthday of Michael Rabinovich, a pioneer of nonlinear science

Preface: Misha Rabinovich and Nonlinear Dynamics in the Last Half-Century

ПОЗНАНИЕ
Узлами скручены спирали,
А между ними судеб путь.
Скопленья звезд спираль порвали,
Хотя б вполглаза мне взглянуть,
Потрогать грань меж Раем - Адом,
Порядок в хаосе понять,
Услышать рев галактик рядом
И... Богом все это назвать.

Михаил Рабинович

COGNITION
Spirals are wrung into knots.
Between them lies the way of the Fates.
Star clusters once tore the spirals.
Oh, if I could glimpse even half of this,
Touch the thin between Heaven and Hell,
Perceive Order in Chaos,
Hear the roar of galaxies next door,
And – to call it God: all of it.

Mikhail Rabinovich
(English translation by Olga Livshin)

This book is comprised of contributions by just a few of a very large cohort of friends, colleagues, and former students of Misha Rabinovich or, as he is often affectionally called after his initials, MIR.[1] In the last half-century, Misha

[1]The word *mir* means both *peace* and *world* in Russian.

Rabinovich has been at the forefront of most major developments in nonlinear dynamics, starting from the development of asymptotic methods for the analysis of nonlinear waves in non-equilibrium media to the present days when he is using nonlinear dynamics to advance the theory of cognition. In these brief introductory notes, we will attempt to sketch Misha's portrait both as a scientist and as a human being and connect it with the contents of this book.

Mikhail Izrailevich Rabinovich was born on April 20, 1941, in Nizhny Novgorod (then Gorky) in the family of Israel Rabinovich, a professor of chemistry at Gorky State University. In his teenage years, Misha was an avid cross-country skier and only got serious about science toward the second half of his university studies. But in science, just as earlier in sports, he was very fast and not only quickly caught up with his peers but left them far behind.

MIR got his first taste of doing research around 1960 during his sophomore university year in the field which now is called robotics. His first scientific mentor Tyoma Alekseev suggested to him a student project to improve the efficiency of a conveyor at Gorky Automobile Plant (GAZ). Before long, Misha developed an automation algorithm and implemented it at one of the plant conveyors. That work not only led to increased efficiency of the conveyor operation and caused some layoffs, but it also became his first published paper. However, MIR decided not to pursue his scientific career in robotics and switched to the nonlinear dynamics and theoretical physics. Nevertheless, this short stint at GAZ possibly played a big role in his future scientific development.[2] While MIR has always been a theorist, he is also known for deep understanding of experiment, often suggesting original ideas to his experimental colleagues.

He was only 26 when in 1967 he defended his candidate of physical and mathematical sciences' dissertation (an analog of Ph.D. thesis in the West), and at 33 he already obtained the doctor of science degree (an analog of habilitation in Germany, exceptionally early by Soviet standards). In 1991, he was elected the corresponding member of the Russian Academy of Sciences. The general direction of his scientific studies from those very early days has remained the theory of nonlinear oscillations and waves. This was quite natural for him, perhaps even unavoidable, because he belongs to the illustrious school of nonlinear dynamicists established by Leonid Mandelshtam and his students Alexander Andronov, Gabriel Gorelik, and Mikhail Leontovich, followed among others by Misha's Ph.D. adviser Andrey Gaponov-Grekhov. This scientific dynasty is widely known not only for their fundamental discoveries in physics of oscillations and waves but also for applying them in practice, ranging from clocks to powerful generators of electromagnetic or acoustic radiation.

One of Misha Rabinovich's first significant contributions to nonlinear dynamics was the theoretical discovery of stable stationary waves in active nonlinear media in the late 1960s. The first experimental studies were performed with the chains of

[2]This early phase of MIR's scientific career is not well documented, and our account of it is mostly based on our many friendly evening conversations.

coupled electronic self-sustained oscillators [6]. These stable nonlinear structures were later found in plasma physics and fluid dynamics and were used in lasers and networks of radio frequency oscillators. At about the same time, he with his Ph.D. student Alexander Rosenblum (who was 15 years older than MIR) proposed an asymptotic method for theoretical analysis of self-sustained oscillations which was a nontrivial extension of the classical Krylov-Bogolyubov asymptotic method to the distributed nonlinear systems [16].

Perhaps it is Misha's extremely active persona that always attracted him to non-equilibrium systems and nonlinear dynamics. In 1972, he discovered a novel phenomenon of explosive instability in the nonlinear interaction of waves in non-equilibrium media. In this situation the amplitudes of all three ways grow super-exponentially and become infinite in a finite time unless higher-order nonlinearities limit their growth [15]. When the new Institute of Applied Physics of the Soviet Academy of Sciences opened its doors in 1977, MIR became the head of the Laboratory for Nonlinear Dynamics and Chaos which quickly became one of the centers of nonlinear dynamics in the Soviet Union. Many of his students and junior collaborators (including editors and a large fraction of authors of this volume) worked in this laboratory and benefited from daily interactions with MIR. However, his role in grooming generations of "nonlinear dynamicists" is even greater. Misha has always been a wonderful teacher and mentor. As a professor of the Gorky State University, he established a yearlong lecture course on the theory of oscillations and waves and taught it himself every year for 30 years. This class was the highlight of our own years at the radiophysics department of the Gorky State University in the 1970s. This nonlinear dynamics course laid the foundation of the popular textbook written by him in collaboration with Dmitry Trubetskov [18]. His another crown achievement was establishing and running famous "Schools on Nonlinear Oscillations and Waves" that were held biannually in a beautiful countryside some 200 km from Gorky. These 2-week-long springtime gatherings brought together famous physicists and mathematicians (V.I. Arnold, B.B. Kadomtsev, Ya.I. Sinai, Ya.B. Zeldovich, and many others) and hundreds of young researchers in a very informal and stimulating atmosphere. This tradition survived even the demise of the Soviet Union and continues to this day; the next school is planned for 2018.

In the 1970–1980s, Misha Rabinovich turned his attention to deterministic chaos and pattern formation. They were nascent fields at the time, and Misha was one of the early pioneers. In his 1979 paper with his former student Anatoly Fabrikant [13], they demonstrated the emergence of low-dimensional deterministic chaos in a spatiotemporal system describing modulational instability of nonlinear waves in dissipative media. They reduced this infinite-dimensional problem to a set of three ordinary differential equations now known as "Rabinovich-Fabrikant equations." In 1980, MIR with Sergey Kiyashko and one of us demonstrated deterministic chaos in a simple electronic circuit [7]. His influential 1978 review in "Soviet

Physics-Uspekhi" called "Stochastic Self-Oscillations and Turbulence" [10][3] and two chapters on chaos and turbulence written for the classic Landau-Lifshitz textbook series on theoretical physics [8] opened up this field and set the agenda for generations of Soviet physicists. One of the most important discoveries made in that field was *synchronization of chaos*. Misha and his group were the first to experimentally observe this new phenomenon by coupling two electronic chaotic oscillators [2]. This early work heralded the beginning of a new rich field with thousands of scientific publications to date. Other significant scientific achievements of that time were the discoveries of stable particle-like localized solutions of nonlinear field equations [4] and the spatiotemporal chaos in the Ginzburg-Landau equation [3]. Very deep studies of turbulence and pattern formation (see his review article [17] and book [12]) followed.

By the mid-1980s, Misha Rabinovich became well known in the West, but only by name and by his influential papers. Soviet authorities never allowed him to leave the country, not even to the "brotherly" socialist countries of the Soviet bloc. The Institute of Applied Physics and the whole city of Gorky were off limits for foreigners as well. However, as soon as the Iron Curtain began to rust, crack, and crumble during the perestroika years, Misha was able to finally meet his Western colleagues in person. It was during this time that Misha established a long-term collaboration with the Institute for Nonlinear Science at UCSD that in a few years became his new scientific home.

At about the same time as MIR made UCSD his home base, he became interested in neuroscience. Early on he sensed that this field and the biology as a whole were becoming new frontiers for applications of nonlinear dynamics. He quickly realized that deterministic chaos must play a major role in complexity and plasticity of neural systems. His pioneering work on deterministic chaos in stomatogastric ganglia of lobsters in collaboration with Henry Abarbanel, Allen Selverston, and others [11] produced not only fresh new insights into the role of nonlinearity in neuroscience but also tasty leftovers that were enthusiastically consumed at INLS parties.

Observing very complex but often reproducible patterns of neural activity, Misha came to the realization that this complexity must have somewhat different origins from by-now-familiar deterministic chaos. Using a simple Lotka-Volterra system as the paradigmatic model, he introduced a new dynamic concept that governs such complex but stable transient phenomenon which is now known as *winnerless competition* [1]. The geometrical image of such dynamics is the so-called *stable heteroclinic channel* which connects a unique sequence of saddle fixed points in the phase space of the corresponding dynamical system [5]. While the original motivation for this work was the olfactory system in locust, Misha and his collaborators uncovered evidence that this principle governs visual and spatial memory, as well as many other neural systems. In the last several years, Misha has been thinking about the role of nonlinear dynamics in cognition [14]. He is

[3]At that time, the word *stochastic* was commonly used in Soviet scientific literature to describe deterministic chaos as opposed to noise that was labeled by the word *random*.

not deterred by the enormous complexity of human brain. Misha's main idea is that different modalities of brain function also represent the saddle points of the global phase space, and brain activity from this vantage point can be represented as sequential switching from one modality to the next, akin to winnerless competition in smaller neural circuits.

Our portrait of Misha would be grossly incomplete if we did not mention his love of poetry. Although he wrote poems since a young age, it became a real passion in the last 15–20 years. He published seven books of poetry to date (December 2016), and by the time this book is in print, this number could well be greater. His poems are often reflections on the philosophy of science and the creative process in general; in his worldview, the creativity does not know the boundaries between exact and ephemeral and science and art.

We already mentioned that Misha's unique personality makes him contagious, in the good sense of the word. Many of his ideas fertilized scores of his junior colleagues which followed his lead into new fields. In this book, we assembled papers of Misha's former students and his past and present colleagues.

Misha Rabinovich's best scientific insights were borne of his physical intuition and experience, but then he often recruits mathematicians to help put his ideas on firm theoretical foundation. On the other hand, his deep insight into the nonlinear phenomena inspired many experimental physicists and biologists. Therefore, in this book, the reader will find quite a broad coverage of modern topics in nonlinear theory of complex systems, from mathematics to experiments. We have organized the contributions into four parts, highlighting also the main milestones in the scientific life of Misha Rabinovich.

Part I "Chaos and Dynamics" is devoted to the field where Misha Rabinovich made seminal contributions. In one of his now classical works published in 1978, he introduced a low-dimensional model for three parametrically coupled waves that exhibited chaotic behavior, now called the Rabinovich system [9]. Two chapters, by Kuznetsov and by Pusuluri et al., report on recent progress in studies of this simple model. Pusuluri with collaborators describe the global organization of the chaotic attractor of that system using a combination of novel analytical and computational techniques. Kuznetsov addresses possible experimental implementation of the Rabinovich system; he demonstrates that this three-wave system can be implemented as a simple electrical circuit. He further shows that this circuit indeed generates chaotic trajectories described by a Lorenz-type quasi-hyperbolic attractor. Three other chapters in this part are devoted to the dynamics on the border of chaos and regularity. The contribution by Pesin et al. introduces a new class of scaled Lyapunov exponents, suitable for quantitative characterization of systems with sub-exponential separation of trajectories. These exponents are used in the chapter by Afraimovich and Neiman for the description of weak transient chaos in the switching dynamics. Another example of dynamics between order and chaos is presented in a chapter by Zaks and Nepomnyashchy, where anomalously slow dynamically generated diffusion is described.

Part II "Synchronization and Networks" reflects Misha's long-term interest in spatially organized models such as coupled map lattices and in synchronization phenomena. It opens with the contribution by Anishchenko et al. devoted to complex states, including chimera-like configurations, in networks of nonlocally coupled chaotic maps. Two chapters deal with phase oscillator networks. V. Belykh et al. consider star networks of phase oscillators and describe regular and chaotic transitions to synchrony. The chapter by Bick describes possible Lotka-Volterra-type dynamics in phase oscillator networks. Effect of symmetry in the network on the appearance of synchronous clusters is analyzed in the contribution by Pecora et al. The chapter by Bunimovich and Webb describes the stability analysis of networks with time delays. Finally, Reimayev and collaborators analyze synchronous states in coupled bursting neurons. This contribution makes a bridge to the following Part III.

The opening chapter of the Part III "Brain" deals with the subject of large-scale brain dynamics, which is the focus of Misha's current interest that has emerged from his previous work on neural dynamics. It is written by Karl Friston, his collaborator and co-editor of the recent book *Principles of Brain Dynamics: Global State Interactions* [14]. In this chapter, Friston proposes a variational principle that casts motor and sensory activity of the brain as an optimization strategy to minimize a particular free energy functional. Chapters by Mangin and Courbage and Rubchinsky et al. deal with synchronization of electrical activity in various biological neural networks. In related contributions, Komarov et al. and Nowotny and Szyszka describe coding of odors in neural activity of olfactory systems.

The final part of the book, "Waves," is devoted to the topic with which MIR started his scientific career. Krinsky and collaborators in their chapter describe how vortices of electrical activity in cardiac tissues can be controlled. Two other chapters deal with conservative problems. Pelinovsky and Shurgalina demonstrate that complexity can emerge even in integrable systems by analyzing statistical properties of a gas of solitons in the classical Korteweg-de Vries (KdV) equation. Stepanyants describes complex patterns emerging in the two-dimensional generalization of the KdV equation, the Kadomtsev-Petviashvili model. This final chapter of the book also contains some personal remarks based on the long-term friendship between the author and Misha Rabinovich.

Acknowledgments: We owe the original idea of this volume to Valentin Afraimovich and Vitaly Vugalter. We are also very grateful to Albert Luo for support at all stages of manuscript preparation.

University Park, PA, USA — Igor S. Aranson
Potsdam, Germany — Arkady Pikovsky
La Jolla, CA, USA — Nikolai F. Rulkov
La Jolla, CA, USA — Lev S. Tsimring
December 2016

References

1. Afraimovich, V., Tristan, I., Huerta, R., Rabinovich, M.I.: Chaos: Interdiscip. J. Nonlinear Sci. **18**(4), 043103 (2008)
2. Afraimovich, V., Verichev, N., Rabinovich, M.I.: Radiophys. Quant. Electron. **29**(9), 795 (1986)
3. Aranson, I., Gaponov-Grekhov, A., Rabinovich, M., Starobinets, I.: Zhurnal Eksperimentalnoi i Teoreticheskoi Fiziki **90**, 1707 (1986)
4. Aranson, I., Gorshkov, K., Lomov, A., Rabinovich, M.: Physica D: Nonlinear Phenom. **43**(2), 435 (1990)
5. Bick, C., Rabinovich, M.I.: Dynam. Syst. **25**(1), 97 (2010)
6. Bogatyrev, Y.K., Rabinovich, M.: Radiophys. Quant. Electron. **10**(12) 978–983 (1967)
7. Kiyashko, S., Pikovsky, A., Rabinovich, M.: Radio Eng. Electron. Phys. **25**, 74 (1980)
8. Landau, L., Lifshitz, E.: Theoretical physics. vol. 6. Hydrodynamics (1986)
9. Pikovski, A.S., Rabinovich, M.I., Trakhtengerts, V.Y.: Appearance of chaos at decay saturation of parametric instability. Sov. Phys. JETP **47**, 715–719 (1978)
10. Rabinovich, M.I.: Sov. Phys. Uspekhi **21**(5), 443 (1978)
11. Rabinovich, M., Abarbanel, H.: Neuroscience **87**(1), 5 (1998)
12. Rabinovich, M.I., Ezersky, A.B., Weidman, P.D.: The Dynamics of Patterns. World Scientific, Singapore (2000)
13. Rabinovich, M., Fabrikant, A.: Zhurnal Eksperimentalnoi i Teoreticheskoi Fiziki **77**, 617 (1979)
14. Rabinovich, M.I., Friston, K.J., Varona, P.: Principles of Brain Dynamics: Global State Interactions. MIT Press, Cambridge, MA (2012)
15. Rabinovich, M., Reutov, V.: Radiophys. Quant. Electron. **16**(6), 625–643 (1973)
16. Rabinovich, M., Rozenblum, A.: Prik. Mat. Meh. (Appl. Math. Mech.) **36**, 330 (1972)
17. Rabinovich, M.I., Sushchik, M.M.: Sov. Phys. Uspekhi **33**(1), 1 (1990)
18. Rabinovich, M., Trubetskov, D.: Vvedenie v Teoriyu Kolebanii i Voln (Introduction to the Theory of Oscillations and Waves). Nauka, Moscow (1984)

Contents

Contributors

Valentin S. Afraimovich Instituto de Investigacion Communicacion Optica, Universidad Autonoma de San Luis Potosi, San Luis Potosi, Mexico

Sungwoo Ahn Department of Mathematics, East Carolina University, Greenville, NC, USA

Vadim S. Anishchenko Department of Physics, Saratov National Research State University, Saratov, Russia

M. Bazhenov Department of Medicine, University of California, San Diego, CA, USA

Vladimir N. Belykh Volga State University of Water Transport, Nizhny Novgorod, Russia

Igor Belykh Department of Mathematics and Statistics and Neuroscience Institute, Georgia State University, Atlanta, GA, USA

Christian Bick Centre for Systems, Dynamics and Control and Department of Mathematics, University of Exeter, Exeter EX4 4QF, UK

Vadim N. Biktashev University of Exeter, Exeter, UK

Maxim I. Bolotov Nizhny Novgorod University, Nizhny Novgorod, Russia

Leonid Bunimovich School of Mathematics, Georgia Institute of Technology, Atlanta, GA, USA

Maurice Courbage Laboratoire Matière et Systèmes Complexes (MSC), Paris Cedex 13, France

Karl Friston The Wellcome Trust Centre for Neuroimaging, University College London, Queen Square, London WC1N 3BG, UK

Aaron M. Hagerstrom Department of Physics, University of Maryland, College Park, MD, USA

Institute for Research in Electronics and Applied Physics, University of Maryland, College Park, MD, USA

M. Komarov Department of Medicine, University of California, San Diego, CA, USA

Valentin I. Krinsky Max Planck Institute DS, BMPG, Göttingen, Germany

INLN, CNRS, Valbonne, France

Sergey P. Kuznetsov Kotel'nikov's Institute of Radio-engineering and Electronics of RAS, Saratov Branch, Zelenaya, Saratov, Russia

Stefan Luther Max Planck Institute DS, BMPG, Göttingen, Germany

Laurence Mangin Laboratoire Matière et Systèmes Complexes (MSC), Paris Cedex 13, France

Physiology Dpt., APHP, Bichat Hospital, Paris 7 University, Paris, France

Thomas E. Murphy Department of Electrical and Computer Engineering, University of Maryland, College Park, MD, USA

Institute for Research in Electronics and Applied Physics, University of Maryland, College Park, MD, USA

Alexander B. Neiman Department of Physics and Astronomy and Neuroscience Program, Ohio University, Athens, OH, USA

Alexander Nepomnyashchy Technion, Haifa, Israel

Thomas Nowotny University of Sussex, Falmer, Brighton BN1 9QJ, UK

Grigory V. Osipov Nizhny Novgorod University, Nizhny Novgorod, Russia

Niels F. Otani Rochester Institute of Technology, Rochester, NY, USA

Choongseok Park Department of Mathematics, North Carolina A&T State University, Greensboro, NC, USA

Louis M. Pecora U. S. Naval Research Laboratory, Washington, DC, USA

Efim Pelinovsky Institute of Applied Physics, Nizhny Novgorod, Russia

Nizhny Novgorod State Technical University, Nizhny Novgorod, Russia

National Research University – Higher School of Economics, Nizhny Novgorod, Russia

Yakov Pesin Department of Mathematics, Pennsylvania State University, University Park, PA, USA

Arkady Pikovsky Institute for Physics and Astronomy, University of Potsdam, Potsdam, Germany

Krishna Pusuluri Neuroscience Institute, Georgia State University, Petit Science Center, Atlanta, GA, USA

Reimbay Reimbayev Department of Mathematics and Statistics, Georgia State University, Atlanta, GA, USA

Rajarshi Roy Department of Physics, University of Maryland, College Park, MD 20742, USA

Institute for Research in Electronics and Applied Physics, University of Maryland, College Park, MD 20742, USA

Institute for Physical Science and Technology, University of Maryland, College Park, MD 20742, USA

Leonid L. Rubchinsky Department of Mathematical Sciences, Indiana University Purdue University Indianapolis, Indianapolis, IN 46202, USA

Stark Neurosciences Research Institute, Indiana University School of Medicine, Indianapolis, IN 46202, USA

Andrey Shilnikov Neuroscience Institute and Department of Mathematics and Statistics, Georgia State University, Petit Science Center, Atlanta, GA, USA

Ekaterina Shurgalina Institute of Applied Physics, Nizhny Novgorod, Russia

Francesco Sorrentino Department of Mechanical Engineering, University of New Mexico, Albuquerque, NM, USA

Yury Stepanyants University of Southern Queensland, Toowoomba, QLD, Australia

M. Stopfer National Institutes of Health, NICHD, Bethesda, MD, USA

Galina I. Strelkova Department of Physics, Saratov National Research State University, Saratov, Russia

Paul Szyszka Department of Biology, University of Konstanz, Konstanz, Germany

Tatiana E. Vadivasova Department of Physics, Saratov National Research State University, Saratov, Russia

Benjamin Webb Department of Mathematics, Brigham Young University, Provo, UT, USA

Michael A. Zaks Humboldt University Berlin, Berlin, Germany

Agnieszka Zelerowicz Department of Mathematics, Pennsylvania State University, University Park, PA, USA

Kun Zhao Department of Mathematics and Statistics, Georgia State University, Atlanta, GA, USA

Yun Zhao Department of Mathematics, Soochow University, Suzhou, Jiangsu, People's Republic of China

Part I
Chaos and Dynamics

Chapter 1
Weak Transient Chaos

Valentin S. Afraimovich and Alexander B. Neiman

1.1 Introduction

Once a new scientific direction arises it influences not only its immediate area, but also is accompanied by the appearance of new ideas and notions in neighboring fields. Our article here can be treated as a manifestation of this general principle. About 15 years ago Mikhail Rabinovich with co-authors started considering specific models with the so-called sequential dynamics, based on the winnerless competition principle [22] (see also [3, 6] and the references therein). It turned out that by using such models one can describe and explain important features of dynamics of neural and cognitive systems. A collection of works of M. Rabinovich in this direction set up a new area in nonlinear dynamics [20, 21]. Mathematically, the key point of these works was the understanding of the important role of heteroclinic networks and heteroclinic channels for sequential dynamics [4, 7]. Trajectories following paths determined by heteroclinic networks may demonstrate behavior that could not be quantitatively described in terms of conventional notions of complexity. Here we describe a regime of weak transient chaos in a model with sequential dynamics and propose a new measure for its quantitative characterization.

There are processes in nature that have adequate mathematical models in the form of dynamical systems (DS) but these models are applicable only for finite intervals of time. For example, a neural network with parameters determined by a given

V.S. Afraimovich (✉)
Institute de Investigacion Communicasion Optica, Universidad Autonoma de San Luis Potosi, A. Obregon 64, San Luis Potosi 78000, Mexico
e-mail: valentin.afraimovich@gmail.com

A.B. Neiman
Department of Physics and Astronomy and Neuroscience Program, Ohio University, Athens, OH, USA
e-mail: neimana@ohio.edu

I.S. Aranson et al. (eds.), *Advances in Dynamics, Patterns, Cognition*, Nonlinear Systems and Complexity 20, DOI 10.1007/978-3-319-53673-6_1

stimulus behaves as a specific DS until the instant when another stimulus arrives. For such systems one should be interested in both attractors and transient motions. In the seminal works [11, 12] C. Grebogi, E. Ott, and J.A. Yorke have discovered the phenomenon of transient chaos, explained its origin, and studied its main features (see [15] for the current state of the transient chaos theory). Roughly speaking, if an initial point belongs to the basin of a regular attractor (e.g., fixed point or a periodic trajectory) and the boundary of the basin contains a chaotic set (e.g., the Smale horseshoe), then the trajectory going through this point behaves chaotically provided that the initial point is close enough to the chaotic set. Of course, if an attractor is chaotic, then almost all trajectories in its basin behave chaotically.

Another type of transient motions is observed in systems operating according to the winnerless competition principle. Such motions can be treated as a sequential switching among metastable sets. In the corresponding phase space metastable sets are represented by invariant sets of the saddle type. For example, saddle equilibrium points or saddle limit cycles may represent metastable sets, while switching is governed by heteroclinic trajectories joining these sets. In this way a heteroclinic network arises and transient motions follow heteroclinic channels around this network's edges. Importantly, such motions can be chaotic during finite intervals of time [5]. However, in contrast to "conventional" transient chaos described by Grebogi–Ott–Yorke [11, 12], this chaotic behavior is not caused by the existence of an unstable chaotic set in the boundary of the basin of an attractor. Hence, this transient dynamics was given the name "finite time chaos" in [5].

The heteroclinic network considered in [5] was not an attractor. That is, each representative point in the heteroclinic channel (that did not belong to the stable manifolds of limit cycles) spent just a finite amount of time inside the channel and then moved away. To make chaotic features of motions more pronounced we consider here the case when a heteroclinic network is an attractor. We consider a master–slave system whereby the master, represented by $\{x\}$-coordinates, has a stable heteroclinic cycle, an attractor in the x-space. Such an attractor consists of a set of saddle equilibrium points with one-dimensional unstable manifolds; saddles are joined by heteroclinic trajectories. The slave, represented by $\{y\}$-coordinates, possesses a stable limit cycle in the absence of the master's drive.

1.2 Detection of Weak Transient Chaos

It turns out that the finite time chaos [5] is weaker than usual chaos since the largest Lyapunov exponent vanishes. To expose such a behavior one could perturb the master by a weak noise, which would bring its sequential heteroclinic dynamics to a steady state, calculate the Lyapunov exponent of the slave and be convinced in chaos, if the Lyapunov exponent is positive. Here we propose an alternative approach: instead of adding noise to the master, we introducing a novel definition of the Lyapunov exponent, which indicates weak transient chaos.

The largest Lyapunov exponent is defined as

$$\lambda_T := \limsup_{t \to \infty} \frac{\ln ||Df^t(p)||}{t}, \tag{1.1}$$

where p is an initial point in the basin and f^t is the flow generated by our master–slave system. $\lambda_T = 0$, if the divergence of trajectories is subexponential in time. In our situation we have exactly such a case because a trajectory of the master tends to the heteroclinic cycle and spends progressively more time in neighborhoods of saddle equilibrium points. While the representative point is near a saddle the dynamics is regular: trajectories in the full phase space do not diverge. The divergence may occur only when the representative point makes transition from one saddle to another.

If one would calculate the topological or the Kolmogorov-Sinai entropy in the case under consideration, one obtains 0. The entropy is defined as the limit of a fraction with time t in the denominator, as $t \to \infty$. The numerator measures an amount of instability accumulated in the system up to time t. If the numerator grows subexponentially, the entropy will be equal to 0. For such situation the so-called sequence entropy was introduced by replacing the time t in the denominator by some function of t which increases slower than t [8, 10, 14, 23]. Similar replacement can be probably done for the Lyapunov exponent, resulting in

$$\lambda_{\text{new}} := \limsup_{t \to \infty} \frac{\ln ||Df^t(p)||}{\rho(t)}, \tag{1.2}$$

see, for instance, [13]. However, the disadvantage of this definition is that the denominator $\rho(t)$ will be the same for all initial points p, while in reality behavior in time for different p can be different.

We propose here another approach based on the works of G.M. Zaslavsky with coauthors [1, 16–18, 24]. It was suggested there to replace (in calculations of complexity functions) time t by the length of a piece of trajectory of temporal length t, or some function of it. It was shown that such a replacement allows to obtain an additional useful information in several interesting situations. Following this lead we introduce here a new Lyapunov exponent,

$$\lambda_S := \limsup_{t \to \infty} \frac{\ln ||Df^t(p)||}{S(p,t)}, \tag{1.3}$$

where $S(p,t)$ is a function of the length of the piece of the trajectory of temporal length t going through an initial point p such that $S \to \infty$ as $t \to \infty$. Let us call λ_S the S-Lyapunov exponent. It shows how the instability evolves according to the lengths of trajectories. Let us remark that some general theory for S-Lyapunov exponents for points, typical for an invariant measure, is presented by Ya. Pesin, A. Zelerowicz, and Y. Zhao in Chap. 3 of this volume [19].

1.3 Master–Slave Model System

We consider a master–slave system in which the master possesses a heteroclinic cycle and drives the slave, Duffing–Van der Pol oscillator. In the absence of noise the system under consideration has the following form:

$$\dot{x}_i = x_i \left(\sigma_i - x_i - \sum_{j \neq i} r_{ij} x_j \right), \quad i, j = 1, 2, 3, \tag{1.4}$$

$$\ddot{y} - k\dot{y}(1 - y^2) + \alpha y^3 + \mu(x) y = 0, \tag{1.5}$$

where parameters $\sigma_i, r_{ij}, \alpha, k$ are positive numbers. The master (1.4) drives the slave (1.5) via the coupling function, $\mu(x)$,

$$\begin{aligned} \mu(x) &= 1 + \frac{\varepsilon}{2} \left[1 + \tanh\left(z(x) - \Delta \right) \right], \\ z(x) &= \left[(x_1 - x_2)^2 + (x_1 - x_3)^2 + (x_3 - x_2)^2 \right], \end{aligned} \tag{1.6}$$

where $\varepsilon > 0$ is the coupling strength and $\Delta > 0$ is a threshold parameter. The master system (1.4) has a heteroclinic cycle consisting of three saddle equilibrium points $O_1 = (\sigma_1, 0, 0), O_2 = (0, \sigma_2, 0)$, and $O_3 = (0, 0, \sigma_3)$ having one-dimensional unstable manifolds, connected by heteroclinic trajectories Γ_{12}, Γ_{23}, and Γ_{31}. The following conditions [2]:

$$\begin{aligned} &\sigma_2 - r_{21}\sigma_1 > 0, \quad \sigma_3 - r_{31}\sigma_1 < 0, \\ &\sigma_3 - r_{32}\sigma_2 > 0, \quad \sigma_1 - r_{12}\sigma_2 < 0, \\ &\sigma_1 - r_{13}\sigma_3 > 0, \quad \sigma_2 - r_{23}\sigma_3 < 0, \end{aligned} \tag{1.7}$$

guarantee that O_1, O_2, and O_3 are saddles with one-dimensional unstable manifolds. Furthermore, we assume that

$$\begin{aligned} -\sigma_1 &< \sigma_3 - r_{31}\sigma_1, \\ -\sigma_2 &< \sigma_1 - r_{12}\sigma_2, \\ -\sigma_3 &< \sigma_2 - r_{23}\sigma_3. \end{aligned}$$

These inequalities imply that the separatrices Γ_{ij} approach saddles O_j along the leading direction transversal to the x_j- axis, $j = 1, 2, 3$. Finally, let

$$\nu_1 := -\frac{\sigma_3 - r_{31}\sigma_1}{\sigma_2 - r_{21}\sigma_1} > 1,$$

$$\nu_2 := -\frac{\sigma_1 - r_{12}\sigma_2}{\sigma_3 - r_{32}\sigma_2} > 1,$$
$$\nu_3 := -\frac{\sigma_2 - r_{23}\sigma_3}{\sigma_1 - r_{13}\sigma_3} > 1. \tag{1.8}$$

Under these assumptions each saddle O_j is dissipative, and the heteroclinic cycle $\Gamma := \cup_{j=1}^{3} O_j \cup (\Gamma_{12} \cup \Gamma_{23} \cup \Gamma_{31})$ is an attractor for the master (1.4).

We will use the length of the projection onto the x-space of the piece of the trajectory of the system (1.4), (1.5) of duration t for calculation of the length function $S(p, t)$ in the definition of S-Lyapunov exponent, λ_S (2). Let us remark that the coupling is effective when the representative point of the master (1.4) is far from the saddles, O_j. In the full phase space of the system (1.4), (1.5) there are three limit cycles $\{O_i\} \times \{L_i\}$, $i = 1, 2, 3$, where L_i is the limit cycle of the slave (1.5) for which the coordinates of O_i are substituted into the coupling function, $\mu(x)$, i.e. $z = 2\sigma_i^2$. Each of these cycles is of the saddle type and has two-dimensional unstable manifolds. While the representative point of the master moves from O_i to O_{i+1}, the trajectories on the unstable manifold of the cycle move from one limit cycle $\{O_i\} \times \{L_i\}$ to the next one. These trajectories thus form a "heteroclinic tube" that for vanishing coupling, $\varepsilon = 0$ is topologically equivalent (even smoothly equivalent, in fact) to the direct product $\Gamma \times S'$, where S' is a circle. However, as the coupling strength increases, the shape of the tube changes. Its intersection with a section $x_i = a_i > 0, a_i \ll 1$, might look as the one in Fig. 1.3b and so the tube is not a topological manifold anymore. Such a tube was called *a bizarre tube* in [5] and it was shown that a complexity function grows faster for the case of the bizarre tube compared to piece-wise smooth tubes.

1.4 Numerical Results

We set the following parameters for the master, $\sigma_1 = 5, \sigma_2 = 7, \sigma_3 = 9, \Gamma_{12} = 1.2243, \Gamma_{13} = 0.0556, \Gamma_{21} = 0.9, \Gamma_{23} = 2.31, \Gamma_{32} = 0.7857$, which satisfy the conditions (1.7)–(1.8) and so the master system possesses a heteroclinic cycle. The slave Duffing–Van der Pol oscillator (1.5) shows chaotic behavior with positive conventional Lyapunov exponent if perturbed by additive white noise [9]. Instead, here it is driven by the master which possesses a heteroclinic cycle. In the following we set $k = 0.5$, $\alpha = 2.5$ for the slave Duffing–Van der Pol oscillator and $\Delta = 8$ for the coupling function, $\mu(x)$.

We compared deterministic dynamics of the full system with the case when the master was perturbed by weak additive Gaussian white noise. In this case the master was governed by stochastic differential equations,

$$\dot{x}_i = x_i \left(\sigma_i - x_i - \sum_{j \neq i} r_{ij} x_j \right) + \sqrt{2D}\, \xi_i(t), \quad i, j = 1, 2, 3, \tag{1.9}$$

where $\xi_i(t)$ are uncorrelated white Gaussian processes and D is their intensity. In the following we set $D = 10^{-6}$.

The length of the phase trajectories was calculated separately for the master, S_M, and for the slave, S_S, as

$$S_M = \int_0^T \left(\sum_{i=1}^{3} \dot{x}_i(t)^2 \right)^{1/2} dt, \quad S_S = \int_0^T \left(\dot{y}(t)^2 + \ddot{y}(t)^2 \right)^{1/2} dt \tag{1.10}$$

Numerical simulations were performed with quadruple precision using 4-th order Runge–Kutta method. To avoid negative values for the master variables we set a reflecting boundary conditions, so that if $x_i(t) < 0$, $x_i(t)$ was replaced by $-x_i(t)$. The largest Lyapunov exponents were calculated for the slave system only over the time span of 10^5 and additionally averaged over a set of 100 randomly chosen initial conditions of the master and slave systems. We then calculated the standard deviation from this mean, which allows putting "errorbars" on the Lyapunov exponent for indication of its dependence on initial conditions. Both, the conventional Lyapunov exponent, λ_T (1.1) and the proposed S-exponent with normalization over the length of the projection of the trajectory on master system, λ_S (1.3), were calculated.

In the absence of noise, $D = 0$, the master shows heteroclinic cycle, slowing down as time progresses. That is, the master system generates long transient motions as Fig. 1.1 (upper trace) indicates. Weak noise accelerates the master when its phase trajectory passes near saddles and results in steady stochastic oscillations shown in Fig. 1.1 (lower trace). As a result, the trajectory length of deterministic slave system shows a limited growth with time, while the length of stochastic slave trajectory exhibits a linear growth, shown in Fig. 1.2a. From this graph it is easy to see that the average speed of the master (i.e., the slope of $S_M(T)$ on Fig. 1.2a) decreases for the deterministic case and is virtually constant for stochastic case.

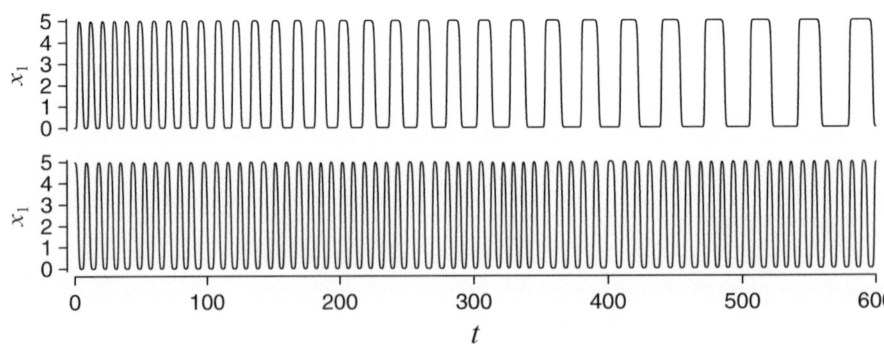

Fig. 1.1 Time series of the master system. *Upper trace*: heteroclinic cycle in the intrinsic case. *Lower trace*: noise perturbed case showing steady state stochastic oscillations

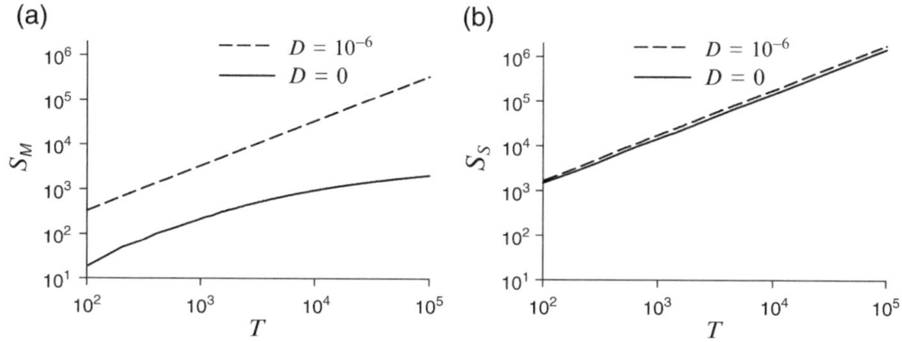

Fig. 1.2 Trajectory lengths vs integration time. Length of the projection on the master system, S_M (**a**), and on the slave Duffing–Van der Pol oscillator, S_S (**b**), are shown for the deterministic case, $D = 0$ (*solid lines*) and for weak noise, $D = 10^{-6}$ (*dashed line*)

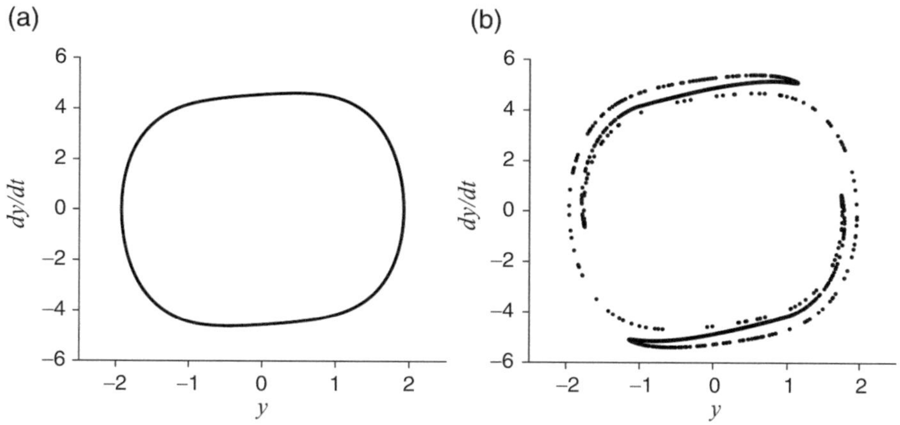

Fig. 1.3 Effect of drive from heteroclinic cycle of the deterministic master system on the ensemble of 10^4 identical Duffing–Van der Pol oscillators. (**a**) Initial distribution of the ensemble. (**b**) Distorted distribution after the heteroclinic drive during time $T = 10^4$ for $\varepsilon = 8$

Figure 1.3 shows results of simulations of an ensemble of 10^4 identical Duffing–Van der Pol slave oscillators subjected to the common drive from the master system. The ensemble was started with random initial condition, so that before the drive from the master was turned on, the slave oscillators were randomly distributed on the stable limit cycle as Fig. 1.3a shows. Under the influence of the heteroclinic sequence in the master system, the slave's limit cycle can be distorted exhibiting multiple folding as shown in Fig. 1.3b. In both, deterministic and stochastic cases, the slave's trajectory length grows linearly with integration time, T, as indicated in Fig. 1.2b.

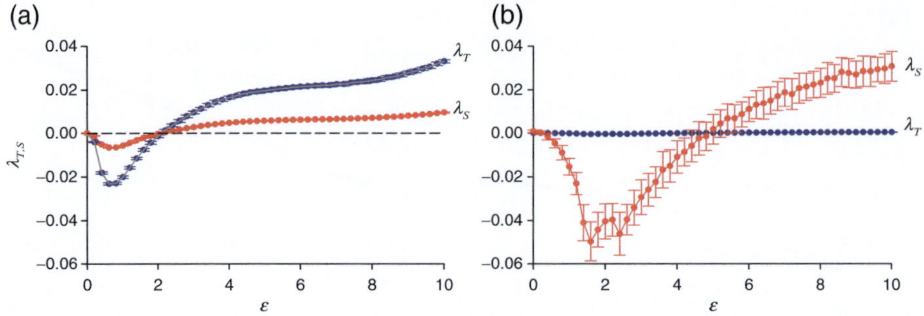

Fig. 1.4 Largest Lyapunov exponents of the slave system normalized to time, λ_T (*blue lines* and *symbols*), and to the length of the master system, λ_S (*red lines* and *symbols*) vs the coupling p parameter, ε. Errorbars indicate standard deviation of the Lyapunov exponent from the mean obtained by averaging over 100 random initial conditions. (**a**) Randomly perturbed master system, $D = 10^{-6}$. (**b**) Deterministic master system, $D = 0$

The results of calculation of the largest Lyapunov exponent are shown in Fig. 1.4. We begin with the noise-perturbed system which reaches a steady state, Fig. 1.4a. In this case both the conventional and S-Lyapunov exponents show qualitatively similar dependence on the coupling parameter, ε: starting with $\varepsilon \approx 2$ both exponents are positive. Noise-induced chaos in this Duffing–Van der Pol oscillator was indeed reported earlier in [9]. Importantly, we notice small errorbars, indicating that the Lyapunov exponents are invariant with respect to initial conditions. The deterministic case shown in Fig. 1.4b is different. The conventional Lyapunov exponent is 0, as expected. However, the S exponent shows positive values for $\varepsilon > 5$, indicating transient chaos. Large errorbars point out the dependence on the initial conditions.

1.5 Concluding Remarks

We studied weak transient chaos in the master–slave system whereby the slave, possessing a stable limit cycle, is driven by the master's heteroclinic cycle. We have shown that if the coupling strength is large enough, then the system manifests a weak transient chaos indicated by positive values of newly introduced S-Lyapunov exponent. We stress that such chaotic behavior is caused neither by the presence of a chaotic unstable set in the boundary of the basin of the attractor (the heteroclinic tube in our case) nor by the chaoticity of the attractor itself. Instead, it is caused by the instability of trajectories in the directions "parallel" to the attractor. Furthermore, on the attractor all trajectories (except for the limit cycles) go from one limit cycle to the next one manifesting trivial non-chaotic behavior. Thus, a weak chaotic behavior is caused by divergence of trajectories going through wandering (transient) points. One can say that the chaos is supported on a subset of wandering points.

This phenomenon, in slightly different interpretations, was discovered in [23] for the sequence entropy. Probabilistic distributions of such initial points and/or measures with supports on the set of these points could not be invariant. From the physical viewpoint these distributions do not correspond to equilibrium or steady-state states. So, in the study of weak transient chaos one should learn how to deal with non-invariant states.

Acknowledgements The authors dedicate this paper to the 75th anniversary of Mikhail Rabinovich. We wish him great health, and maintenance of his remarkable enthusiasm and energy, which generated and will generate many exciting ideas and fundamental works. The authors thank T. Young for useful discussions and Ya. Pesin for allowing them to read the manuscript of his chapter [19] in this volume. The authors acknowledge support by the RSF grant 14-41-00044 of the Russian Science Foundation during their stay at Nizhny Novgorod University.

References

1. Afraimovich, V., Zaslavsky, G.: Space-time complexity in Hamiltonian dynamics. Chaos: Interdiscip. J. Nonlinear Sci. **13**(2), 519 (2003)
2. Afraimovich, V., Zhigulin, V., Rabinovich, M.: On the origin of reproducible sequential activity in neural circuits. Chaos: Interdiscip. J. Nonlinear Sci. **14**(4), 1123 (2004)
3. Afraimovich, V., Tristan, I., Huerta, R., Rabinovich, M.I.: Winnerless competition principle and prediction of the transient dynamics in a Lotka-Volterra model. Chaos: Interdiscip. J. Nonlinear Sci. **18**(4), 043103 (2008)
4. Afraimovich, V., Ashwin, P., Kirk, V.: Robust heteroclinic and switching dynamics. Dyn. Syst. **25**(3), 285 (2010)
5. Afraimovich, V., Cuevas, D., Young, T.: Sequential dynamics of master-slave systems. Dyn. Syst. **28**(2), 154 (2013)
6. Afraimovich, V., Tristan, I., Varona, P., Rabinovich, M.: Transient dynamics in complex systems: heteroclinic sequences with multidimensional unstable manifolds. Discontinuity, Nonlinearity and Complexity **2**(1), 21 (2013)
7. Ashwin, P., Coombes, S., Nicks, R.: Mathematical frameworks for oscillatory network dynamics in neuroscience. J. Math. Neurosci. **6**(1), 1 (2016)
8. Canovas, J.: A guide to topological sequence entropy. In: Progress in Mathematical Biology Research, pp. 101–139. Nova Science Publishers, New York (2008)
9. Goldobin, D.S., Pikovsky, A.: Synchronization and desynchronization of self-sustained oscillators by common noise. Phys. Rev. E **71**(4), 045201 (2005)
10. Goodman, T.: Topological sequence entropy. Proc. Lond. Math. Soc. **29**(3), 331 (1974)
11. Grebogi, C., Ott, E., Yorke, J.A.: Crises, sudden changes in chaotic attractors, and transient chaos. Physica D: Nonlinear Phenom. **7**(1), 181 (1983)
12. Grebogi, C., Ott, E., Yorke, J.A.: Critical exponent of chaotic transients in nonlinear dynamical systems. Phys. Rev. Lett. **57**(11), 1284 (1986)
13. Klages, R.: From Hamiltonian Chaos to Complex Systems, pp. 3–42. Springer, Berlin (2013)
14. Kushnirenko, A.G.: On metric invariants of entropy type. Russ. Math. Surv. **22**(5), 53 (1967)
15. Lai, Y.C., Tél, T.: Transient Chaos: Complex Dynamics on Finite Time Scales, vol. 173. Springer Science & Business Media, Dordrecht (2011)
16. Leoncini, X.: Hamiltonian Chaos Beyond the KAM Theory, pp. 143–192. Springer, Berlin/Heidelberg (2010)
17. Leoncini, X., Zaslavsky, G.M.: Jets, stickiness, and anomalous transport. Phys. Rev. E **65**(4), 046216 (2002)

18. Leoncini, X., Agullo, O., Benkadda, S., Zaslavsky, G.M.: Anomalous transport in Charney-Hasegawa-Mima flows. Phys. Rev. E **72**(2), 026218 (2005)
19. Pesin, Y., Zelerowicz, A., Zhao, Y.: Time rescaling of Lyapunov exponents. In: Aranson, I.S., Pikovsky, A., Rulkov, N.F., Tsimring, L. (eds.) Advances in Dynamics, Patterns, Cognition. Springer, Cham (2017)
20. Rabinovich, M.I., Simmons, A.N., Varona, P.: Dynamical bridge between brain and mind. Trends Cogn. Sci. **19**(8), 453 (2015)
21. Rabinovich, M.I., Tristan, I., Varona, P.: Hierarchical nonlinear dynamics of human attention. Neurosci. Biobehav. Rev. **55**, 18 (2015)
22. Rabinovich, M., Volkovskii, A., Lecanda, P., Huerta, R., Abarbanel, H., Laurent, G.: Dynamical encoding by networks of competing neuron groups: winnerless competition. Phys. Rev. Lett. **87**(6), 068102 (2001)
23. Szlenk, W.: On weakly conditionally compact dynamical systems. Stud. Math. **66**(1), 25 (1979)
24. Zaslavsky, G., Edelman, M.: Polynomial dispersion of trajectories in sticky dynamics. Phys. Rev. E **72**(3), 036204 (2005)

Chapter 2
Lorenz Type Attractor in Electronic Parametric Generator and Its Transformation Outside the Parametric Resonance

Sergey P. Kuznetsov

2.1 Introduction

This chapter is inspired by seminal works of M.I. Rabinovich and his collaborators in 70-th devoted to complex dynamics of parametric oscillators [34, 36, 42], mainly, by the article of Pikovsky et al. [34].

It was shown in [34] that in a case of parametric excitation of two modes by the pump at the sum frequency and with the energy transfer to decaying third mode at the difference frequency, chaotic dynamics can occur. Particularly, the authors of [34] considered the situation in application to waves in magnetized plasma. Assuming a quadratic nonlinearity, they formulated amplitude equations, which in the case of a fixed phase relation were reduced to a set of three differential equations of the first order possessing the Lorenz type attractor. The same mechanism of chaos generation can be implemented with the parametric interaction of waves or oscillatory modes in various physical objects, such as mechanical, electronic, optical, acoustic systems [2, 3, 21, 28, 32].

Lorenz attractor is a popular example of a strange chaotic attractor [20, 27, 40], which was originally discovered in a model system of three first order differential equations for the problem of fluid convection in a layer heated from below. It belongs to a class of singular hyperbolic (quasi-hyperbolic) attractors [4, 7] and generates robust chaos [5, 11] in the sense that the chaotic behavior is not destroyed by a small variation of the system parameters. In the years after the Lorenz publication it became clear that this type of attractor may be related to many different natural systems, including laser dynamics [15, 30, 31], mechanical systems based on the rigid body rotations [8, 10, 14, 18], and others [1, 9, 13, 16, 29, 33, 35, 38].

S.P. Kuznetsov (✉)
Kotel'nikov's Institute of Radio-Engineering and Electronics of RAS, Saratov Branch, Zelenaya 38, Saratov 410019, Russia
e-mail: spkuz@yandex.ru

I.S. Aranson et al. (eds.), *Advances in Dynamics, Patterns, Cognition*, Nonlinear Systems and Complexity 20, DOI 10.1007/978-3-319-53673-6_2

Some aspects of the dynamics of the model of Pikovsky, Rabinovich, and Trakhtengerts were analyzed later by other authors [23–25, 43]. In particular, it concerns the dynamics, accompanied by violation of the phase locking [43], comparison of the model with a ten-dimensional system for the problem of the interaction of waves in plasma [24], and mathematical analysis of global dynamics [26]. In [23] an electronic parametric oscillator was considered based on three resonant LC circuits, where the parametric excitation and the interaction of modes take place due to the presence of a varactor diode. It has been found that with accurate description of the nonlinear characteristic of the diode, the equations for slow amplitudes are essentially represented in complex variables, so that in the dynamics of the excited oscillatory modes the phases are relevant, and the attractor ceases to be quasi-hyperbolic. In particular, this is expressed by appearance of windows of regularity in the parameter space, where the periodic dynamics occur instead of chaos, and attractors are represented by limit cycles.

Here we analyze the parametric oscillator circuit similar to that in [23], but using a specially designed nonlinear reactance element composed on a basis of an operational amplifier and an analog multiplier, with the characteristic exactly given by a quadratic function. Because of this it is possible to realize the dynamics of the Pikovsky–Rabinovich–Trakhtengerts model in pure form. With exact parametric resonance conditions it manifests the Lorenz type attractor. In the case of violation of the exact frequency relations a situation occurs similar to that observed in [23]. Namely, the phase dynamics becomes relevant; the attractor ceases to be quasi-hyperbolic, and windows of regular dynamics appear in the parameter space besides the chaotic regions.

2.2 Parametric Oscillator Circuit Diagram and the Basic Equations

Consider the circuit diagram of Fig. 2.1a. It is composed of three resonant circuits: L_1–C_1, L_2–C_2, and L_3–C_3. Parametric excitation is provided by the pump from the AC voltage source V_1 in presence of the quadratic nonlinear reactance $\tilde{C}$.

The nonlinear element circuit diagram is shown separately in panel (b). When a voltage U is applied to the input of the element with respect to the ground, the potential U takes place on the both input terminals of the operational amplifier OA. Since the input impedance of the operational amplifier ideally is infinite, the presence of the current U/R through the resistor R, which has a grounded outlet, implies the same current through the second resistor R connected to the previous one, and thus the voltage at the input of the analog multiplier AM must be equal to $2U$. Hence we have the voltage $4KU^2$ at its output. The currents through the one and the other capacitors C_0 are $C_0 \frac{dU}{dt}$ and $\frac{d}{dt}\left(4KU^2 - C_0 U\right)$; in amount, they comprise the current through the nonlinear element $\frac{d}{dt}\left(4KU^2\right)$.

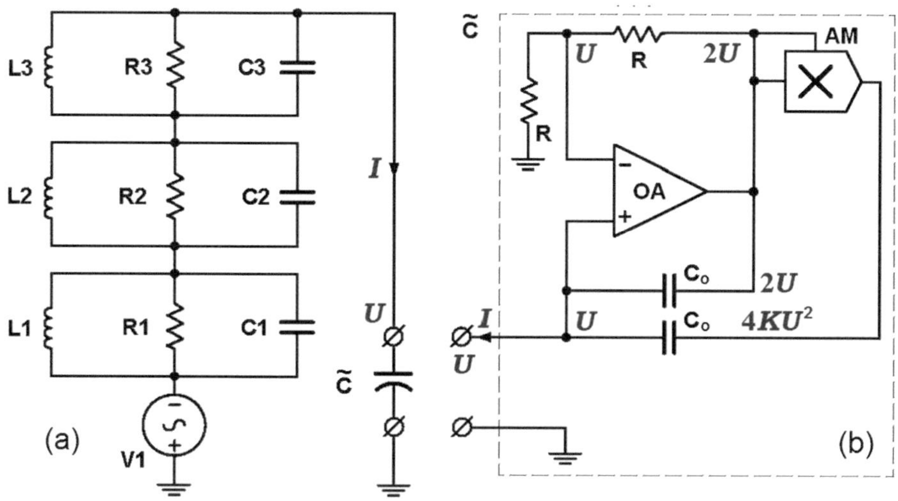

Fig. 2.1 The circuit diagram of the chaotic parametric oscillator (**a**). The pumping is provided by the voltage source V_1. The symbol $\tilde{C}$ denotes a two-pole reactance element with quadratic nonlinearity, the scheme of which is shown separately on the panel (**b**)

The natural frequencies of the LC resonant circuits (without taking into account the dissipation) will be assumed to satisfy, at least approximately, the parametric resonance conditions

$$\Omega_0 \approx \Omega_1 + \Omega_2, \quad \Omega_3 \approx \Omega_1 - \Omega_2. \tag{2.1}$$

Figure 2.2 shows plots of voltages on the capacitors C_1, C_2, and C_3 obtained from the virtual oscilloscope in the course of simulation using the Multisim environment of the circuit with the component values indicated in the figure caption. After the transient decay, a sustained regime of nonlinear oscillations persists. In the scale of the figure, the high-frequency filling is indistinguishable, but one can observe clearly the irregular, apparently chaotic behavior of the amplitudes.

In a framework of the circuit simulation in Multisim it is difficult to get information concerning some of the essential features of the dynamics including the expected presence of the Lorenz type attractor and to determine relevant characteristics such as the Lyapunov exponents. Therefore, in the following sections we will discuss the equations describing the system, and analyze some results of their numerical integration.

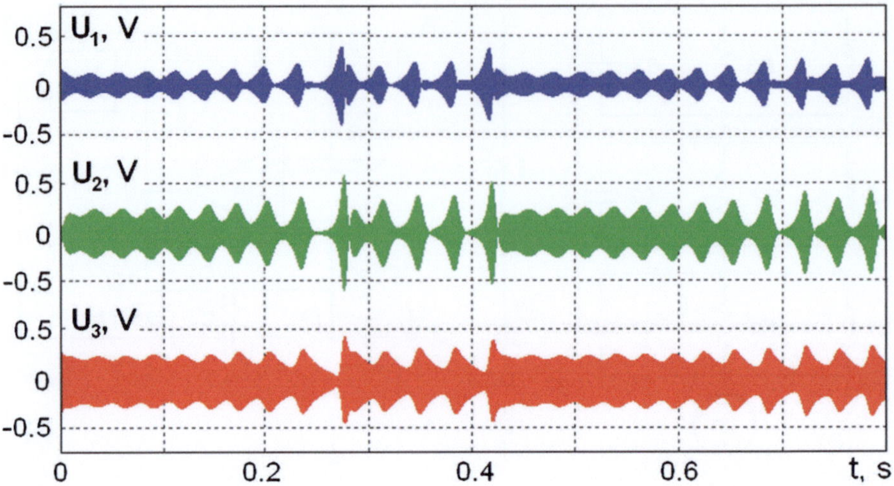

Fig. 2.2 Typical waveforms for the voltages across the capacitors C_1, C_2, C_3 obtained by simulation of the circuit of Fig. 2.1 using the Multisim environment. The component values: $C_1 = C_2 = C_3 = 40$ nF, $L_1 = 1.583$ mH, $L_2 = 4.398$ mH, $L_3 = 9.895$ mH, $R_1 = 250\,\text{k}\Omega$, $R_2 = 62.5\,\text{k}\Omega$, $R_3 = 250\,\text{k}\Omega$. The natural frequencies of the LC circuits are $f_1 = 20$ kHz, $f_2 = 12$ kHz, $f_3 = 8$ kHz. Pumping is carried out by the voltage source V_1 with amplitude of 0.245 V at the frequency $f_0 = 32$ kHz. The capacitors in the circuit diagram of the nonlinear reactance element are of capacitance $C_0 = 2$ nF; the transmission coefficient of the analog multiplier AM is $K = 1/8\,\text{V}^{-1}$

2.3 Basic Equations of the Parametric Oscillator

Suppose that U_1, U_2, U_3 are voltages on the capacitors C_1, C_2, C_3, and I_1, I_2, I_3 are currents through the inductors L_1, L_2, L_3. Assuming equality of the capacities $C = C_1 = C_2 = C_3$ for simplicity, write down the Kirchhoff equations as follows:

$$\begin{aligned}
L_1\dot{I}_1 &= U_1,\\
L_2\dot{I}_2 &= U_2,\\
L_3\dot{I}_3 &= U_3,\\
C\dot{U}_1 + U_1/R_1 + I_1 &= -I,\\
C\dot{U}_2 + U_2/R_2 + I_2 &= -I,\\
C\dot{U}_3 + U_3/R_3 + I_3 &= -I.
\end{aligned} \tag{2.2}$$

Here I is the current through the non-linear element defined by the expression

$$I = \frac{d}{dt}4KC_0U^2 = C\varepsilon\frac{d}{dt}\frac{U^2}{2}, \tag{2.3}$$

where

$$\varepsilon = 8KC_0/C, \quad U = U_1 + U_2 + U_3 + U_0, \quad U_0 = -\kappa \sin \omega_0 t, \tag{2.4}$$

and the values of κ and ω_0 designate the amplitude and frequency of the pump signal. Using the normalized dimensionless time $t' = t/(2R_3C)$, the equation can be rewritten as

$$\ddot{X}_k + 2\nu_k \dot{U}_k + \Omega_k^2 U_k = 0, \quad X_k = U_k + \frac{1}{2}\varepsilon U^2, \quad k = 1, 2, 3, \tag{2.5}$$

where

$$\nu_{1,2} = \frac{R_3}{R_{1,2}}, \quad \nu_3 = 1, \quad \Omega_{1,2,3} = 2R_3\sqrt{\frac{C}{L_{1,2,3}}}. \tag{2.6}$$

For the numerical integration it is convenient to reformulate the problem representing it by the set of the first order differential equations

$$\dot{Y}_k = -\Omega_k(X_k - \frac{1}{2}\varepsilon U^2), \quad \dot{X}_k = \Omega_k Y_k - 2\nu_k, \quad k = 1, 2, 3,$$
$$U = \frac{-1 + \sqrt{6\varepsilon(X_1 + X_2 + X_3 - \kappa \sin \Omega_0 t') + 1}}{3\varepsilon}, \quad U_i = X_k - \frac{1}{2}\varepsilon U^2. \tag{2.7}$$

Figure 2.3 shows the time dependences for the quantities U_1, U_2, U_3, obtained by numerical integration of the equations (2.7) with component values indicated

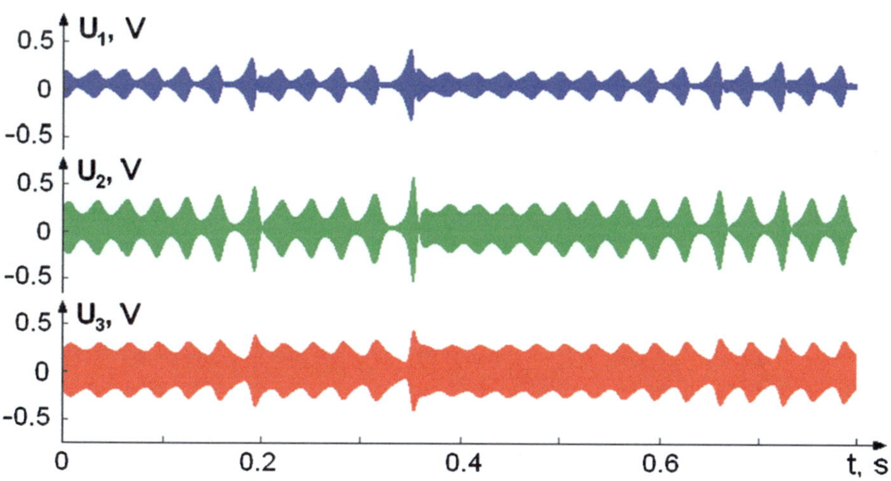

Fig. 2.3 Typical waveforms of voltages across the capacitors C_1, C_2, C_3, obtained from numerical integration of the equations (2.7) for the circuit shown in Fig. 2.1, with component values indicated in the caption of Fig. 2.2

in the caption of Fig. 2.2. Comparing the plots in Figs. 2.2 and 2.3, we can see a good agreement between the observed dynamics. In both cases we have chaotic waveforms containing similar fragments, and compliance in characteristic scales of times and voltages. (One should not expect to see exact correspondence of the waveforms because of the inherent sensitivity of the chaotic dynamics to small perturbations of the initial conditions.)

2.4 Equations for Slow Amplitudes

To obtain equations in a form that allows comparison with the Lorenz and Pikovsky–Rabinovich–Trakhtengerts models, it is necessary to apply the method of slow varying amplitudes. First, it is convenient to rewrite the equations considering only those terms, which can contribute to the resonant interaction of the modes corresponding to the relation of the frequencies (2.1). In the first, the second, and the third equations (2.5) one can set, respectively,

$$\begin{aligned} \frac{1}{2}U^2 &\approx U_0U_2 + U_2U_3, \\ \frac{1}{2}U^2 &\approx U_0U_1 + U_1U_3, \\ \frac{1}{2}U^2 &\approx U_1U_2. \end{aligned} \tag{2.8}$$

In addition, replace the operation of the second derivative of the nonlinear terms in the equations by the multiplier $(-\Omega_k^2)$. The result is

$$\begin{aligned} \ddot{U}_1 + 2\nu_1\dot{U}_1 + \Omega_1^2U_1 &= \varepsilon\Omega_1^2(U_0U_2 + U_2U_3), \\ \ddot{U}_2 + 2\nu_2\dot{U}_2 + \Omega_2^2U_2 &= \varepsilon\Omega_2^2(U_0U_1 + U_1U_3), \\ \ddot{U}_3 + \quad 2\dot{U}_3 + \Omega_3^2U_3 &= \varepsilon\Omega_3^2U_1U_2. \end{aligned} \tag{2.9}$$

Omitting for brevity the prime at the time variable, we seek a solution in the form

$$\begin{aligned} U_1 &= A_1e^{i\omega_1t} + A_1^*e^{-i\omega_1t}, & \dot{U}_1 &= i\omega_1A_1e^{i\omega_1t} - i\omega_1A_1^*e^{-i\omega_1t}, \\ U_2 &= A_2e^{i\omega_2t} + A_2^*e^{-i\omega_2t}, & \dot{U}_2 &= i\omega_2A_2e^{i\omega_2t} - i\omega_2A_2^*e^{-i\omega_2t}, \\ U_3 &= -iA_3e^{i\omega_3t} + iA_3^*e^{-i\omega_3t}, & \dot{U}_3 &= \omega_3A_3e^{i\omega_3t} + \omega_3A_3^*e^{-i\omega_3t}, \end{aligned} \tag{2.10}$$

where the reference frequencies $\omega_{1,2,3}$ are defined as

$$\omega_1 = \frac{\Omega_1 - \Omega_2 + \Omega_0}{2}, \quad \omega_2 = \frac{-\Omega_1 + \Omega_2 + \Omega_0}{2}, \quad \omega_3 = \Omega_1 - \Omega_2. \tag{2.11}$$

Being close to the values $\Omega_{1,2,3}$ they satisfy the resonance conditions precisely:

$$\omega_3 = \omega_1 - \omega_2, \quad \Omega_0 = \omega_1 + \omega_2. \tag{2.12}$$

Use of (2.10) implies fulfillment of the additional equalities for the amplitudes:

$$\dot{A}_1 e^{i\omega_1 t} + \dot{A}_1^* e^{-i\omega_1 t} = 0, \quad \dot{A}_2 e^{i\omega_2 t} + \dot{A}_2^* e^{-i\omega_2 t} = 0, \quad \dot{A}_3 e^{i\omega_3 t} - \dot{A}_3^* e^{-i\omega_3 t} = 0. \tag{2.13}$$

Next, according to (2.4), we have

$$U_0 = -\kappa \sin \Omega_0 t = \frac{1}{2} i\kappa e^{i\Omega_0 t} - \frac{1}{2} i\kappa e^{-i\Omega_0 t}. \tag{2.14}$$

Substitution of (2.10) into Eqs. (2.9) yields

$$\begin{aligned}
\dot{A}_1 + \nu_1 A_1 - i\Delta A_1 &= \frac{1}{2}\varepsilon\Omega_1\left(\frac{1}{2}\kappa A_2^* - A_2 A_3\right),\\
\dot{A}_2 + \nu_2 A_2 - i\Delta A_2 &= \frac{1}{2}\varepsilon\Omega_2\left(\frac{1}{2}\kappa A_1^* + A_1 A_3^*\right),\\
\dot{A}_3 + A_3 - i\delta A_3 &= \frac{1}{2}\varepsilon\Omega_3 A_1 A_2^*,
\end{aligned} \tag{2.15}$$

where

$$\Delta \approx \Omega_1 - \omega_1 = \Omega_2 - \omega_2 = \frac{\Omega_1 + \Omega_2 - \Omega_0}{2}, \quad \delta \approx \Omega_3 - \omega_3 = \Omega_3 - \Omega_1 + \Omega_2.$$

With normalization

$$A_1 = \frac{2}{\varepsilon\sqrt{\Omega_2\Omega_3}} a_1, \quad A_2 = \frac{2}{\varepsilon\sqrt{\Omega_3\Omega_1}} a_2, \quad A_3 = \frac{2}{\varepsilon\sqrt{\Omega_1\Omega_2}} a_3, \quad h = \frac{\varepsilon\kappa\sqrt{\Omega_1\Omega_2}}{4}, \tag{2.16}$$

we obtain the equations exactly corresponding to [34]:

$$\dot{a}_1 + \nu_1 a_1 - i\Delta a_1 = h a_2^* - a_2 a_3, \quad \dot{a}_2 + \nu_2 a_2 - i\Delta a_2 = h a_1^* + a_1 a_3^*, \quad \dot{a}_3 + a_3 - i\delta a_3 = a_1 a_2^*. \tag{2.17}$$

Here κ is the dimensionless parameter of the pumping amplitude, the value of Δ can be adjusted by varying the pumping frequency, and δ by varying the inductance L_3.

2.5 Precise Parametric Resonance: Lorenz Type Attractor

In the absence of detuning, $\Delta = 0$, $\delta = 0$, using the substitution

$$a_1 = x e^{i\varphi}, \quad a_2 = y e^{-i\varphi}, \quad a_3 = z e^{2i\varphi} \tag{2.18}$$

with $\varphi = \text{const}$ we arrive at the equations

$$\dot{x} = hy - \nu_1 x - yz, \quad \dot{y} = hx - \nu_2 y + xz, \quad \dot{z} = -z + xy, \tag{2.19}$$

which may be considered in domain of real variables.

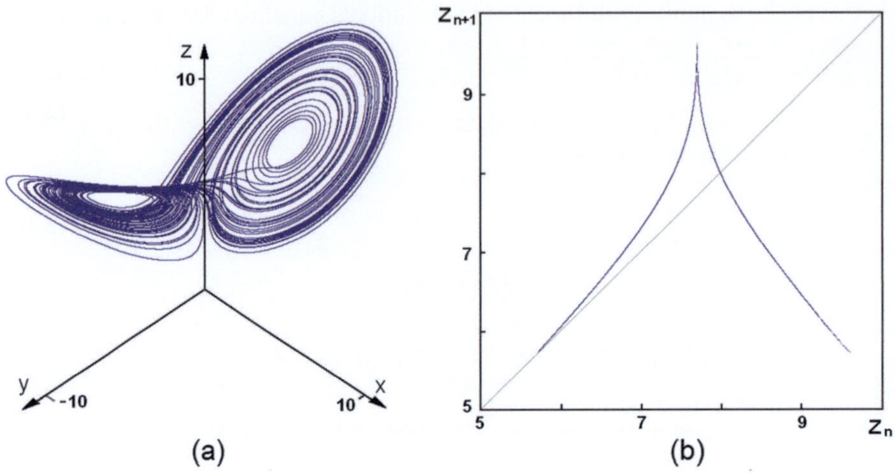

Fig. 2.4 Attractor in the three-dimensional state space of system (2.19) (**a**) and plot of the map for successive maxima of the variable z in the course of temporal evolution (**b**); parameters are $\nu_1 = 1,\ \nu_2 = 4,\ h = 5.962$

According to [34], the model (2.19) has attractor of Lorenz type in the three-dimensional phase space of the variables x, y, z, which is true, particularly, in the case $\nu_1 = 1$, $\nu_2 = 4$, $h = 5.962$. Figure 2.4 shows a portrait of the attractor according to the results of numerical integration of the equations (2.19). Also, the plot is shown obtained with the procedure applied by Lorenz in his work [27]: on the axes are the values of the maxima of the variable z achieved sequentially during the time evolution of the system. The view of the plot with a sharp peak, which resembles a classic "saw tooth" map [20, 27, 34, 40], indicates that the attractor is quasi-hyperbolic, just like the classic Lorenz attractor.

The Lyapunov exponents calculation by joint numerical solution of equations (2.19) and the corresponding variational equations is based on a known algorithm with Gram–Schmidt orthonormalization of perturbation vectors [6, 20] and yields[1]

$$\lambda_1 = 0.394 \pm 0.001,\ \ \lambda_2 = 0.0000 \pm 0.0001,\ \ \lambda_3 = -6.394 \pm 0.001. \qquad (2.20)$$

The presence of a positive Lyapunov exponent indicates occurrence of chaos, characterized by the exponential growth of the deviation from the reference trajectory on the attractor under small perturbations of the initial conditions. The second exponent is zero up to a calculation error; it is associated with a perturbation of a shift along the reference phase trajectory. The third exponent is negative

[1] Lyapunov exponents' calculations were performed on time intervals of duration of 50,000 with the counting of the average values and standard deviations on 20 samples. As an error, the standard deviations are indicated.

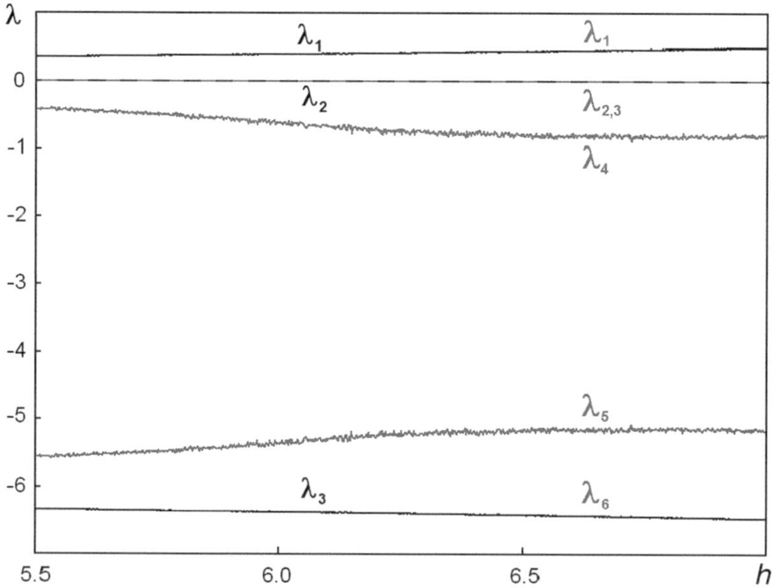

Fig. 2.5 Dependence of Lyapunov exponents on the parameter h. The exponents of the model (2.19) are marked as λ_{1-3} on the *left side* of the diagram, and those for the model (2.17) at $\Delta = 0$, $\delta = 0$ are labeled as λ_{1-6} in the *right part* of the figure. Other parameters: $\nu_1 = 1$, $\nu_2 = 4$

and is responsible for the approach of the trajectories to the attractor. The fact that the sum of the exponents is negative indicates the volume compression in the three-dimensional phase space. It is consistent with the analytical calculation of the divergence of the vector field defined by the right sides of the equations (2.19): $\mathrm{div}\mathbf{F} = \partial_x f_x + \partial_y f_y + \partial_z f_z = -\nu_1 - \nu_2 - 1 = -6$. Estimate of the dimension of the attractor from the well-known formula of Kaplan–Yorke provides $D = 2 + \lambda_1/|\lambda_3| \approx 2.06$.

Figure 2.5 shows the three Lyapunov exponents of the model (2.19) versus the parameter h. The smooth nature of the dependence and the lack of notable dips (regularity windows) for the senior exponent in the graph indicate the robustness of chaos in the three-dimensional system (2.19) and correspond to the conclusion that the nature of the attractor is quasi-hyperbolic as motivated by the view of the graph displayed in Fig. 2.5b.

It is interesting to compare the Lyapunov exponents calculated at the same parameters for equations in real and complex amplitudes. For the system (2.17) at $\Delta = 0$, $\delta = 0$ we have

$$\begin{aligned} &\lambda_1 = 0.394 \pm 0.001, \quad \lambda_2 = 0.0000 \pm 0.0002, \quad \lambda_3 = 0.0000 \pm 0.0003, \\ &\lambda_4 = -0.618 \pm 0.05, \quad \lambda_5 = -5.381 \pm 0.05, \quad \lambda_6 = -6.394 \pm 0.003. \end{aligned} \tag{2.21}$$

In this list, there are two zero exponents, one of which is associated with the shift perturbation along the reference phase trajectory, the second with the phase shift in the variable ϕ [see (2.18)]. The exponents $\lambda_{1,2,6}$ are in agreement with the exponents $\lambda_{1,2,3}$ from the list (2.20). Additional exponents $\lambda_{4,5}$ correspond obviously to the relaxation of phases to the situation described by equations for real amplitudes. Figure 2.5 shows the plot for six Lyapunov exponents of the model (2.17), three of which are indistinguishable from those of the model (2.19).

If we talk about the system without reduction to the slow amplitudes (2.7) and about the model with complex amplitudes (2.17), it would be incorrect to relate to them the conclusion concerning robustness of the Lorenz type attractor from the three-dimensional real model (2.19). A formal sign pointing to a possible violation of the robustness in this sense is occurrence of an additional zero Lyapunov exponent in the complex system (2.17). In particular, introduction of the frequency detuning leads to a disruption of the phase relations (2.18) and to a change in the nature of the attractor.

2.6 Chaotic and Regular Dynamics in the Parametric Oscillator in Presence of Frequency Detuning

We now turn to the situation when the conditions of parametric resonance are not accurately fulfilled, and one has to take into account the frequency detuning of the pump from the sum of natural frequencies of the first and second oscillators, and detuning of the difference frequency from the frequency of the third oscillator. This corresponds to non-zero parameters Δ, δ in the complex amplitude equations (2.17), which cannot be reduced now to a three-dimensional system for real amplitudes.

In this situation, there is a problem of graphical presentation of attractors allowing a visual comparison with the Lorenz attractor. We proceed with assumption that the approximate correspondence of instantaneous phases of the complex variables a_1 and a_2 to the formulas (2.18) roughly persists for the most part, although the value of φ, generally speaking, will not be constant in time. For graphical representation it is natural to use the variables $x = \mathrm{Re}(a_1 e^{-i\varphi})$, $y = \mathrm{Re}(a_2 e^{i\varphi})$, where ϕ is chosen each time to minimize the value $[\mathrm{Im}(a_1 e^{-i\varphi})]^2 + [\mathrm{Im}(a_2 e^{i\varphi})]^2$. As the third variable we use $z = |a_3|$.

Note that the same method is applicable to processing numerical data for the original Kirchhoff equations (2.7) if we convert the vectors defined by the voltages $\mathbf{U} = (U_1, U_2, U_3)^T$ to complex amplitudes as

$$a_{1,2,3} = \frac{1}{2}(U_{1,2,3} - i\omega_{1,2,3}^{-1}\dot{U}_{1,2,3}). \tag{2.22}$$

Figure 2.6a shows a portrait of the attractor, drawn using the results of the numerical integration of equations (2.17) in a mode corresponding to a slight shift of the pump frequency from the exact parametric resonance at $h = 5.962$, $\nu_1 = 1$, $\nu_2 = 4$, $\Delta = 0.3$, $\delta = 0$. (In the original system, this corresponds to the pumping

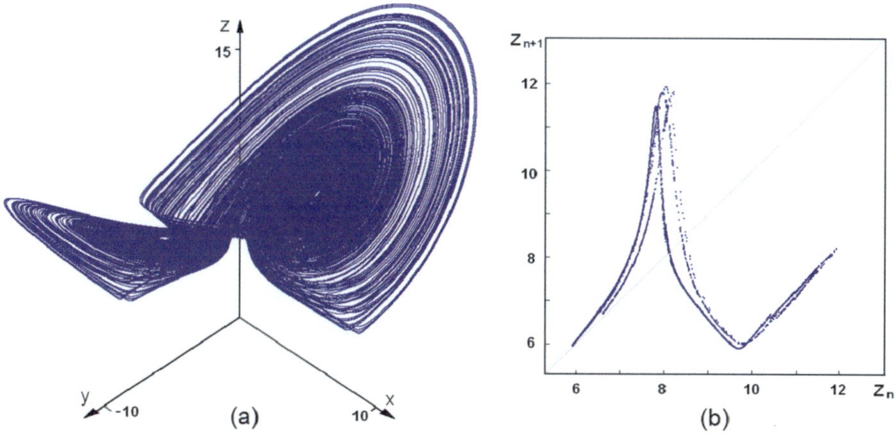

Fig. 2.6 Three-dimensional portrait of the attractor drawn using the techniques described in the text (**a**), and a plot of the map for successive maxima of the variable $z = |a_3|$ (**b**) for the system (2.17) at $h = 5.962$, $\nu_1 = 1$, $\nu_2 = 4$, $\Delta = 0.3$, $\delta = 0$

frequency of 31998.8 Hz at amplitude of 62.25 mV setting resistances $R_1 = 1\,\text{M}\Omega$, $R_2 = 250\,\text{k}\Omega$, $R_3 = 1\,\text{M}\Omega$, and the remaining components correspond to the caption of Fig. 2.2.) The phase portrait looks like Lorenz type attractor: it has two "wings," each of which corresponds to orbits spiraling from the central blank area, with successive transitions from one wing to the other, and the committed numbers of turns vary from once again chaotically.

Figure 2.6b shows a plot of the map, where the axes correspond to successive maxima of the variable $z = |A_3|$ achieved in the course of the temporal evolution of the system. The picture is significantly different from the map corresponding to the Lorenz type attractor in Fig. 2.4. Firstly, the graph looks composed not of a single curve, but a set of curves, i.e. it possesses a transverse fractal structure expressed much stronger than that for the Lorenz type attractor, where it is visually indistinguishable. Secondly, the curves representing the mapping manifest smooth quadratic maxima instead of a sharp tip at the top. In this connection, in this case one cannot speak of robust quasi-hyperbolic attractor. Rather, the properties of chaotic dynamics should be similar to attractors in the Hénon map [17] and Rössler model [37], which in mathematical works are interpreted within the concept of quasi-attractor [1, 39].

Figure 2.7 shows the dependence of six Lyapunov exponents of the model (2.17) on parameters of the frequency detuning Δ and δ. Note the symmetry of one and the other graph in Fig. 2.7; it occurs due to the fact that the equations transform into themselves under complex conjugation together with the sign change of Δ and δ.

Unlike the case of exact resonance, the graph for the senior Lyapunov exponent manifests dips (the regularity windows), which are also accompanied by tips or dips in the graphs of other exponents. As one can verify by the numerical integration of the equations, these windows correspond to the emergence of attracting limit cycles

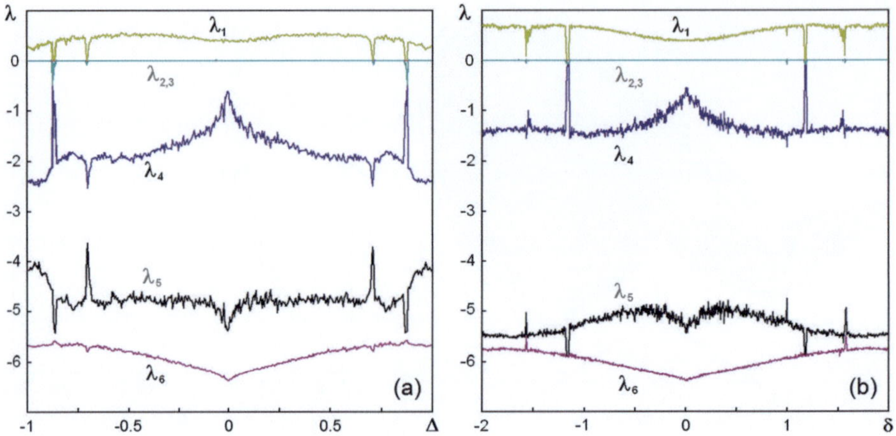

Fig. 2.7 Lyapunov exponents of the model (2.17) versus parameter Δ at $\delta = 0$ (**a**) and versus parameter δ at $\Delta = 0$. (**b**) Other parameters: $\nu_1 = 1$, $\nu_2 = 4$

of (2.17), i.e., instead of chaos periodic oscillations of the amplitude variables arise. It is this kind of the plots for Lyapunov exponents intrinsic particularly to one-dimensional maps with quadratic extremum and for many other dissipative systems, including the Hénon map and the Rössler model, which are associated with the concept of quasi-attractor [1, 17, 37, 39].

Let us turn to the chart of dynamical regimes in the parameter plane Δ, δ. The procedure consists of scanning parameter space area in two dimensions over the grid nodes with some small step. At each point about 10^3 iterations are performed for the Poincaré map defined for the system (2.17) via the section surface $S = |a_3| - h + \sqrt{\nu_1 \nu_2} = 0$ in the phase space (in the direction of passage of the orbits with decreasing S). According to the latest recorded data of iterations, the analysis is carried out for the presence or absence of a repetition period of the states in the Poincaré section from 1 to 14 (with some accepted small level of errors). When the periodicity is detected, the corresponding pixel in the chart is indicated by some color depending on the period, and the procedure proceeds with analyzing next point in the parameter plane. At the new point, as the initial conditions it is reasonable to assign the state resulting in the end of iterations at the previous point ("scanning with inheritance") to accelerate the convergence to the steady state dynamics.

In the center of Fig. 2.8 the parameter plane chart is shown for the system (2.17), and on the periphery the portraits of attractors are depicted corresponding to some representative points of the parameter plane (Δ, δ). Attractors in panels (a), (c), (h), (e) are limit cycles of the amplitude equations, i.e. relate to periodic modes for oscillations of amplitudes in the colored parameter plane areas. On the other hand, attractors in the diagrams (b), (g), (f) are chaotic corresponding to not colored regions where the periodicity is not detected. Diagram (f) relates to the origin on the chart where the Lorenz type attractor occurs, which was discussed in the previous section. Dark blue areas in "north-west" and "south-east" parts of the chart

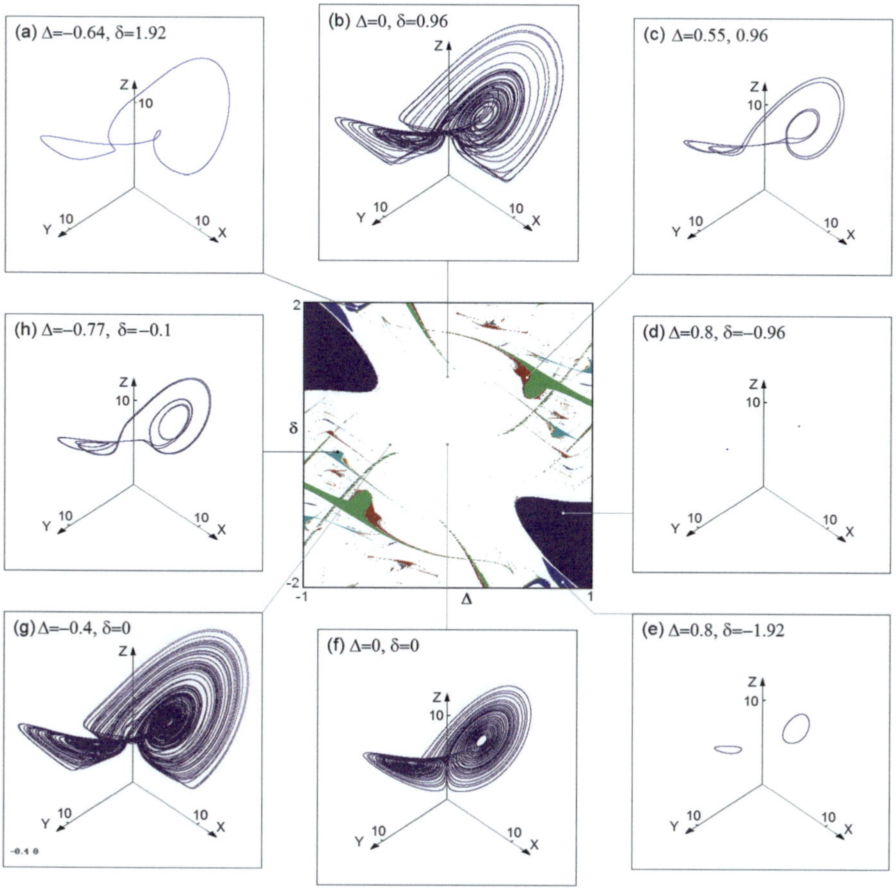

Fig. 2.8 A chart for the system (2.17) in the plane of parameters Δ and δ and portraits of attractors corresponding to representative points of the parameter plane. Other parameters: $h = 5.962$, $\nu_1 = 1$, $\nu_2 = 4$. For explanation of the method of the graphical presentation see in the text

correspond to the fixed point attractor of the Poincaré map [panel (d)] associated with a stable regime of stationary oscillations of constant amplitude in the initial equations.

Similar regimes are observed in numerical simulations of the original system of Kirchhoff's equations (2.7). Figure 2.9 shows attractors related to the system with the pump of amplitude of 62.25 mV, and the resistances $R_1 = 1\,\text{M}\Omega$, $R_2 = 250\,\text{k}\Omega$, $R_3 = 1\,\text{M}\Omega$. The frequency of the pump and inductance L_3 were selected to provide the parameters Δ and δ indicated in the inscriptions, and the other parameter values correspond to the caption of Fig. 2.2. The pictures clearly resemble those on the periphery of Fig. 2.8 with the difference that the trajectories on the attractors look

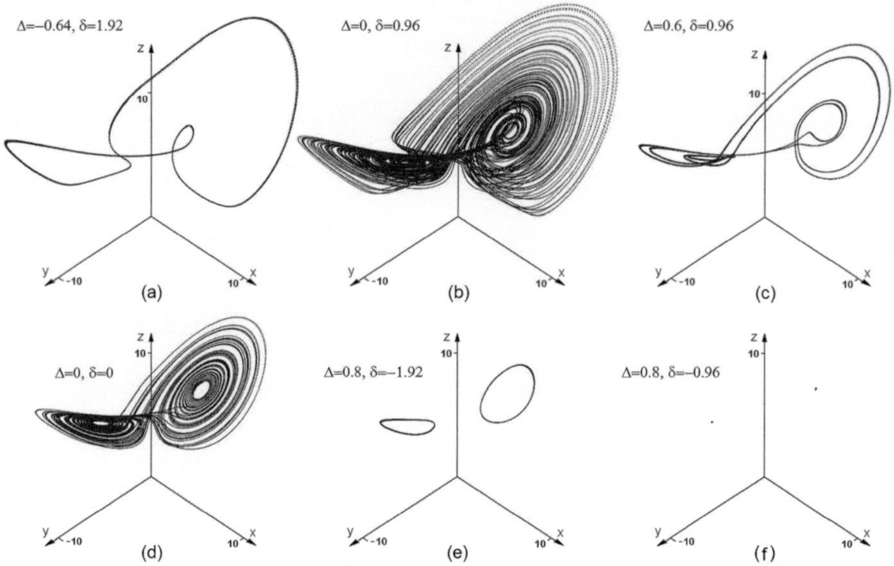

Fig. 2.9 Portraits of attractors of the original Kirchhoff's equations (2.3) at the resistances $R_1 = 1\,\text{M}\Omega$, $R_2 = 250\,\text{k}\Omega$, $R_3 = 1\,\text{M}\Omega$ and the pumping amplitude 62.25 mV. The pump frequency and the inductance L_3 were set to provide the parameter values Δ and δ: -0.64, 1.92 (**a**); 0, 0.96 (**b**); 0.6, 0.96 (**c**); 0, 0 (**d**); 0.8, -1.92 (**e**); 0.8, -0.96 (**f**). The remaining parameters correspond to the caption of Fig. 2.2

a little "fluffed," which is obviously connected with the presence of non-resonant components of relatively low amplitude contributing to the instantaneous values of dynamic variables, which were excluded in the amplitude equations.[2]

2.7 Conclusion

In this chapter we presented the analysis of the chaotic parametric oscillator composed of three resonant circuits with pumping provided by periodic variation in voltage on a quadratic nonlinear element. The methodological value of this model is that it allows a pure realization and exploration of the mechanism of parametric generation of chaos when two modes are excited due to the pump at the sum frequency, and the energy extraction is carried out by the mode at the difference frequency [34]. This circuit may serve as an analog simulator for systems of different nature where the same mechanism of parametric oscillations takes place, or, more widely, for systems where similar equations occur on some reason [19, 22].

[2] The lack of a perfect match in the parameters for Fig. 2.9 in comparison to Fig. 2.8 is due to the approximate nature of description in terms of slow amplitudes.

In the case of deviation from the exact parametric resonance, instead of the quasi-hyperbolic Lorenz type attractor, one obtains the dynamics although resembling those of Lorenz, but lacking robustness: by varying parameters destruction of chaos is possible with the emergence of regular motions. This conclusion has been illustrated particularly by the chart of the plane of the detuning frequency parameters, which gives a visual representation of location of areas of chaotic and regular dynamics.

References

1. Afraimovich, V.S.: Strange attractors and quasiattractors. Nonlinear and turbulent processes in physics. In: Proceedings of the Second International Workshop, Kiev, vol. 1, pp. 1133–1138 (1984)
2. Akhmanov, S.A., Khokhlov, R.V.: Parametric amplifiers and generators of light. Physics-Uspekhi **9**(2), 210–222 (1966)
3. Akulenko, L.D.: Parametric control of the oscillations and rotations of a physical pendulum (swing). Prikladnaya Matematika i Mekhanika **57**(2), 82–91 (1993)
4. Anishchenko, V.S.: Attractors of dynamical systems. Izvestija VUZ. Appl. Nonlinear Dynam. **5**(1), 109–127 (1997)
5. Banerjee, S., Yorke, J.A., Grebogi, C.: Robust chaos. Phys. Rev. Lett. **80**, 3049–3052 (1998)
6. Benettin, G., Galgani, L., Giorgilli, A., Strelcyn, J.-M.: Lyapunov characteristic exponents for smooth dynamical systems and for Hamiltonian systems: a method for computing all of them. Meccanica **15**, 9–20 (1980)
7. Bonatti, C., Diaz, L.J., Viana, M.: Dynamics beyond uniform hyperbolicity. A global geometric and probabilistic perspective. In: Encyclopedia of Mathematical Sciences, vol. 102. Springer, Berlin, Heidelberg, New York (2005)
8. Chen, H.K., Lee, C.I.: Anti-control of chaos in rigid body motion. Chaos Solitons Fractals **21**, 957–965 (2004)
9. Cuomo, K.M., Oppenheim, A.V.: Circuit implementation of synchronized chaos with applications to communications. Phys. Rev. Lett. **71**, 65–68 (1993)
10. Doroshin, A.V.: Modeling of chaotic motion of gyrostats in resistant environment on the base of dynamical systems with strange attractors. Commun. Nonlinear Sci. Numer. Simul. **16**, 3188–3202 (2011)
11. Elhadj, Z., Sprott, J.C.: Robust Chaos and its Applications. World Scientific, Singapore (2011)
12. Fowler, A.C., Gibbon, J.D., McGuinness, M.J.: The complex Lorenz equations. Physica D: Nonlinear Phenom. **4**(2), 139–163 (1982)
13. Gibbon, J.D., McGuinness, M.J.: The real and complex Lorenz equations in rotating fluids and lasers. Physica D: Nonlinear Phenom. **5**(1), 108–122 (1982)
14. Glukhovskii, A.B.: Nonlinear systems that are superpositions of gyrostats. Sov. Phys. Doklady **27**, 823–827 (1982)
15. Haken, H.: Analogy between higher instabilities in fluids and lasers. Phys. Lett. A **53**(1), 77–78 (1975)
16. Hemail, N.: Strange attractors in brushless DC motor. IEEE Trans. Circuits Syst.-I: Fundam. Theory Appl. **41**(1), 40–45 (1994)
17. Hénon, M.: A two-dimensional mapping with a strange attractor. Commun. Math. Phys. **50**, 69–77 (1976)
18. Kolář, M., Gumbs, G.: Theory for the experimental observation of chaos in a rotating waterwheel. Phys. Rev. A **45**, 626–637 (1992)

19. Kozlov, V.V.: On the problem of fall of a rigid body in a resisting medium. Moscow Univ. Mech. Bull. **45**(1), 30–35 (1990)
20. Kuznetsov, S.P.: Dynamical Chaos, 2nd edn. Moscow, Fizmatlit (2006)
21. Kuznetsov, S.P.: Dynamical chaos and uniformly hyperbolic attractors: from mathematics to physics. Physics-Uspekhi **54**(2), 119–144 (2011)
22. Kuznetsov, S.P.: Plate falling in a fluid: regular and chaotic dynamics of finite-dimensional models. Regul. Chaot. Dynam. **20**(3), 345–382 (2015)
23. Kuznetsov, S.P.: Parametric chaos generator operating on a varactor diode with the instability limitation decay mechanism. Tech. Phys. **61**(3), 436–445 (2016)
24. Letellier, C., Aguirre, L.A., Maquet, J., Lefebvre, B.: Analogy between a 10D model for nonlinear wave–wave interaction in a plasma and the 3D Lorenz dynamics. Physica D: Nonlinear Phenom. **179**(1), 33–52 (2003)
25. Liu, Y., Yang, Q., Pang, G.: A hyperchaotic system from the Rabinovich system. J. Comput. Appl. Math. **234**(1), 101–113 (2010)
26. Llibre, J., Messias, M., da Silva, P.R.: On the global dynamics of the Rabinovich system. J. Phys. A: Math. Theor. **41**(27), 275210 (2008)
27. Lorenz, E.N.: Deterministic nonperiodic flow. J. Atmos. Sci. **20**(2), 130–141 (1963)
28. Louisell, W.H.: Coupled Mode and Parametric Electronics. Wiley, New York (1960)
29. Mahmoud, G.M., Ahmed, M.E., Mahmoud, E.E.: Analysis of hyperchaotic complex Lorenz systems. Int. J. Mod. Phys. C **19**, 1477–1494 (2008)
30. Oraevskii, A.N.: Masers, lasers, and strange attractors. Quant. Electron. **11**(1), 71–78 (1981)
31. Oraevsky, A.N.: Dynamics of single-mode lasers and dynamical chaos. Izvestija VUZ. Appl. Nonlinear Dynam. **4**, 3–13 (1996)
32. Ostrovskii, L.A., Papilova, I.A., Sutin, A.M.: Parametric ultrasound generator. Sov. Phys. JETP Lett. **15**, 322–323 (1972)
33. Peters, F., Lobry, L., Lemaire, E.: Experimental observation of Lorenz chaos in the Quincke rotor dynamics. Chaos: Interdiscip. J. Nonlinear Sci. **15**, 013102 (2005)
34. Pikovski, A.S., Rabinovich, M.I., Trakhtengerts, V.Y.: Appearance of chaos at decay saturation of parametric instability. Sov. Phys. JETP **47**, 715–719 (1978)
35. Poland, D.: Cooperative catalysis and chemical chaos: a chemical model for the Lorenz equations. Physica D: Nonlinear Phenom. **65**, 86–99 (1993)
36. Rabinovich, M.I., Fabrikant, A.L.: Stochastic wave self-modulation in nonequilibrium media. Sov. Phys. JETP **50**, 311–317 (1979)
37. Rössler, O.E.: Continuous chaos: four prototype equations. Ann. N. Y. Acad. Sci. **316**, 376–392 (1979)
38. Rucklidge, A.M.: Chaos in magnetoconvection. Nonlinearity **7**, 1565–1591 (1994)
39. Shilnikov, L.P.: Bifurcations and strange attractors. Vestnik Nizhegorodskogo Universiteta **4**(2), 364–366 (2011)
40. Sparrow, C.: The Lorenz Equations: Bifurcations, Chaos, and Strange Attractors. Springer, NY, Heidelberg, Berlin (1982)
41. Tucker, W.: A rigorous ODE solver and Smale's 14th problem. Comp. Math. **2**, 53–117 (2002)
42. Vyshkind, S.Ya., Rabinovich, M.I.: The phase stochastization mechanism and the structure of wave turbulence in dissipative media. Sov. Phys. JETP **44**, 292–299 (1976)
43. Wang, P.K.C., Masui, K.: Intermittent phase unlocking in a resonant three-wave interaction with parametric excitation. Phys. Lett. A **81**(2), 97–101 (1981)

Chapter 3
Time Rescaling of Lyapunov Exponents

Yakov Pesin, Agnieszka Zelerowicz, and Yun Zhao

3.1 Introduction

Lyapunov exponents are classical characteristics of instability of trajectories and in the presence of nonzero Lyapunov exponents the system is expected to exhibit a certain level of chaotic behavior. This is indeed the case if the system preserves a smooth measure or more generally a Sinai–Ruelle–Bowen (SRB) measure. This is one of the manifestations of the classical non-uniform hyperbolicity theory (see [1]). For system preserving SRB measures the Kolmogorov–Sinai (metric) entropy of the system can be computed using Pesin's entropy formula: the entropy is the mean over the phase space of the system of the sum of positive Lyapunov exponents. In particular, the entropy of the system is positive.

On the other hand, if a measure invariant under the system has all its Lyapunov exponent zero, then by the Margulis–Ruelle inequality, the entropy of the measure is zero. When entropy of the measure is positive it characterizes the complexity of the system (with respect to this measure) but in the case when entropy is zero little if any meaningful information about the complexity can be recovered.

This observation is crucial, since there are many examples of physical systems which exhibit sub-exponential instability of trajectories and hence, have zero Lyapunov exponents with respect to some "natural" invariant measures. Such systems include some models with sequential dynamics studied by Rabinovich [9]

Y. Pesin (✉) • A. Zelerowicz
Department of Mathematics, Pennsylvania State University, University Park, PA 16802, USA
e-mail: pesin@math.psu.edu; axz157@psu.edu

Y. Zhao
Department of Mathematics, Soochow University, Suzhou, Jiangsu 215006, People's Republic of China
e-mail: zhaoyun@suda.edu.cn

I.S. Aranson et al. (eds.), *Advances in Dynamics, Patterns, Cognition*, Nonlinear Systems and Complexity 20, DOI 10.1007/978-3-319-53673-6_3

in connection to his work on dynamics of neural and cognitive systems, see [10, 11]. It also includes a model of weak transient chaos considered by V. Afraimovich and A. Neiman (see their paper in this volume).

To study models with sub-exponential instability of trajectories, one may introduce a more appropriate sub-exponential scale in which Lyapunov exponents should be computed. This is a "better adapted" or "internal" scale of the system. The classical notion of Lyapunov exponent is based on the exponential scale and if it happens to be the "internal" scale of the system which has positive entropy, one obtains positive Lyapunov exponents. Otherwise, one should switch to a different scale (e.g., the polynomial scale) with respect to which the Lyapunov exponents and entropy may become positive. This would allow one to evaluate the level of complexity of the system.

Finding an internal scale for a given system or proving that it exists may be difficult if at all possible. However, if such a scale is found one hopes to use the corresponding *scaled Lyapunov exponents* to recover at least some part of non-uniform hyperbolicity theory. In particular, one hopes to establish a version of the Margulis-Ruelle inequality (or in some cases even Pesin's entropy formula) to connect an appropriately rescaled metric entropy with scaled Lyapunov exponents.

In [8] using the general Carathéodory construction as described in [7], the concept of scaled entropy was introduced in both topological and metric settings. While the standard approach to topological entropy defines it as the exponential growth rate of the number of periodic points, the definition of the scaled topological entropy allows asymptotic rates of the general form $e^{\alpha a(n)}$, where $\alpha > 0$ is a parameter and $a(n)$ is a *scaled sequence*. Similar idea was used in [8] in defining scaled metric entropy.

Measures with zero Lyapunov exponents often appear as infinite invariant measures for dynamical systems on compact phase spaces. A classical example is the Manneville–Pomeau map $x \to x+x^{1+\alpha}(\pmod 1)$, where α controls the degree of intermittency at the neutral fixed point. If $\alpha \in (0, 1)$, then the systems preserve a finite measure which is absolutely continuous with respect to the Lebesgue measure. However, for $\alpha > 1$ this measure becomes infinite and the corresponding Lyapunov exponents are zero. After the rescaling $t \to t^{\alpha}$ the scaled Lyapunov exponent becomes positive and one recovers the "rescaled" version of Pesin's entropy formula as well, see [3, 4].

The goal of this article is to examine the dependence of Lyapunov exponents on the scale in which they are computed and to outline the abstract theory of scaled Lyapunov exponents. We will do this in the general setting of cocycles over dynamical systems.

3.2 Scaled Lyapunov Exponents for Cocycles

In this section, we describe the general theory of scaled Lyapunov exponents for cocycles over dynamical systems. Consider an invertible measurable transformation f of a measure space X.

3.2.1 Linear Multiplicative Cocycles

Let $f : X \to X$ be an invertible measurable transformation of a measure space X. We call the function $\mathcal{A} : X \times \mathbb{Z} \to GL(d, \mathbb{R})$ a *linear multiplicative cocycle over* f or simply a *cocycle* if it has the following properties:

(1) $\mathcal{A}(x, 0) = Id$ for every $x \in X$ and $\mathcal{A}(x, m+n) = \mathcal{A}(f^n(x), m)\mathcal{A}(x, n)$ for all $m, n \in \mathbb{Z}$;
(2) the function $\mathcal{A}(\cdot, n) : X \to GL(d, \mathbb{R})$ is measurable for each $n \in \mathbb{Z}$.

Every cocycle is generated by a measurable function $A : X \to GL(d, \mathbb{R})$, which is called the *generator*. In fact, every such function determines a cocycle by the formula

$$\mathcal{A}(x, n) = \begin{cases} A(f^{n-1}(x)) \cdots A(f(x))A(x) & \text{if } n > 0, \\ Id & \text{if } m = 0, \\ A(f^n(x))^{-1} \cdots A(f^{-2}(x))^{-1} A(f^{-1}x)^{-1} & \text{if } n < 0. \end{cases}$$

On the other hand, a cocycle $\mathcal{A}$ is generated by the matrix function $A = \mathcal{A}(\cdot, 1)$.

A simpler way to describe a cocycle $\mathcal{A}$ over f is by considering a *linear extension* $F : X \times \mathbb{R}^d \to X \times \mathbb{R}^d$ of f that is induced by the cocycle (for simplicity of presentation we only consider the trivial bundle in this paper). It is given by

$$F(x, v) = (f(x), A(x)v).$$

If $\pi : X \times \mathbb{R}^d \to X$ is the projection defined by $\pi(x, v) = x$, then it is easy to see that $\pi \circ F = f \circ \pi$.

If f is a differentiable map of a compact phase space M, then f generates a *differential* cocycle $\mathcal{A}(x, n)$ over f whose generator is $A(x) = df_x$ and it acts on the tangent bundle TM of M. Therefore, the results in this and next sections are applicable to smooth maps.

3.2.2 Definition of Scaled Lyapunov Exponents

Let $\mathcal{A}$ be a cocycle over an invertible measurable transformation f of a measure space X. Using the notion of the standard Lyapunov exponent (see [1] for details), we introduce the notion of the scaled Lyapunov exponent for the cocycle.

Given $x \in X$, we call a sequence of positive numbers $\mathbf{a} = \{a(x,n)\}_{n\geq 1}$ a *scaled sequence* if

1. it is monotonically increasing to infinity, e.g., $a(n) = n^\alpha$, $\log n$, etc.;
2. for each n the function $a(x,n)$ is Borel;
3. $a(f(x),n) = a(x,n+1)$, in other words the function $a(x,n)$ depends on the entire trajectory of the point x.

Given a point $x \in X$, a scaled sequence $\mathbf{a} = \{a(x,n)\}$ and a vector $v \in \mathbb{R}^d$, we call the following quantity

$$\chi(x,v,\mathbf{a}) = \limsup_{n\to+\infty} \frac{1}{a(x,n)} \log \|\mathcal{A}(x,n)(v)\|$$

the *scaled Lyapunov exponent* of (x,v) (with respect to the scaled sequence $\mathbf{a}$ and the cocycle $\mathcal{A}$). With the convention that $\log 0 = -\infty$ this extends the definition of the *standard Lyapunov exponent* corresponding to the scaled sequence $a(n) = n$ for each n.

3.2.3 Choices of Scaled Sequences

The above definition of the scaled Lyapunov exponent allows any scaled sequence $\mathbf{a} = \{a(x,n)\}_{n\geq 1}$, which grows slower than n. Depending on the choice of $\mathbf{a}$ the value of the scaled Lyapunov exponent can be positive, negative, or zero. It can also be $\pm\infty$.

Given a point $x \in X$, consider the sequence of positive numbers

$$\mathbf{b} = \left\{ b(x,n) = \max_{0\leq k\leq n} |\log \|\mathcal{A}(x,k)\||\right\}.$$

This sequence is non-decreasing and if it is bounded, then for any scaled sequence $\mathbf{a}$ the corresponding values of the Lyapunov exponent are all zero. However, if the sequence $\mathbf{b}$ is unbounded, then the corresponding values of the Lyapunov exponent are all finite (except for $v = 0$).

Consider the collection of vectors for which $\chi(x,v,\mathbf{b}) = 0$. By the properties of the scaled Lyapunov exponents described in the next section, this collection is a linear subspace of $\mathbb{R}^d$, which we denote by $\mathbb{R}^{d_1}$. Consider the "restricted" cocycle $\mathcal{A}_1(x,n)$ with values in $GL(d_1,\mathbb{R})$. Repeating the above argument we find the non-decreasing sequence of positive numbers $\mathbf{b_1}$ and if this sequence is unbounded then we can use it to rescale the Lyapunov exponents for vectors $v \in \mathbb{R}^{d_1}$. The above procedure produces a filtration

$$\mathbb{R}^d = \mathbb{R}^{d_0} \supset \mathbb{R}^{d_1} \supset \cdots \supset \mathbb{R}^{d_k},$$

a collection of scaled sequences $\mathbf{b_i}$ and the corresponding collection of scaled Lyapunov exponents $\chi(x, v, \mathbf{b_i})$, $i = 1, \ldots, k-1$ such that $\chi(x, v, \mathbf{b_i}) \neq 0$ for every $v \in \mathbb{R}^{d_{i-1}} \setminus \mathbb{R}^{d_i}$ and $i = 1, \ldots, k-1$. If the sequence $\mathbf{b_k}$ is unbounded, then $\mathbb{R}^{d_k} = 0$.

3.3 Basic Properties of Scaled Lyapunov Exponents

The function χ has the following basic properties of a Lyapunov exponent, see [1, Chap. 2] or [2, Proposition 1] for detailed proofs.

Proposition 3.1 *For each $x \in X$, $v, w \in \mathbb{R}^d$ and $c \in \mathbb{R} \setminus \{0\}$,*

(1) $\chi(x, cv, \mathbf{a}) = \chi(x, v, \mathbf{a})$;
(2) $\chi(x, 0, \mathbf{a}) = -\infty$;
(3) $\chi(x, v + w, \mathbf{a}) \leq \max\{\chi(x, v, \mathbf{a}), \chi(x, w, \mathbf{a})\}$.

It follows from the abstract theory of Lyapunov exponents (see [1, Theorem 2.1]) that:

1. $\chi(x, v + w, \mathbf{a}) = \max\{\chi(x, v, \mathbf{a}), \chi(x, w, \mathbf{a})\}$ for any $v, w \in \mathbb{R}^d$ whenever $\chi(x, v, \mathbf{a}) \neq \chi(x, w, \mathbf{a})$;
2. if for some nonzero vectors $v_1, \ldots, v_m \in \mathbb{R}^d$, the numbers $\chi(x, v_1, \mathbf{a}), \ldots, \chi(x, v_m, \mathbf{a})$ are distinct, then these vectors are linearly independent;
3. the function $\chi(x, \cdot, \mathbf{a})$ attains only finitely many values on $\mathbb{R}^d \setminus \{0\}$, which we denote by $\chi_1(x, \mathbf{a}) < \cdots < \chi_{s(x,\mathbf{a})}(x, \mathbf{a})$, where $s(x, \mathbf{a}) \leq d$; note that, in general, χ_1 may be $-\infty$ and $\chi_{s(x,\mathbf{a})}$ may be $+\infty$.

Further we denote by $\mathcal{V}_\mathbf{a}(x)$ the *filtration* of $\mathbb{R}^d$ associated with $\chi(x, \cdot, \mathbf{a})$:

$$\{0\} = V_\mathbf{a}^0(x) \subsetneq V_\mathbf{a}^1(x) \subsetneq \cdots \subsetneq V_\mathbf{a}^{s(x,\mathbf{a})}(x) = \mathbb{R}^d,$$

where $V_\mathbf{a}^i(x) = \left\{v \in \mathbb{R}^d \colon \chi(x, v, \mathbf{a}) \leq \chi_i(x, \mathbf{a})\right\}$ for $i = 1, \ldots, s(x, \mathbf{a})$. It is easy to see that

$$\chi_i(x, v, \mathbf{a}) = \chi_i(x, \mathbf{a}) \text{ for all } v \in V_\mathbf{a}^i(x) \setminus V_\mathbf{a}^{i-1}(x).$$

The number $k_i(x, \mathbf{a}) = \dim V_\mathbf{a}^i(x) - \dim V_\mathbf{a}^{i-1}(x)$ is the *multiplicity* of the value $\chi_i(x, \mathbf{a})$. We have that

$$\sum_{i=1}^{s(x,\mathbf{a})} k_i(x, \mathbf{a}) = d.$$

The collection of pairs

$$\operatorname{Sp}\chi_{\mathbf{a}}(x) = \{(\chi_i(x,\mathbf{a}), k_i(x,\mathbf{a})) \colon 1 \leq i \leq s(x,\mathbf{a})\}$$

forms the *Lyapunov spectrum* of the scaled Lyapunov exponent χ at the point $x \in X$ with respect to the scaled sequence $\mathbf{a} = \{a(x,n)\}$.

Definition 3.1 A basis $\mathbf{v} = (v_1, \cdots, v_d)$ of $\mathbb{R}^d$ is said to be *subordinate* to the filtration $\mathcal{V}_{\mathbf{a}}(x)$ if for every $1 \leq i \leq s(x,\mathbf{a})$ there exists a basis of $V^i_{\mathbf{a}}(x)$ composed of d_i vectors from $(v_1, \ldots, v_d)$. A subordinate basis $\mathbf{v}$ is *ordered* if for every $1 \leq i \leq s(x,\mathbf{a})$ the vectors $v_1, \ldots, v_{d_i}$ form a basis of $V^i_{\mathbf{a}}(x)$.

Using the same arguments in [1, Chap. 12], we can find that there always exists a subordinate basis for a filtration.

Proposition 3.2 *Given a filtration $\mathcal{V}_{\mathbf{a}}(x)$ of $\mathbb{R}^d$, there exists a basis $\mathbf{w} = (w_1, \ldots, w_d)$ of $\mathbb{R}^d$ such that*

$$\inf\left\{\sum_{j=1}^{d} \chi(x, v_j, \mathbf{a}) \colon \mathbf{v} = (v_1, \ldots, v_d) \text{ is a basis of } \mathbb{R}^d\right\} = \sum_{j=1}^{d} \chi(x, w_j, \mathbf{a}).$$

3.4 The Lyapunov and Perron Regularity Coefficients

Consider the dual matrix $B(x) = (A(x)^*)^{-1}$ at each point $x \in X$. Given a scaled sequence $\mathbf{a} = \{a(x,n)\}_{n\geq 1}$ and a point $(x, v^*) \in X \times \mathbb{R}^d$, the *dual scaled Lyapunov exponent* is given by the formula:

$$\widetilde{\chi}(x, v^*, \mathbf{a}) = \limsup_{m\to+\infty} \frac{1}{a(x,m)} \log \|\mathcal{B}(x,m)v^*\|$$

where $\mathcal{B}(x,m) = B(f^{m-1}(x)) \cdots B(f(x))B(x)$ for $m > 0$. In fact, choose dual bases $(v_1, \ldots, v_d)$ and $(v_1^*, \ldots, v_d^*)$, i.e., $\langle v_i, v_j^* \rangle = \delta_{ij}$ for each i and j (here δ_{ij} is the Kronecker symbol), and set $v_{i,m} = \mathcal{A}(x,m)v_i$ and $v_{i,m}^* = \mathcal{B}(x,m)v_i^*$. For each $m \in \mathbb{N}$ we have

$$\langle v_{i,m}, v_{i,m}^* \rangle = \langle \mathcal{A}(x,m)v_i, (\mathcal{A}(x,m)^*)^{-1}v_i^* \rangle = 1.$$

Hence, $1 \leq \|\mathcal{A}(x,m)v_i\| \cdot \|\mathcal{B}(x,m)v_i^*\|$ and the exponents χ and $\widetilde{\chi}$ are *dual* at the point x, i.e., $\chi(x, v_i, \mathbf{a}) + \widetilde{\chi}(x, v_i^*, \mathbf{a}) \geq 0$ for each $1 \leq i \leq d$.

Arguing as above one can show that for each $v^* \in \mathbb{R}^d$, the Lyapunov exponent $\widetilde{\chi}$ can only attain finitely many values on $\mathbb{R}^d \setminus \{0\}$. We denote them by $\widetilde{\chi}_{r(x,\mathbf{a})}(x,\mathbf{a}) < \cdots < \widetilde{\chi}_1(x,\mathbf{a})$ for some integer $r(x,\mathbf{a}) \leq d$. Let $\widetilde{\mathcal{V}}_{\mathbf{a}}(x) = \{\widetilde{V}^i_{\mathbf{a}}(x) : i = 1, \ldots, r(x,\mathbf{a})\}$ be the filtration associated with $\widetilde{\chi}$. Note that, in general, $\widetilde{\chi}_{r(x,\mathbf{a})}$ may be $-\infty$ and $\widetilde{\chi}_1$ may be $+\infty$ and therefore, from now on we assume that the sums $\chi_1 + \widetilde{\chi}_1$ and $\chi_{s(x,\mathbf{a})} + \widetilde{\chi}_{r(x,\mathbf{a})}$ are well defined, that is $|\chi_1|$ and $\widetilde{\chi}_1$ are not both $+\infty$ and similar for $\chi_{s(x,\mathbf{a})}$ and $|\widetilde{\chi}_{r(x,\mathbf{a})}|$.

We call the quantity

$$\gamma_{\mathbf{a}}(x, \chi, \widetilde{\chi}) = \min\max\left\{\chi(x, v_i, \mathbf{a}) + \widetilde{\chi}(x, v_i^*, \mathbf{a}) \colon 1 \le i \le d\right\}$$

the *regularity coefficient* of the pair of scaled Lyapunov exponents χ and $\widetilde{\chi}$ at the point x (with respect to the scaled sequence $\mathbf{a} = \{a(x, n)\}$). Here the minimum is taken over all pairs of dual bases $(v_1, \ldots, v_d)$ and $(v_1^*, \ldots, v_d^*)$ of $\mathbb{R}^d$. We say that a point $x \in X$ is *regular* (with respect to the pair of scaled Lyapunov exponents $(\chi, \widetilde{\chi})$) if $\gamma_{\mathbf{a}}(x, \chi, \widetilde{\chi}) = 0$.

Now we let

$$\chi_1'(x, \mathbf{a}) \le \cdots \le \chi_d'(x, \mathbf{a}) \text{ and } \widetilde{\chi}_1'(x, \mathbf{a}) \ge \cdots \ge \widetilde{\chi}_d'(x, \mathbf{a})$$

be, respectively, the values of χ and $\widetilde{\chi}$ at the point x counted with their multiplicities. Define the *Perron coefficient* of the pair χ and $\widetilde{\chi}$ at $x \in X$ (with respect to the scaled sequence $\mathbf{a} = \{a(x, n)\}$) by

$$\pi_{\mathbf{a}}(x, \chi, \widetilde{\chi}) = \max\{\chi_i'(x, \mathbf{a}) + \widetilde{\chi}_i'(x, \mathbf{a}) \colon 1 \le i \le d\}.$$

The following theorem can be proven by the same arguments as in the proof of Theorem 2.8 in [1].

Theorem 4.1 *For a point $x \in X$, if $\chi_1 + \widetilde{\chi}_1$ and $\chi_{s(x,\mathbf{a})} + \widetilde{\chi}_{r(x,\mathbf{a})}$ are well defined, then*

$$0 \le \pi_{\mathbf{a}}(x, \chi, \widetilde{\chi}) \le \gamma_{\mathbf{a}}(x, \chi, \widetilde{\chi}) \le d\pi_{\mathbf{a}}(x, \chi, \widetilde{\chi}).$$

It follows that a point $x \in X$ is regular if and only if $\pi_{\mathbf{a}}(x, \chi, \widetilde{\chi}) = 0$ and also if and only if $\chi_i'(x, \mathbf{a}) = -\widetilde{\chi}_i'(x, \mathbf{a})$ for $i = 1, \ldots, d$.

Theorem 4.2 *If a point $x \in X$ is regular, then the filtrations $\mathcal{V}_{\mathbf{a}}(x)$ and $\widetilde{\mathcal{V}}_{\mathbf{a}}(x)$ are orthogonal, that is, $s(x, \mathbf{a}) = r(x, \mathbf{a}) := s$, $\dim V_{\mathbf{a}}^i(x) + \dim \widetilde{V}_{\mathbf{a}}^{s-i}(x) = d$ and $\langle v, v^* \rangle = 0$ for every $v \in V_{\mathbf{a}}^i(x)$ and $v^* \in \widetilde{V}_{\mathbf{a}}^{s-i}(x)$.*

Reversing the time, we can introduce the scaled Lyapunov exponents for negative time. The above result provides a basis to study Lyapunov–Perron regularity for scaled Lyapunov exponents. The ultimate goal is to find out whether various regularity criteria that hold in the case of the standard scale can be extended to general scaled sequences and to what extent the Multiplicative Ergodic theorem may hold for scaled Lyapunov exponents.

3.5 Examples

In this section we present two examples that illustrate that there are systems for which the Lyapunov exponents can be rescaled to achieve non-zero values but that in general this should not be expected.

3.5.1 *Existence of Scaled Lyapunov Exponents*

Consider the elliptic matrix $A = \begin{bmatrix} 1 & 1 \\ 0 & 1 \end{bmatrix}$ whose only eigenvalue is 1. Hence, the two Lyapunov exponents for the constant cocycle generated by A are zero. In addition, the vector $v = \begin{bmatrix} 1 \\ 0 \end{bmatrix}$ is the eigenvector for A, so that $\|A^n v\| = 1$ for all integers n where $\|\cdot\|$ is the Euclidean norm. Consequently, the scaled Lyapunov exponent, $\chi(v, \mathbf{a})$ is zero for any scaled sequence $\mathbf{a}$. On the other hand, taking the vector $w = \begin{bmatrix} 0 \\ 1 \end{bmatrix}$ we observe that $A^n w = \begin{bmatrix} n \\ 1 \end{bmatrix}$ so that the norm $||A^n w||$ grows linearly with $|n|$. Choosing $a(n) = \log n$ we obtain that

$$\chi(w, \mathbf{a}) = \lim_{n \to +\infty} \frac{1}{a(n)} \log \|\mathcal{A}(n)(w)\| = 1,$$

and the limit exists. Similar observation can be made for any non-diagonalizable elliptic matrix. In fact, the following is true:

Proposition 5.1 ([6]) *Suppose all eigenvalues of a linear map $A : \mathbb{R}^n \to \mathbb{R}^n$ have absolute value one. Then there exist an invariant subspace $C = C(A) \subset \mathbb{R}^n$ and a norm in $\mathbb{R}^n$ such that A acts in C as an isometry and for every vector $v \in \mathbb{R}^n \setminus C$ the norm $\|A^n v\|$ grows polynomially as $|n| \to \infty$.*

To see how nonzero scaled Lyapunov exponents can be obtained consider a diffeomorphism $f : S^2 \to S^2$ of the unit sphere in $\mathbb{R}^3$ which fixes the South and North poles and moves every other point along the meridian from the North pole toward the South pole. In the spherical coordinates we may write $f(\theta, \varphi, r) = (f_1(\theta, \varphi), \varphi, r)$, where for every fixed angle $\varphi \in [-\pi, \pi)$, the function $g_\varphi(\theta) := f_1(\theta, \varphi)$ satisfies: (1) $g_\varphi(\frac{\pi}{2}) = \frac{\pi}{2}$ and $g_\varphi(-\frac{\pi}{2}) = -\frac{\pi}{2}$; (2) $g_\varphi(\theta) < \theta$ for all $-\frac{\pi}{2} < \theta < \frac{\pi}{2}$.

We have that $f^n(\theta, \varphi, r) = (f_n(\theta, \varphi), \varphi, r)$, where $f_n(\theta, \varphi) = g^n_\varphi(\theta)$. Consequently,

$$df^n = \begin{bmatrix} \frac{dg^n_\varphi}{d\theta} & \frac{df_n}{d\varphi} & 0 \\ 0 & 1 & 0 \\ 0 & 0 & 1 \end{bmatrix}. \tag{3.1}$$

Observe that:

- the direction spanned by the vector $v^s = \begin{bmatrix} 1 \\ 0 \\ 0 \end{bmatrix}$ is df-invariant,
- the subspace spanned by $w^s = \begin{bmatrix} 0 \\ 1 \\ 0 \end{bmatrix}$ is not df-invariant,
- the vector $v^0 = \begin{bmatrix} 0 \\ 0 \\ 1 \end{bmatrix}$ is df-invariant, that is $df^n v^0 = v^0$ for all n.

Consider the special case, when $g_\varphi(\theta)$ has the following form on the interval $[-\frac{\pi}{2}, -\frac{\pi}{2} + \delta]$ for some small $\delta > 0$:

$$g_\varphi(\theta) = \begin{cases} \left(1 + (\theta + \frac{\pi}{2})^{-\frac{1}{\alpha}}\right)^{-\alpha} - \frac{\pi}{2} & \text{for } \theta \neq -\frac{\pi}{2} \\ g_\varphi(-\frac{\pi}{2}) = -\frac{\pi}{2} & \text{for } \theta = -\frac{\pi}{2}, \end{cases}$$

where $\alpha = \alpha(\varphi) > 0$ is a smooth function such that $\alpha'(\varphi) > 0$.

One can see that $g'_\varphi(-\frac{\pi}{2}) = 1$ and hence all Lyapunov exponents (in the standard scale) for all points (except maybe for the North pole) are zero. On the other hand, for $-\frac{\pi}{2} < \theta < -\frac{\pi}{2} + \delta$ we can see that

$$g^n_\varphi(\theta) = \left(n + \left(\theta + \frac{\pi}{2}\right)^{-\frac{1}{\alpha}}\right)^{-\alpha} - \frac{\pi}{2}.$$

Having an explicit formula for $g^n_\varphi(\theta)$ allows us to compute $\frac{dg^n_\varphi}{d\theta}$ and $\frac{df_n}{d\varphi}$ in (3.1) and conclude that:

- The norm $||df^n v^s||$ decays polynomially with exponent $-(1 + \alpha(\varphi))$ as n grows to infinity.
- The norm $||df^n w^s||$ decays polynomially with exponent $-\alpha(\varphi)$ as n grows to infinity.
- The angle $\angle(df^n(w^s), w^s)$ goes to zero as n grows to infinity.

Consequently taking the scaled sequence $\mathbf{a} = \log n$, we obtain the following values of the corresponding scaled Lyapunov exponents:

- $\chi(x, v^s, \mathbf{a}) = -(1 + \alpha(\varphi(x)))$
- $\chi(x, w^s, \mathbf{a}) = -\alpha(\varphi(x))$
- $\chi(x, v^0, \mathbf{a}) = 0.$

We therefore obtained two distinct non-zero values of the scaled Lyapunov exponents that vary with x.

3.5.2 Non-existence of the Scaled Limit

In general, one cannot expect the existence of the limit in the definition of scaled Lyapunov exponents. To see this consider a diffeomorphism f for which there exists an invariant family of one dimensional subspaces $E(x)$. Denoting by v_x a unit vector in $E(x)$ we have that

$$\log ||df^n(x)(v_x)|| = \sum_{k=0}^{n-1} \log ||df(f^k(x))(v_{f^k(x)})||.$$

Considering a function $\varphi(x) := \log ||df(x)(v_x)||$ we can rewrite the above sum as $\sum_{k=0}^{n-1} \varphi \circ f^k(x)$. Assume that the standard Lyapunov exponent, $\chi(x, v_x)$ is zero almost everywhere with respect to some ergodic measure μ. It means that

$$\lim_{n\to+\infty} \frac{1}{n} \sum_{k=0}^{n-1} \varphi \circ f^k = 0 \; \mu - a.e. \tag{3.2}$$

In that case one cannot obtain a nonzero finite limit by rescaling as the following result shows.

Theorem 5.1 ([5, 12]) *Let $(X, \mathfrak{B}, \mu)$ be a probability space, $f : X \to X$ a measure preserving ergodic transformation, and let $\varphi \in L^1(X, \mathfrak{B}, \mu)$ be such that (3.2) holds. If $g \in L^\infty(X, \mathfrak{B}, \mu)$ is such that for some scaled sequence $\{a(x, n)\}_{n\geq 1}$*

$$\frac{1}{a(x,n)} \sum_{m=0}^{n-1} (\varphi \circ f^m)(x) \to g(x) \; a.e.$$

as $n \to \infty$, then $g = 0$ a.e.

This result is an immediate corollary of the following stronger and more general statement.

Theorem 5.2 ([5, 12]) *Let $(X, \mathfrak{B}, \mu)$ be a probability space, $f : X \to X$ a measure preserving ergodic transformation, and let $\varphi \in L^1(X, \mathfrak{B}, \mu)$ be such that (3.2) holds. Then for any scaled sequence $\{a(x, n)\}_{n\geq 1}$ and almost every $x \in X$,*

$$\liminf_{n\to\infty} \frac{1}{a(x,n)} \sum_{k=0}^{n-1} (\varphi \circ f^k)(x) \leq 0$$

while

$$\limsup_{n\to\infty} \frac{1}{a(x,n)} \sum_{k=0}^{n-1} (\varphi \circ f^k)(x) \geq 0.$$

We finish this section with a positive result on shift spaces which can be easily extended to hyperbolic diffeomorphisms using their symbolic representations. Under some additional assumptions on the shift space and the function φ one can construct a set of positive Hausdorff dimension $\dim_H$ on which the scaled sum converges to an arbitrary real number a.

Theorem 5.3 ([12]) *Let (X, σ) be the full shift on the space of double sided infinite sequences on a finite alphabet. Let also φ be a Hölder continuous potential not cohomologous to a constant and such that (3.2) holds for $f = \sigma$. Finally, let μ be the unique equilibrium state for φ, then the Hausdorff dimension*

$$\dim_H \left\{ x \in X | \lim_{n\to\infty} \frac{1}{b(n)} \sum_{m=1}^{n-1} \varphi(\sigma^m(x)) = a \right\} \geq \dim_H \mu$$

for any $a \in \mathbb{R}$ and for any invertible, strictly increasing, continuous, positive function $b(R)$ satisfying:

1. $\lim_{R\to\infty} b(R) = \infty$;
2. $\lim_{R\to\infty} \frac{b(R)}{R} = 0$;
3. $\alpha_n/\beta_n \to 1$ *implies that* $b(\alpha_n)/b(\beta_n) \to 1$ *for any two sequences* $\{\alpha_n\}, \{\beta_n\} \subset \mathbb{N}$.

Acknowledgements Ya. P. and A. Z. are partially supported by NSF grant DMS-1400027. Y. Z. is partially supported by NSFC 1137127. A part of this work was done when the authors visited Brown University Institute for Computational and Experimental Research in Mathematics (ICERM). We would like to thank the institute for their hospitality. We are also grateful to V. Afraimovich for useful discussions and comments.

References

1. Barreira, L., Pesin, Y.: Introduction to smooth ergodic theory. In: GSM, vol. 148. AMS, Providence, RI (2013)
2. Barreira, L., Valls, C.: Growth rates and nonuniform hyperbolicity. Discret. Contin. Dynam. Syst. **22**(3), 509–528 (2008)
3. Corabel, N., Barkai, E.: Pesin-type identity for intermittent dynamics with zero Lyapunov exponent. Phys. Rev. Lett. **102**, 050601 (2009)
4. Gaggero-Sager, L.M., Pujals, E.R., Sotolongo-Costa, O.: Infinite ergodic theory and non-extensive entropies. Braz. J. Phys. **41**, 297–303 (2011)
5. Halász, G.: Remarks on the remainder in Birkhoff's ergodic theorem. Acta Math. Acad. Sci. Hungar. **28**(3–4), 389–395 (1976)
6. Katok, A., Hasselblatt, B.: Introduction to the modern theory of dynamical systems. In: Encyclopedia of Mathematics and Its Applications, vol. 54. Cambridge University Press, Cambridge (1995)
7. Pesin, Y.: Dimension theory in dynamical systems. In: Contemporary Views and Applications. University of Chicago Press, Chicago (1997)
8. Pesin, Y., Zhao, Y.: Scaled entropy for dynamical systems. J. Stat. Phys. **158**, 447–475 (2015)

9. Rabinovich, M., Volkovskii A., Lecanda, P., Huerta, R., Abarbanel, H.: Dynamical encoding by networks of competing neuron groups: winnerless competition. Phys. Rev. Lett. **87**, 068102-1-068102-4 (2001)
10. Rabinovich, M., Simmons, A., Varona, P.: Dynamical bridge between brain and mind. Trends Cogn. Sci. **19**, 453–461 (2015)
11. Rabinovich, M., Tristan, I., Varona, P.: Hierarchical nonlinear dynamics of human attention. Neurosci. Biobehav. Rev. **55**, 18–35 (2015)
12. Zelerowicz, A.: Time Rescaling of Birkhoff sums, Preprint, PSU (2016)

Chapter 4
Unraveling the Chaos-Land and Its Organization in the Rabinovich System

Krishna Pusuluri, Arkady Pikovsky, and Andrey Shilnikov

4.1 Introduction

Nonlinear wave and mode interactions often result in complex dynamics. Remarkably, already elementary systems of two and three weakly interacting waves can demonstrate chaotic behavior [6, 12, 16, 28, 29, 39, 41]. One such simple model is the so-called Rabinovich system, describing wave interaction with complex dynamics in a system of three resonantly coupled waves, comprised of two parametrically excited waves, and another wave that is in synchronism with this pair [27]. It can exhibit the following states while remaining phase locked:

- the trivial static stabilization of parametric instability at low pump fields, which corresponds to a fixed point with zero wave amplitudes in the phase space;
- the static and cyclic stabilizations of parametric instability elimination, corresponding to stable fixed points with non-zero wave amplitudes, and cyclic oscillations of wave amplitudes, respectively;
- eventually, Lorenz like chaotic behavior with a stochastic stabilization region of self-oscillations of wave amplitudes, at higher pump fields.

K. Pusuluri (✉)
Neuroscience Institute, Georgia State University, Petit Science Center,
100 Piedmont Av., Atlanta, GA 30303, USA
e-mail: pusuluri.krishna@gmail.com

A. Pikovsky
Institute for Physics and Astronomy, University of Potsdam, Karl-Liebknecht-Str 24/25, Bld. 28,
Potsdam 14476, Germany
e-mail: pikovsky@uni-potsdam.de

A. Shilnikov
Neuroscience Institute and Department of Mathematics and Statistics, Georgia State University,
Petit Science Center, 100 Piedmont Av., Atlanta, GA 30303, USA
e-mail: ashilnikov@gsu.edu

I.S. Aranson et al. (eds.), *Advances in Dynamics, Patterns, Cognition*,
Nonlinear Systems and Complexity 20, DOI 10.1007/978-3-319-53673-6_4

The physical motivation for the Rabinovich system proposed in [27] is as follows. A whistler wave with wave vector **q** and frequency ω_q propagates along a magnetic field **H** in a non-isothermal magnetoactive plasma. This wave parametrically excites a plasma wave ($\mathbf{k}$, ω_k) and an ion sound (κ, ω_κ), provided that the resonance conditions $\mathbf{k} + \kappa = \mathbf{q}$, $\omega_k + \omega_\kappa = \omega_q$ are fulfilled. These two parametrically excited waves are resonant with the plasma wave ($\mathbf{k}_1$, ω_{k_1}) where $\mathbf{k}_1 = \mathbf{k} - \kappa$, $\omega_{k_1} = \omega_k - \omega_\kappa$, which is synchronous to the produced pair. As a result, one obtains a closed set of amplitude equations for the three waves ω_k, ω_κ, and ω_{k_1}, where the energy that comes from the constant pump ω_q is distributed between the waves due to nonlinear resonant coupling, and eventually is dissipated due to linear damping.

The simplified set of equations governing this resonantly coupled wave triplet system—the Rabinovich system—is given by the following equations:

$$\begin{aligned} \dot{x} &= hy - \nu_1 x - yz \\ \dot{y} &= hx - \nu_2 y + xz \\ \dot{z} &= -z + xy\,. \end{aligned} \tag{4.1}$$

Here, x, y and z correspond to the amplitudes of the three resonantly coupled waves—the parametrically excited plasma wave **k**, the parametrically excited ion sound κ and the synchronous plasma wave $\mathbf{k}_1$, respectively. Quantities h, ν_1, and ν_2 are the parameters of the system: the value of h is proportional to the pump field, whereas ν_1 and ν_2 are the normalized damping decrements in the parametrically excited waves **k** and κ, respectively. After the original investigation in [27], the studies of this system have further been continued in [9–11, 15, 17, 18, 20–26, 37, 38, 40], see also the contribution by S. Kuznetsov in this volume [19].

Although initial numerical simulations have revealed the presence of a Lorenz-like chaotic behavior in the Rabinovich system, the exact boundaries of static, periodic, and chaotic dynamics in the parametric space have not been identified. The underlying structures governing the organization of chaos in the system, such as the various homoclinic and heteroclinic connections, and codimension two bifurcation points called the Bykov terminal points (T-points), with characteristic spirals typical for Lorenz-like systems [2, 4, 8, 13, 34, 35], have not been disclosed either. Even as of now, there is only a limited set of computational tools that can be employed to detect such structures in the parametric space of a system, and especially, in Lorenz-like models. In particular, tools based on Lyapunov exponents are computationally effective to sweep, and find regions of stationary (equilibria states), periodic, and chaotic dynamics. However, they fail to reveal details of fine or any constructions of homoclinic and heteroclinic structures in the parametric spaces, that are the basic and imperative building blocks of structurally unstable, deterministic chaos in most systems. While parameter continuation techniques let some such structures to be revealed, one has to possess specific skills and enormous patience to perform a painstaking reconstruction of the bifurcation unfolding of the system in question, in its 2D parameter plane, by separately following a few dozens of principal bifurcation curves, one after the other [34, 35].

One of the aims of the current study is to discuss and demonstrate a recent advance in the field that became possible with the development of a suite of computational tools utilizing symbolic representations of simple and chaotic dynamics. This allows for fast and effective identification of bifurcation structures underlying, and governing, deterministic chaos in systems with the Lorenz strange attractors, as well as those with spiral chaos with the Shilnikov saddle-focus [5, 42–44]. Moreover, the latest advances in GPU and parallel computing techniques have empowered us to achieve a tremendous degree of parallelization to reconstruct bi-parametric sweeps, at a fraction of the time taken for traditional serial computational approaches, for a comparable analysis. In this paper, we employ this computational toolkit to disclose the bifurcation features of complex dynamics in the Rabinovich system.

In the following sections, we will analytically describe, and numerically simulate, the solutions of the Rabinovich system. We will then describe the symbolic apparatus and the computational techniques, and apply them to study this system. Next, we will present our results and identify various important structures that provide a framework to organize the complex dynamics arising in the system. A brief description of the methods used in the study is presented towards the end.

4.2 Solutions of the Rabinovich System

The system (1) is $\mathbf{Z}_2$-symmetric—i.e., invariant under the involution $(x, y, z) \leftrightarrow (-x, -y, z)$. All of its trajectories are confined within an ellipsoid given by $u \leq 9h^2k^{-1}$, where

$$
\begin{aligned}
u &= 2x^2 + y^2 + (z - 3h)^2 \\
\dot{u} &\leq -ku + 9h^2
\end{aligned}
\tag{4.2}
$$

At low pump amplitudes $h < (\nu_1\nu_2)^{\frac{1}{2}}$, the system has just a single equilibrium state $O(0, 0, 0)$, which is a global attractor of the system, pulling all trajectories inwards. In this state, the system is below the threshold of parametric instability. This equilibrium state O undergoes a pitchfork bifurcation at $h = (\nu_1\nu_2)^{\frac{1}{2}}$, that gives rise to two more equilibrium states $C^{\pm}$ $(\pm(z^0 l)^{\frac{1}{2}}, \pm(z^0 l)^{\frac{1}{2}}, z^0)$ for larger pump fields $h > (\nu_1\nu_2)^{\frac{1}{2}}$, where $z^0 = (h^2 - \nu_1\nu_2)^{\frac{1}{2}}$ and $l = (h - z^0)\nu_1^{-1}$. After the bifurcation, i.e., beyond the parametric instability threshold, the zero equilibrium state O becomes unstable, resulting in parametric instability elimination. The fixed points $C^{\pm}$ can be either stable or unstable, depending upon the parameters of the system h, ν_1 and ν_2. Stable equilibria $C^{\pm}$ correspond to a static stabilization of parametric instability elimination (Fig. 4.1a, b and d). Equilibria $C^{\pm}$ lose stability through an Andronov–Hopf bifurcation that gives rise to a pair of stable periodic orbits, corresponding to stable cyclic self-oscillations of wave amplitudes (Fig. 4.1f). Besides periodic oscillations, the Rabinovich system may possess a Lorenz-like strange attractor

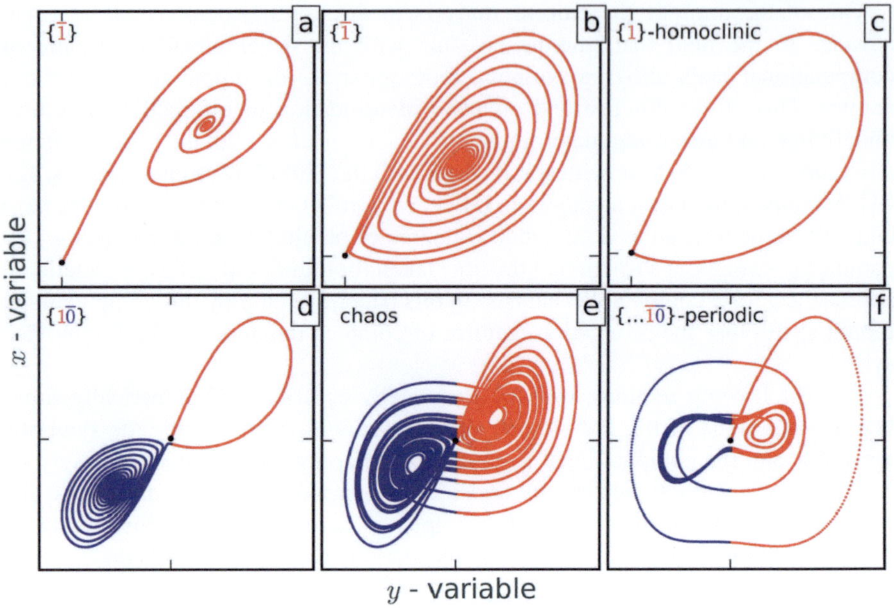

Fig. 4.1 Snapshots of the dynamics of the right unstable separatrix Γ_1 of the saddle O (*black dot*) in the Rabinovich system, in the (x, y)-projection. Sections of Γ_1 are color-coded in *red* when $y > 0$, and in *blue* when $y < 0$, for the sake of clarity. Symbolic representations are as described in Sect. 4.3. A bar in the symbolic representation denotes a repetitive sequence. The parameters are set as $\nu_1 = 1, \nu_2 = 4$. At $h = 3$, Γ_1 converges to the stable focus C^+ in (**a**), and comes close to the saddle O at $h = 3.95$ in (**b**), generating a persistent sequence $\{111\ldots\bar{1}\}$. After the primary homoclinic orbit of the saddle at the origin at $h \simeq 3.9998$ in (**c**), Γ_1 converges to the stable focus C^- at $h = 4.1$ to generate the sequence $\{100\ldots\bar{0}\}$ in (**d**). Chaotic attractor is seen in the system at $h = 8$ in (**e**), and convergence to the periodic attractor $\{\overline{10}\}$ at $h = 18$ in (**f**).

within the finite-size ellipsoid, with stochastic variations of wave amplitudes, which is associated with chaotic saturation of the parametric instability (Fig. 4.1e). Here, the origin is a saddle with a two-dimensional stable manifold and a pair of one-dimensional unstable separatrices, while the equilibrium states $C^{\pm}$ are saddle-foci with one-dimensional incoming separatrices and two-dimensional unstable manifolds. In addition, like the original Lorenz model, the Rabinovich system can be bi-stable with coexisting stable equilibrium states $C^{\pm}$ and the strange attractor, see [27]. Also, note that the three resonating waves remain phase locked while their amplitudes exhibit the above-mentioned complex behaviors (see [40] for a further exploration of this phase locking).

A homoclinic bifurcation occurs in the system when both outgoing separatrices of the saddle (or of either saddle-focus) come back to it along the 2D stable manifold. Figure 4.1c illustrates a single "right" separatrix of the saddle at the homoclinic bifurcation. Before and after the primary homoclinic bifurcation, the separatrix spirals converge towards either C^+ or C^- (Fig. 4.1b, d). Similarly, a one-way heteroclinic connection occurs when the outgoing separatrix of the saddle O connects with either of the saddle-foci $C^{\pm}$ by merging with a 1D incoming separatrix (see Fig. 4.6). Note that such connections always come in pairs, due to $\mathbf{Z}_2$-symmetry.

4.3 Symbolic Representation

The fundamental feature of the Lorenz attractor is that it is both dynamically and structurally unstable [3, 14]. A trademark of any Lorenz-like system is the strange attractor of the iconic butterfly shape, as the one shown in Fig. 4.1e. The "wings" of the butterfly are marked with two symmetric "eyes" containing equilibrium states C^+ and C^-, stable or not, isolated from the trajectories of the Lorenz attractor. This attractor is structurally unstable as it bifurcates constantly as the parameters are varied. The primary cause of structural and dynamic instability of chaos in the Lorenz equations and similar models is the singularity at the origin—a saddle with two one-dimensional outgoing separatrices. Both separatrices densely fill the two spatially symmetric wings of the Lorenz attractor in the phase space [2]. The Lorenz attractor undergoes a homoclinic bifurcation when the separatrices of the saddle change the alternating pattern of switching between the butterfly wings centered around two other symmetric equilibria, which can be stable foci or saddle-foci depending on the parameter values. At such a change, the separatrices come back to the saddle, thereby causing a homoclinic explosion in phase space. The computational approach that we employ for studying Lorenz-like and similar systems capitalizes on the key property of deterministic chaos—the sensitive dependence of solutions on variations of control parameters. In particular, for the Lorenz-type attractors, chaotic dynamics are characterized by unpredictable flip-flop switching between the two spatial wings of the strange attractor, separated by the saddle singularity at the origin. This is the main reason why the saddle O is the primary source of instability in such systems, including the Rabinovich system. The ideas of this computational research are greatly inspired by, and deeply rooted in, the pioneering studies of L.P. Shilnikov [30, 31, 36]. His extensive knowledge of homoclinic bifurcations helped to transform the theory of strange attractors into a mathematical marvel [1, 2, 32, 33]. The reader may find more detailed information about the Lorenz-like systems and symbolic computations in the original papers [5, 42–44].

In order to identify regions with topologically identical dynamics in the parametric space, we follow the time progression of a single trajectory—the right outgoing separatrix Γ_1 of the saddle at the origin. We convert the flip-flopping patterns of this trajectory around $C^{\pm}$ (see Fig. 4.2) into a binary symbolic sequence $\{k_n\}$ obeying the following rule:

$$k_n = \begin{cases} 1, & \text{when the separatrix } \Gamma_1 \text{ turns around } C^+, \\ 0, & \text{when the separatrix } \Gamma_1 \text{ turns around } C^-. \end{cases}$$

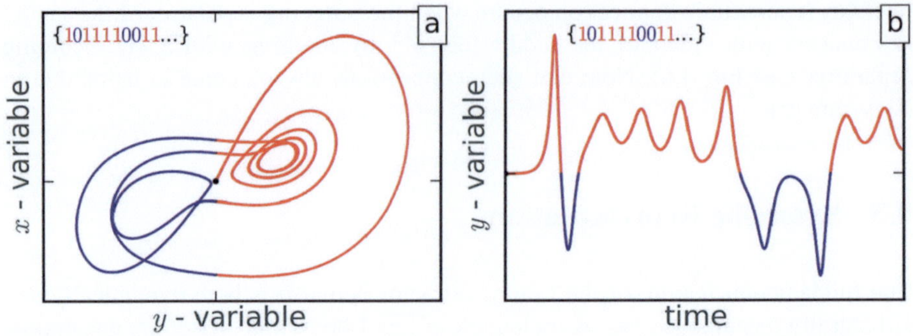

Fig. 4.2 Symbolic representation of the right separatrix Γ_1 at the parameter values $\nu_1 = 1$, $\nu_2 = 4$, and $h = 9$. (**a**) Evolution of the trajectory in the phase space, projected onto the (xy)-plane. (**b**) Time evolution of the y-variable. The principle of symbolic encoding: each portion of the trajectory turning around the equilibrium state C^+ (*red section*) is represented by 1, whereas each portion looping around C^- (*blue section*) is represented by 0. The trajectory shown is thus converted into the symbolic sequence $\{1011110011\ldots\}$

Alternatively, one can detect releva00nt events of $\frac{dy}{dt} = 0$, provided that $\frac{d^2y}{dt^2}$ is negative or positive for 1 or 0, respectively, see the sampled trace $y(t)$ in Fig. 4.2b. We use an overbar symbol to represent repetitive sequences. For example, the periodic orbit turning once around C^+, then once around C^- and so on, generates an infinite repetitive sequence $\{1010101010\ldots\}$, or $\{\overline{10}\}$ for short.

For a sequence $\{k_n\}$ of length N, starting with some $(j+1)$-th symbol (the very first j-transients are skipped), we define a formal power series as follows:

$$P(N) = \sum_{n=j+1}^{j+N} \frac{k_n}{2^{(N+j+1)-n}} \tag{4.3}$$

This series is convergent, with its limit ranging between 0 and 1. Whenever the followed separatrix, or any other such trajectory, after some initial transient dynamics, orbits only around C^+, so that its $y(t)$-coordinate always remains positive, the corresponding binary symbolic sequence contains only 1s, i.e. $k_n = 1$, and therefore, $P(N) = 1$ in the limit as $N \to \infty$. In the case where the trajectory continuously orbits only around C^- after some transient, so that $y(t) < 0$, $k_n = 0$ and $P(N) = 0$. Otherwise, periodic or aperiodic flip-flopping between and around equilibria C^+ and C^- generate either regular or chaotic sequences of 1s and 0s, so that $0 \leq P \leq 1$. This power series provides a way to uniquely quantify the dynamics of the system for a given set of parameters, making it a dynamic invariant. Two different sets of parameters with the same dynamic invariant value show topologically identical behavior. The way we define this power series is slightly different from how it was previously defined, with the current definition giving increasingly higher weights to symbols towards the end of the sequence, rather than the beginning, see [5, 42, 43]. This lets us achieve a greater contrast in the bi-parametric scans, and thereby, revealing greater dynamical details, between neighboring regions of largely similar dynamics, that differ only in the last symbol in their binary sequences due to homoclinic curves separating such regions (see Sect. 4.5). Alternatively, one can also convert the binary sequences into a decimal representation.

4.4 Bi-parametric Scans with Symbolic Computations

In order to obtain bi-parametric scans, we keep one of the three parameters of the system—ν_1, ν_2, or h, constant, while varying the other two. For each set of parameters in the bifurcation plane, we always follow the positive unstable separatrix Γ_1 of the saddle at the origin in the Rabinovich system (1). Note that, as the system is Z_2-symmetric, our results stay the same even if the left separatrix of the saddle is followed, provided there is the swapping of the symbols $0 \rightleftharpoons 1$, resulting in the same symbolic sequence $\{k_n\}$, and the corresponding invariant $P(N)$ values ranging within [0, 1], by the above construction. These invariant values are then projected on to the 2D parametric space, using a colormap that can uniquely identify up to 2^{24} different values of $P(N)$, via the whole spectrum of colors. This results in the desired bi-parametric scans, such as the ones sampled in Figs. 4.3, 4.5, 4.7, 4.8, 4.9, and 4.10. Regions corresponding to similar dynamics, that generate identical symbolic sequences of a given length, and therefore, carry the same dynamic invariant values $P(N)$, are identified by the same colors in the bi-parametric bifurcation sweep of appropriate resolution.

4.4.1 Emergence of Chaos via Homoclinic Explosion

In this section, we demonstrate how the symbolic computations technique can gradually reveal the complex organization of dynamics, and the underlying non-local bifurcations in the system. Figure 4.3 presents a series of (ν_2, h)-bi-parametric sweeps, $\nu_2 \in [2, 10]$ and $h \in [7, 12]$ and at fixed $\nu_1 = 1$, with increasing length/depth of symbolic sequences, from 3 through 8, along with the case of n: 105–128 range. As such, at every point in the given region, the symbolic dynamics remain identical up to the first two symbols $\{10\}$, because the right separatrix always makes its first loop around C^+ with $y > 0$, followed by a loop around C^- with $y < 0$. As the length of the trajectories, and therefore, of the sequences generated, is increased, initially up to three symbols, i.e. n: $1, 2, 3$, the scan detects a secondary homoclinic bifurcation curve that separates the two sub-regions shown in blue and red in Fig. 4.3a. In the red sub-region, the separatrix makes the third loop around C^- (Fig. 4.4a), whereas in the blue sub-region, its third loop is around C^+ (Fig. 4.4c). The borderline separating these two sub-regions is a bifurcation curve (arguably of codimension-1), corresponding to the secondary homoclinic orbit of the saddle at

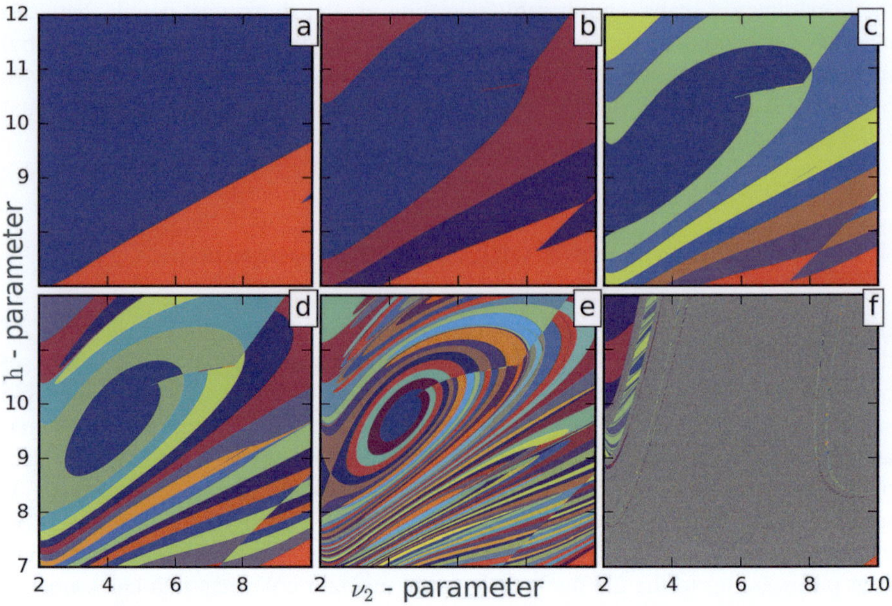

Fig. 4.3 Emergent chaos via homoclinic explosion in bi-parametric (ν_2, h)-scans with a fixed $\nu_1 = 1$. (**a**) Bi-parametric scan of length 3, n: 1–3. The secondary homoclinic curve encoded as {10} (see Fig. 4.4b) divides the parametric space into two distinct subregions: {100} (*red*) (see Fig. 4.4a) and {101} (*blue*) (see Fig. 4.4c); (**b**) a longer scan with the first four symbols, n: 1–4, reveals an additional homoclinic curve, {100}, dividing the *red region* {100} (*red*) of (**a**) into two subregions: {1000} and {1001}, while the *blue region* {101} of (**a**) is partitioned into two subregions {1010} and {1011} by the homoclinic curve {101}; sweeps with (**c**) n: 1–5, (**d**) n: 1–6, and (**e**) n: 1–8 gradually disclose finer underlying structures of homoclinic bifurcation unfolding; (**f**) multi-colored, "noisy" sweep with a long sequence of n: 105–128 is indicative of a region with structurally unstable, chaotic dynamics in the system (see Sect. 4.4.3.1)

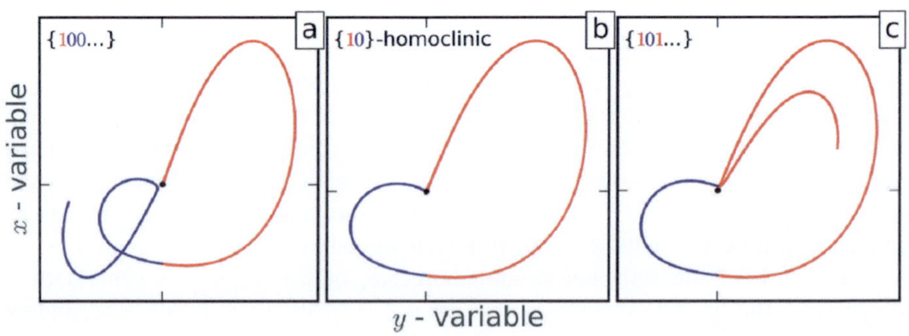

Fig. 4.4 Progressive snapshots of the right 1D separatrix of the saddle in different sub-regions of Fig. 4.3a with a fixed $\nu_1 = 1$, decoded with the first three symbols. (**a**) Dynamical and symbolic representation {100} at $\nu_2 = 4$, $h = 7.5$; (**b**) secondary homoclinic orbit {10} at $\nu_2 = 4$, $h \simeq 7.6$; (**c**) at $\nu_2 = 4$, $h = 7.7$, the trajectory corresponds to the symbolic sequence {101}

the origin (Fig. 4.4b). This homoclinic bifurcation curve, marked by {10}, divides the parametric space in Fig. 4.3a into two distinct sub-regions corresponding to the sequences starting with {100...} and {101...}, respectively. Note that both the homoclinic curve and the entire region of Fig. 4.3a are identified with the same symbolic sequence. As we include one more symbol, i.e., as n runs from 1 to 4 in the computation of $P(N)$, both sub-regions of Fig. 4.3a, red and blue, are further subdivided by bifurcation curves representing longer homoclinic orbits of the saddle, see Fig. 4.3b. The homoclinic curve {100} divides the red sub-region {100} of Fig. 4.3a into two sub-regions coded by {1000} and {1001} in Fig. 4.3b. Similarly, the blue sub-region {101} in Fig. 4.3a is subdivided into two sub-regions coded by {1010} and {1011} in Fig. 4.3b, by the homoclinic bifurcation curve {101}. Adding more symbols to the computation of the bi-parametric sweep increases its depth, lets us detect more complex homoclinic bifurcations, and gradually reveals the underlying structures that result in the complexity of the system (Fig. 4.3c–e). In this case, the complexity is organized around a central point called a terminal point (T-point) (Fig. 4.3e), which will be discussed further in the next section. For very long sequences with n: 105–128 range, the bi-parametric scan indicates that the system continuously undergoes a plethora of homoclinic bifurcations as the parameters are varied, and exhibits structurally unstable dynamics due to these uncontrollable homoclinic explosions (Fig. 4.3f).

4.4.2 Heteroclinic Connections and Bykov T-Points

Fine organization of the structure of the chaotic region with the primary T-point is revealed in greater detail in Fig. 4.5. It demonstrates the complex universality and self-similarity of characteristic spirals typical for most Lorenz-like systems. Here, the primary T-point is marked $T1$. At this codimension-2 point, the 1D outgoing (unstable) separatrix of the saddle, after the first two loops {10} (common to the entire parametric space under consideration here), merges with the 1D incoming (stable) separatrix of the saddle-focus C^+, thus forming a one-way heteroclinic connection (Fig. 4.6a, e). Note that both the saddle and the saddle-focus have stable and unstable manifolds that transversally intersect in the 3D phase space of the Rabinovich system. This makes the heteroclinic connection closed as $t \to \pm\infty$. Thus, at the T-point $T1$, the separatrix makes an infinite number of revolutions around C^+ before it comes back to the saddle. As such, its symbolic representation is given by the sequence $\{10\overline{1}\}$, where the overbar represents a repetitive subsequence. As we move away from $T1$ along the adjacent spiral in the parameter space, the number of revolutions of the separatrix around C^+ keeps decreasing, and becomes finite. It is known that a Lorenz-like system, near a T-point, exhibits a multiplicity of secondary T-points with increasing complexity, called as Bykov T-points [2, 4, 7, 8, 13, 34, 35]. The short parametric scan in Fig. 4.5 detects several notable secondary T-points marked as $T2$, $T2'$, $T3$, $T3'$, $T4$, and $T4'$, and the spiral structures associated with them. At $T4$, the outgoing separatrix of the

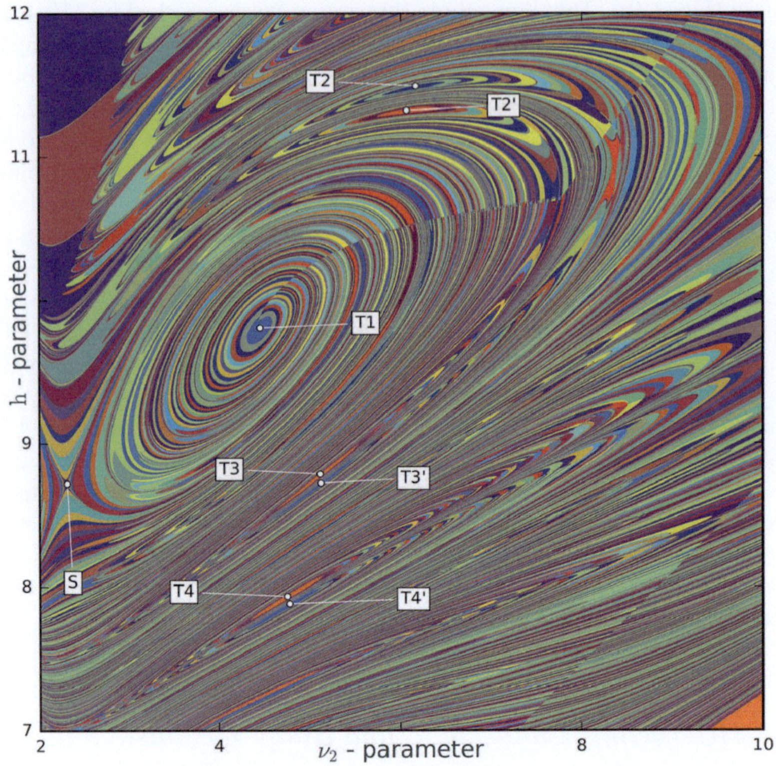

Fig. 4.5 Self-similar organization of Bykov T-Points and heteroclinic connections: bi-parametric (ν_2, h)-sweep at $\nu_1 = 1$, with n: 5–12 range, detects—primary T-point: $T1$—$\{10\overline{1}\}$; secondary T-points: $T2$—$\{10110\overline{1}\}$, $T2'$—$\{10111\overline{0}\}$, $T3$—$\{1011\overline{0}\}$, $T3'$—$\{1010\overline{1}\}$, $T4$—$\{101\overline{0}\}$, and $T4'$—$\{100\overline{1}\}$. A primary saddle in the parametric space is marked with the symbol S. Heteroclinic connections with corresponding y-progressions at some of these T-points are presented in Fig. 4.6

saddle O, after the initial two loops $\{10\}$, makes one more loop towards C^+ and then merges with the 1D incoming, stable separatrix of the other saddle-focus C^-. That is why, this heteroclinic connection is symbolically represented as $\{101\overline{0}\}$ (Fig. 4.6b, f). Similarly at $T4'$, the outgoing separatrix, after the initial two loops $\{10\}$, makes a loop towards C^-, and then hits C^+. Thus, this heteroclinic connection is represented as $\{100\overline{1}\}$ (Fig. 4.6c, g). At $T2$, the complexity of the heteroclinic connection further increases: after the two initial loops $\{10\}$, the next two loops are around C^+, followed by one loop around C^-, and then the separatrix comes back to C^+; its coding is given by $\{10110\overline{1}\}$ (Fig. 4.6d, h). The T-points depicted in Fig. 4.5 can be summarized as follows:

- $T1$: $\{10\ \overline{1}\}$—Two loops followed by heteroclinic connection to C^+
- $T4$: $\{101\ \overline{0}\}$—Three loops followed by heteroclinic connection to C^-
- $T4'$: $\{100\ \overline{1}\}$—Three loops followed by heteroclinic connection to C^+

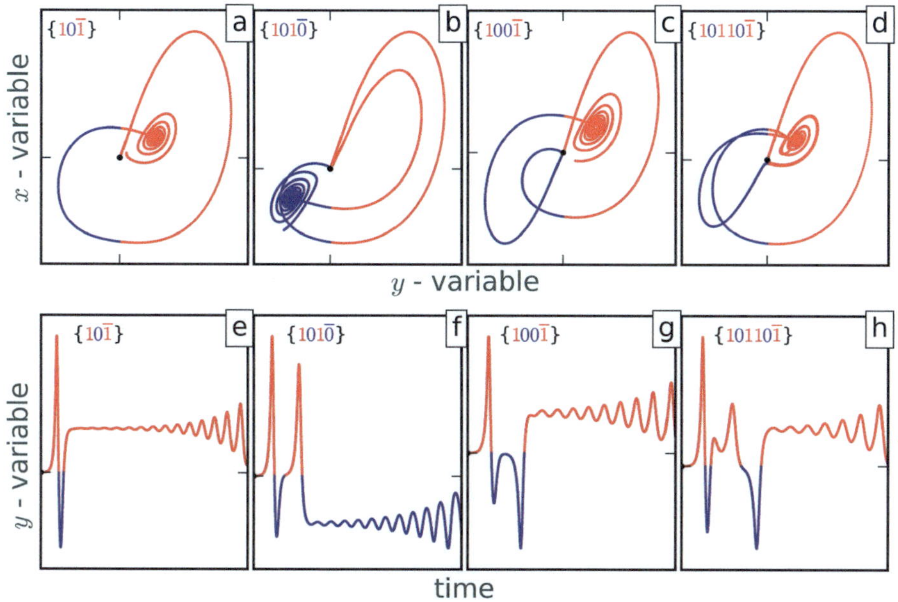

Fig. 4.6 T-point configurations (**a**)–(**d**) and matching time progressions (**e**)–(**h**): heteroclinic connections and the y-variable time evolutions at: (**a**), (**e**) $T1$—{$10\bar{1}$}; (**b**), (**f**) $T4$—{$101\bar{0}$}; (**c**), (**g**) $T4'$—{$100\bar{1}$}; and (**d**), (**h**) $T2$—{$10110\bar{1}$}, as depicted in the bifurcation diagram in Fig. 4.5

- $T3$: {$1011\ \bar{0}$}—Four loops followed by heteroclinic connection to C^-
- $T3'$: {$1010\ \bar{1}$}—Four loops followed by heteroclinic connection to C^+
- $T2$: {$10110\ \bar{1}$}—Five loops followed by heteroclinic connection to C^+
- $T2'$: {$10111\ \bar{0}$}—Five loops followed by heteroclinic connection to C^-

4.4.3 Global Bifurcations and Organization of Chaos

In this section, we study the global organization of chaos using bi-parametric sweeps of the (ν_2, h)-parameter plane (Figs. 4.7 and 4.8) and of the (ν_1, ν_2)-parameter plane (Figs. 4.9 and 4.10). We begin this discussion with Fig. 4.7, showing the (ν_2, h)-sweep with n: 5–12 range, at $\nu_1 = 1$. It detects several low-order T-point-like structures, labeled by $T1$ through $T7$, as well as characteristic spirals—bifurcation curves of homoclinic orbits and separating saddles. A small sub-region of this diagram nearby $T1$ is magnified in Fig. 4.5 of Sect. 4.4.2. Figure 4.8a presents a deeper/longer bi-parametric sweep with n: 105–128 range. Here, we skip a relatively long initial transient of the separatrix, to reveal the long-term dynamics of the Rabinovich system. The underlying idea here is a sweep utilizing a typical trajectory of the Rabinovich system, which does not necessarily have to be the separatrix

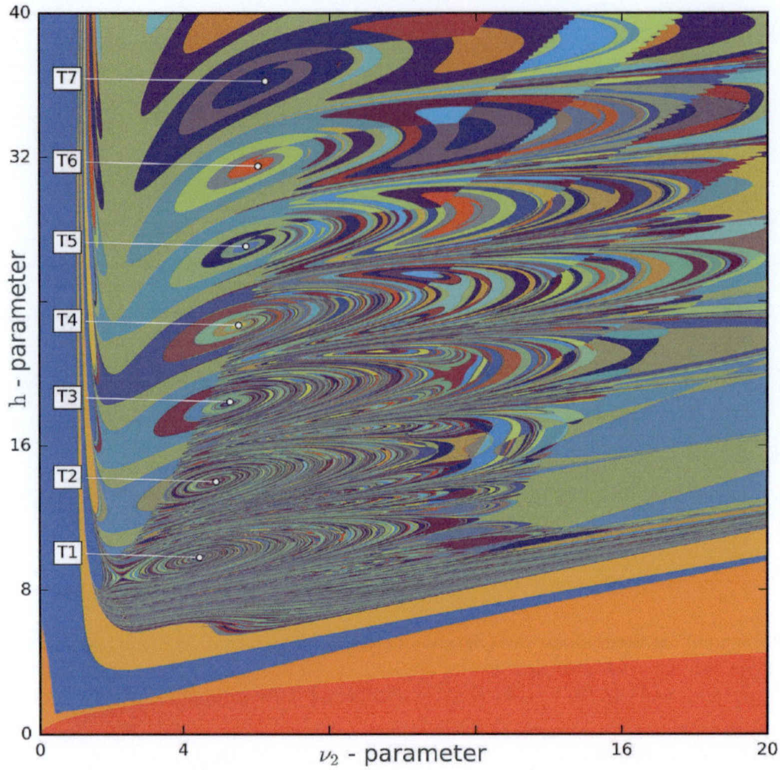

Fig. 4.7 Bi-parametric (ν_2, h)-sweep at $\nu_1 = 1$, with n: 5–12 range, discloses a recursive series of Bykov T-point-like structures marked as T1–T7, saddles, as well as regions of stable periodic dynamics (*solid colors*) of the system, in the bifurcation diagram

that is employed for the purpose of homoclinic structures. With this new approach, we can reveal the occurrence of chaotic, structurally unstable dynamics emerging through homoclinic explosions, and detect these regions in the parameter space. Such a region in the parameter plane appears to look like a noisy region, due to the interference of multiple colors corresponding to constantly changing $P(N)$, due to homoclinic bifurcations that densely fill in, as the control parameters are varied. On the other hand, regions, corresponding to structurally stable (normally hyperbolic) dynamics due to Lyapunov stable equilibria and periodic orbits, i.e., the so-called *Morse-Smale* systems, are coded with solid colors. Note that the same color throughout a region or across regions, corresponds to topologically identical dynamics, by construction.

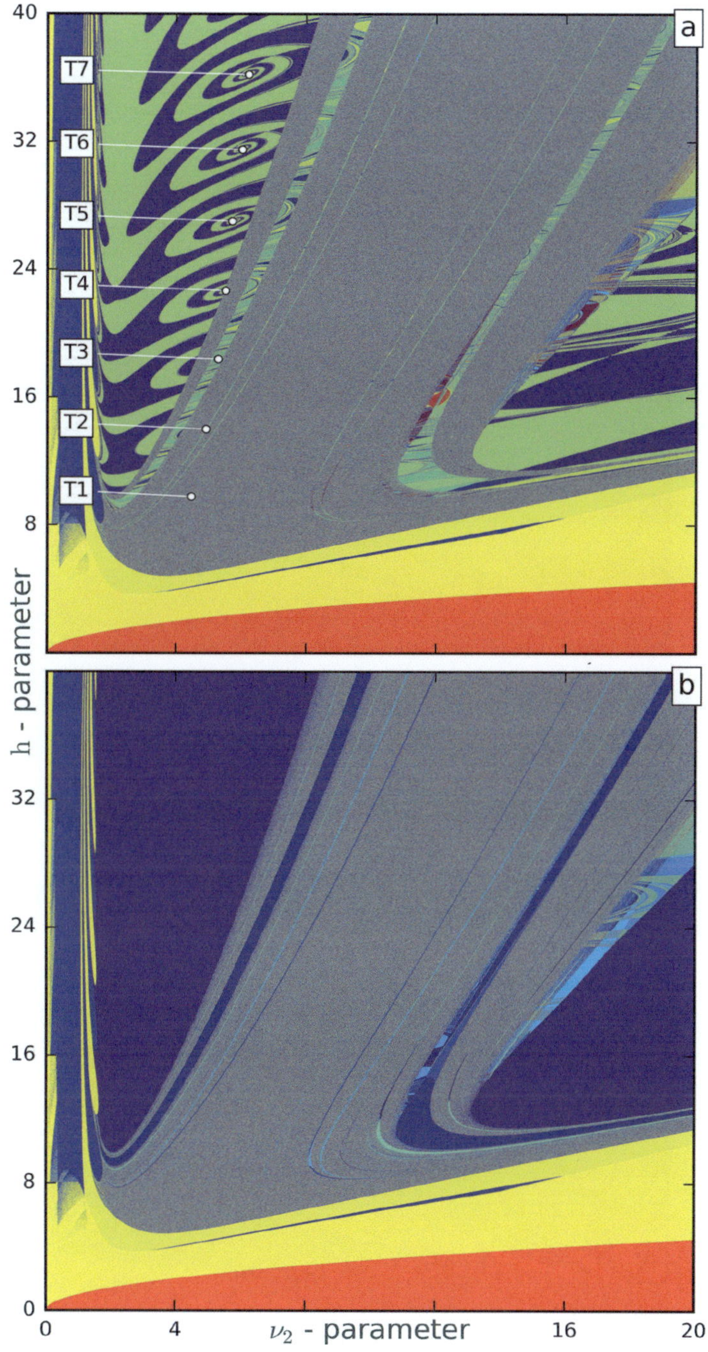

Fig. 4.8 Deterministic Chaos Prospector in action: bi-parametric (ν_2, h)-sweep at $\nu_1 = 1$ with n: 105–128 to study long-term dynamics of the system: (**a**) sweep without periodicity correction reveals spiraling artifacts, pseudo T-points, as well as Bykov T-points (labeled by $T1$–$T7$). (**b**) Sweep enhanced with periodicity correction eliminates spiraling artifacts due to transient dynamics in the existence region of stable periodic orbits. It also detects multiple stability windows (parallel bands of *solid colors*) representing stable dynamics within the otherwise chaotic regions

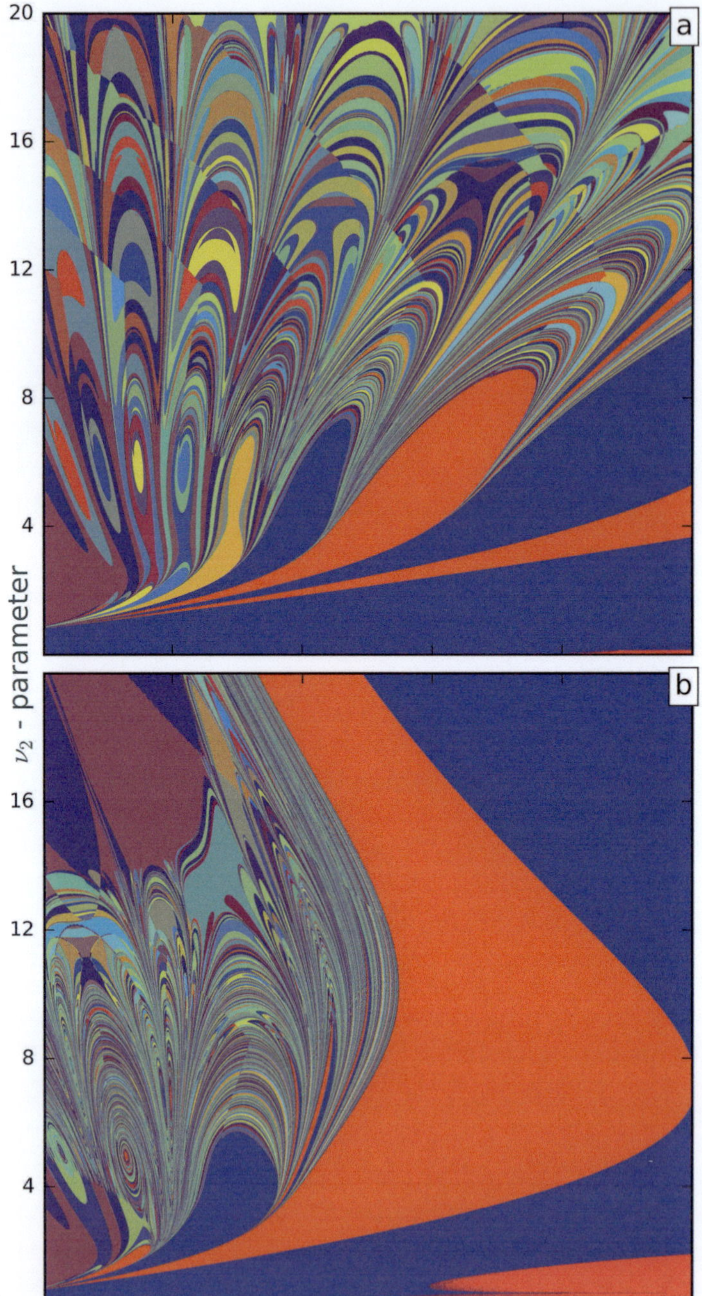

Fig. 4.9 Bi-parametric (ν_1, ν_2)-sweeps with n: 5–12 range at (**a**) $h = 35$, and (**b**) $h = 15$, reveal universality and organization of the chaos-land, featuring characteristic spirals and saddles that are embedded into the *solid-color* regions of stable dynamics

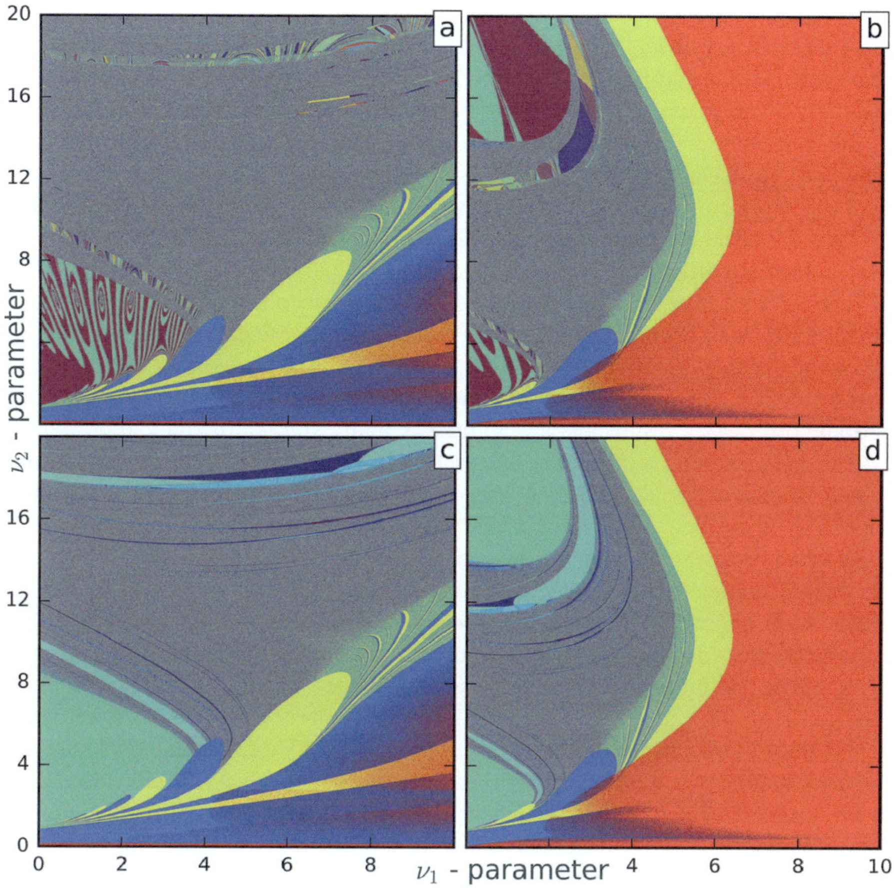

Fig. 4.10 Deterministic Chaos Prospector in action: bi-parametric (ν_1, ν_2)-sweeps with n: 105–128 range at $h = 35$ in panels (**a**), (**c**) and at $h = 15$ in panels (**b**), (**d**), detect regions of chaotic, structurally unstable dynamics, and simple, Morse-Smale dynamics due to the existence of stable equilibria and periodic orbits, in the "noisy" and *solid-colored regions* in the parameter plane, respectively. Artificial spiral structures appearing in (**a**), (**b**) are eliminated with the periodicity correction technique used in (**c**), (**d**). Panels (**c**), (**d**) reveal multiple stability windows of simple dynamics in the otherwise chaotic—"noisy" regions

4.4.3.1 Deterministic Chaos Prospector Using Periodicity Correction

A discernible problem of consequence with the symbolic representation of stable periodic orbits of complex configurations existing in the Morse-Smale systems is their shift-symmetry or shift-circularity feature. For example, the following four sequences: $\{\overline{0110}\}$, $\{\overline{0011}\}$, $\{\overline{1001}\}$, and $\{\overline{1100}\}$ represent the same [stable] periodic orbit, which can be either symmetric or asymmetric in the phase space. To compare whether two such orbits are topologically conjugate or not, at least in their symbolic representation, one has to come up with a consistent rule to sort out and normalize all of their corresponding binary sequences. This issue becomes principal when one deals with skipping arbitrarily long initial transient of typical trajectories converging to a periodic attractor of the system, at the given parameter values.

Let us re-iterate: the transient history of two orbits, before they settle down on the same, or topologically same, periodic orbit can be quite different. Consider a transient trajectory converging to a figure-8 periodic orbit that alternatively loops around C^+ and C^- back and forth. This orbit admits two symbolic descriptions: $\{\overline{10}\}$ and $\{\overline{01}\}$, which are differently color-mapped on to the parameter space. The existence of such an orbit is detected by the 105–128-range sweep shown in Fig. 4.8a; here, we skip 104 initial symbols. Depending on the transient behavior and whether the following 105th symbol is either "0" or "1", the point in the corresponding region is color-coded differently, even though the global attractor is the same figure-8 periodic orbit. As a result, this sweep detects faulty spiral structures, representing $\{\overline{10}\}$ or $\{\overline{01}\}$ sequences, shown in blue and light green colors, respectively, around centers labeled as $T5$, $T6$, and $T7$ in Fig. 4.8a, along with some faulty saddles, which happen to be artifacts of the implemented simulation approach.

In order to overcome this issue, we developed the technique of "Periodicity Correction" to detect periodic orbits, and to determine their periods using a circular permutation approach. This allows us to consistently choose the same symbolic representations for similar periodic orbits, correctly compute the corresponding $P(N)$-value, and colormap it to the parameter plane. As with the above example, identical periodic sequences $\{\overline{10}\}$ and $\{\overline{01}\}$ are normalized to $\{\overline{01}\}$ to determine its P-value. Similarly, the three representations—$\{\overline{101}\}$, $\{\overline{110}\}$, and $\{\overline{011}\}$ for the same (or topologically similar) periodic orbit(s) of periodicity 3, are normalized to the smallest valued binary sequence $\{\overline{011}\}$ to evaluate the $P(N)$-value, and the corresponding color-code (see Sect. 4.5).

Figure 4.8b presents the sweep of the same resolution and depth as Fig. 4.8a, but using periodicity correction. One can see that, with this technique, the diagram is free of the aforementioned spiral artifacts in the region (dark blue) of existence of the stable figure-8 periodic orbit corresponding to the sequences $\{\overline{10}\}$ or $\{\overline{01}\}$. Nevertheless, the presence of those spirals in the scan without periodicity correction (Fig. 4.8a) indicates the existence of heteroclinic connections close to the centers of those pseudo T-point spirals—$T5, T6$, and $T7$. Indeed, the points labeled by $T1$ and $T2$ appear to be genuine T-points with quite complex long-term dynamics, whereas the points labeled by $T3$, $T5$, $T6$, and $T7$ correspond to stable heteroclinic connections between the saddle-foci C^+ and C^-, that the transitioning figure-8 periodic orbit approaches in the limiting case. The point $T4$ is located next to the boundary between stable and chaotic regions, and it is rather difficult to evaluate its contribution at this resolution. In addition, Fig. 4.8b reveals multiple parallel bands of constant colors with gradually decreasing widths, within the otherwise noisy regions. This is indicative of the presence of stability windows corresponding to regular, periodic dynamics alternating with chaotic behaviors. Bands of such constant colors are not detected in Fig. 4.8a, which suggests that, in these regions, though the long-term dynamics are ultimately identical, convergence along distinct paths creates a "masquerading" effect that the developed technique of periodicity correction exposes. Overall, we call this technique "Deterministic Chaos Prospector," since it readily identifies regions of simple (Morse-Smale) and chaotic structurally unstable dynamics in the parametric plane.

Figure 4.9 presents the (ν_1, ν_2)-parametric sweeps of the Rabinovich system for two different values of h: $h = 35$ in panel (a) and $h = 15$ in panel (b); in both cases, a short symbolic scanning is done with n: 5–12 range. Both sweeps disclose a stunning complexity of the organization of the bifurcation unfolding of the system, with a plethora of Bykov T-points with characteristic spirals, and separating saddles in the chaotic region, that greatly stand in contrast to the Morse-Smale regions of simple and stable dynamics. To conclude, long sweeps with the scanning range n: 105–128 to expose the long-term dynamics of the Rabinovich system at $h = 35$ and at $h = 15$ are presented in Fig. 4.10a and b, respectively. Both indicate the occurrence of chaotic, structurally unstable dynamics, clearly depicted by the seemingly noisy regions. As seen above, periodicity correction gets rid of some of the spiral structures around the pseudo T-points (Fig. 4.10c, d). With the enhanced technique, we can also easily identify multiple bands or stability windows corresponding to periodic attractors, within the chaos-land of the system under consideration.

4.5 Methods

Computations of bi-parametric sweeps are performed on a workstation with Intel Xeon(R) 3.5 GHz 12-core CPU and 32 GB RAM, with an NVidia Tesla K40 GPU for parallelization using CUDA. A bi-parametric sweep over a grid of 5000×5000 mesh points, with a scanning depth n: 5–12 (Fig. 4.5) takes about 4.45 s, whereas, for long-term dynamics with n: 105–128 (Fig. 4.3f), it is about 42 s. With the periodicity correction algorithm employed, these numbers increase to approximately 4.68 s and 43.7 s, respectively. Visualizations of the sweeps and trajectories are done using Python. The colormap is constructed by discretizing the range of computed P-values, i.e., [0,1], into 2^{24} distinct levels and assigning them RGB (Red-Green-Blue)-color values which are arranged in the following order—increasing values of Blue color from 0 through 1, decreasing values of Red color from 1 through 0, and randomly assigned values of Green color between 0 and 1. As such, $P = 1$ is associated with a bluish color, while $P = 0$ with a reddish color. With this colormap, we can identify distinct topological dynamics of up to 24 symbols long. As seen in Fig. 4.3, since two neighboring regions differ in the last symbol, the corresponding P-value is defined in such a way that the weight of the last symbol is the highest, so that the two neighboring regions fall in either half of the color map range [0, 1], and thus, have the greatest contrast. In regions where the equilibrium state at the origin is stable, and is the only global attractor of the system, the P value is complementarily set to -0.1, which is outside the normalized range of all computed P-values. Similarly, adjacent regions in the parametric space close to the pitch-fork bifurcation, through which the stable origin becomes a saddle with 2D stable and 1D unstable manifold, and also gives rise to a couple of saddle foci $C^{\pm}$, are assigned a P value -0.05, which is also outside the normalized computed range. In both cases, this is done due to monotone convergence to the steady state or slow transients, which do not admit proper partitioning of the phase space of the system to generate symbolic representation of its solutions.

In order to construct sweeps with periodicity correction, we first detect periodic orbits in each sequence. For a sequence of length N, we check for periodic orbits of periodicity up to $\frac{N}{2}$, starting from 2. If a periodic orbit is detected, it is normalized to the principal cyclic permutation with the lowest numerical value in the sorted list of all of its cyclic permutations. For example, the periodic orbits $\{\overline{110}\}$, $\{\overline{101}\}$, $\{\overline{011}\}$ are all normalized to $\{\overline{011}\}$, which is then used to fill up the symbolic sequence of length N for the computation of the corresponding $P(N)$-value.

4.6 Conclusions

In this study we have shown that

- Symbolic representation is an effective tool to reveal the bifurcation origins of complex, chaotic dynamics in the Rabinovich system, and in other similar systems.
- Bi-parametric scans disclose fine organizational features of deterministic chaos due to complex, self-similar assemblies of homoclinic and heteroclinic bifurcations with a plethora of accompanying T-points, spiral structures, and separating saddles in the fractal, self-similar regions in the parameter plane, that correspond to complex chaotic dynamics. There is no other current computational technique that can reveal the complexity of chaos in the parameter plane with such stunning clarity and completeness.
- The technique of periodicity correction gets rid of a variety of artifacts due to shift-cycling in symbolic representation of complex periodic orbits, in addition to effectively detecting stability windows of regular dynamics that are embedded within the chaotic regions.
- We developed "Deterministic Chaos Prospector"—the paradigm based on the concept of structural and dynamic instability, which presents a novel and highly efficient approach to identify the regions of chaotic and stable dynamics in the parameter space of a system under consideration.
- Massively parallel multi-parametric sweeping based on symbolic representation, using general purpose GPU-computing, presents the new generation, optimal time computational method to study a chaotic system that admits a proper partition of its phase space.

Acknowledgements AS acknowledges the financial support from RSF grant 14-41-00044 at the Lobachevsky University of Nizhny Novgorod, as well as NSF BIO-DMS grant IOS-1455527. We thank the acting members of the NEURDS (Neuro Dynamical Systems) lab at GSU for helpful discussions and proof-reading the manuscript. We are very thankful to Sunitha Basodi for her deep insights into the computational aspects of GPU programming, and the Periodicity Correction algorithm. We gratefully acknowledge the support of NVIDIA Corporation with the donation of the Tesla K40 GPU used in this research.

References

1. Afraimovich, V., Shilnikov, L.: Strange attractors and quasiattractors. In: Nonlinear Dynamics and Turbulence. Interaction of Mechanics and Mathematics Series, pp. 1–34. Pitman, Boston (1983)
2. Afraimovich, V.S., Bykov, V., Shilnikov, L.P.: Akademiia Nauk. SSSR Doklady **234**, 336–339 (1977)
3. Afraimovich, V., Bykov, V., Shilnikov, L.P.: Trans. Moscow Math. Soc. **2**, 153 (1983)
4. Barrio, R., Blesa, F., Serrano, S., Shilnikov, A.: Phys. Rev. E. **84**(3), 035201 (2011)
5. Barrio, R., Shilnikov, A., Shilnikov, L.P.: Int. J. Bifurc. Chaos **22**(04), 1230016 (2012)
6. Bussac, M.N.: Phys. Scr. **T2**(SI) (1982). http://iopscience.iop.org/article/10.1088/0031-8949/1982/T2A/014/meta
7. Bykov, V.V.: On the structure of bifurcations sets of dynamical systems that are systems with a separatrix contour containing saddle-focus. In: Methods of Qualitative Theory of Differential Equations, pp. 44–72. Gorky University (1980) (in Russian)
8. Bykov, V.: Physica D: Nonlinear Phenom. **62**(1–4), 290 (1993)
9. Chen, C., Cao, J., Zhang, X.: Nonlinearity **21**(2), 211 (2008)
10. Dmitrieva, T.V., Ryskin, N.M.: J. Exp. Theor. Phys. **89**(5), 1015 (1999)
11. Emiroglu, S., Uyaroğlu, Y.: Sci. Res. Essays **5**(21), 3298 (2010)
12. Ezersky, A.B., Rabinovich, M.I., Stepanyants, Y.A., Shapiro, M.F.: ZETP **76**(3), 991 (1979)
13. Gallas, J.A.:Int. J. Bifurc. Chaos **20**(02), 197 (2010)
14. Guckenheimer, J., Williams, R.F.: Publications Mathématiques de l'IHÉS **50**, 59 (1979)
15. Kocamaz, U.E., Uyaroğlu, Y., Kizmaz, H.: Int. J. Adapt. Control Signal Process. **28**(12), 1413 (2014)
16. Kostrykin, S.V., Romanova, N.N., Yakushkin, I.G.: On stochastic stabilization of the Kelvin-Helmholtz instability by three-wave resonant interaction. CHAOS **21**, 043117 (2011). doi: http://dx.doi.org/10.1063/1.3656800
17. Krishchenko, A., Starkov, K.: Phys. Lett. A **367**(1–2), 65 (2007)
18. Kuznetsov, S.P.: Tech. Phys. **61**(3), 436 (2016)
19. Kuznetsov, S.P.: Lorenz Type Attractor in Electronic Parametric Generator and Its Transformation Outside the Parametric Resonance. In: Aranson, I., Pikovsky, A., Rulkov, N., Tsimring, L. (eds.) Advances in Dynamics, Patterns, Cognition. Nonlinear Systems and Complexity, vol. 20. Springer, New York (2017)
20. Letellier, C., Aguirre, L., Maquet, J., Lefebvre, B.: Physica D **179**(1–2), 33 (2003)
21. Liu, A., Li, L.: Nonlinear Dynam. **81**(4), 2141 (2015)
22. Liu, Y., Yang, Q., Pang, G.: J. Comput. Appl. Math. **234**(1), 101 (2010)
23. Liu, Y., Li, L., Wang, X.: Int. J. Geom. Methods Mod. Phys. **12**(09), 1550092 (2015)
24. Llibre, J., Valls, C.: J. Geom. Phys. **58**(12), 1762 (2008)
25. Llibre, J., Messias, M., da Silva, P.R.: J. Phys. A: Math. Theor. **41**(27), 275210 (2008)
26. Neukirch, S.: Phys. Rev. E **63**, 036202 (2001)
27. Pikovsky, A., Rabinovich, M., Trakhtengerts, V.: Sov. Phys.—JETP **47**(4), 715 (1978)
28. Porter, J., Knobloch, E.: Physica D **143**(1–4), 138 (2000)
29. Rabinovich, M.I., Fabrikant, A.L.: ZETP **77**(2), 617 (1979)
30. Shilnikov, L.P.: Sov. Math. **6**, 163 (1965)
31. Shilnikov, L.P.: Matematicheskii Sbornik **119**(3), 461 (1968)

32. Shilnikov, L.: Bifurcation theory and the Lorenz model. In: Marsden, J., McCraken, M. (eds.) Appendix to Russian Edition of The Hopf Bifurcation and Its Applications, pp. 317–335. Mir Publications (1980) (in Russian)
33. Shilnikov, L.: The theory of bifurcations and quasiattractors. Uspekhi Math. Nauk **36**, 240–242 (1981) (in Russian). http://www.mathnet.ru/php/journal.phtml?jrnid=rm&option_lang=rus
34. Shilnikov, A.: Bifurcations and chaos in the Morioka–Shimizu model. Part I. In: Methods of Qualitative Theory of Differential Equations, pp. 180–193. Gorky University (1986) (in Russian)
35. Shilnikov, A.L.: Physica D: Nonlinear Phenom. **62**(1–4), 338 (1993)
36. Shilnikov, L.P., Shilnikov, A.L., Turaev, D.V., Chua, L.O.: Methods of Qualitative Theory in Nonlinear Dynamics. World Scientific, Singapore (1998)
37. Tudoran, R.A.: Int. J. Geom. Methods Mod. Phys. **09**(05), 1220008 (2012)
38. Tudoran, R.M., Gîrban, A.: Discret. Contin. Dynam. Syst. Ser. B **15**(3), 789 (2011)
39. Vishkind, S.Y., Rabinovich, M.I.: ZETP **71**(8), 557 (1976)
40. Wang, P., Masui, K.: Phys. Lett. A **81**(2), 97 (1981)
41. Wersinger, J.M., Finn, J.M., Ott, E.: Phys. Fluids **23**(6), 1142 (1980)
42. Xing, T., Barrio, R., Shilnikov, A.: Int. J. Bifurc. Chaos **24**(08), 1440004 (2014)
43. Xing, T., Wojcik, J., Barrio, R., Shilnikov, A.: International Conference on Theory and Application in Nonlinear Dynamics (ICAND 2012), pp. 129–140. Springer, Berlin (2014)
44. Xing, T., Wojcik, J., Zaks, M., Shilnikov, A.L.: In: Nicolis, G., Basios, V. (eds.) Chaos, Information Processing and Paradoxical Games: The Legacy of J.S. Nicolis. World Scientific Publishing, Singapore (2015)

Chapter 5
Anomalous Transport in Steady Plane Viscous Flows: Simple Models

Michael A. Zaks and Alexander Nepomnyashchy

5.1 Introduction

Traditionally, the enhancement of particle transport by advection is described in terms of an *effective diffusivity*: the mean square displacement of a particle satisfies the relation

$$\langle r^2(t)\rangle = 2dD_{\rm eff}\, t, \tag{5.1}$$

where d is the spatial dimension of the flow region, and the effective diffusion coefficient $D_{\rm eff}$ can be significantly larger than the molecular diffusion coefficient D. Relation (5.1) stays valid for a wide class of flows, from a slow laminar flow in a channel [18] to turbulent flows [17]; for review, see [9]. Deviations from the relation (5.1), called "*anomalous diffusion*," usually originate in the geometry of certain flow patterns, and can be observed within the restricted time range, as transient phenomena [2, 13, 20]. Permanent anomalous diffusion is known to be a characteristic feature of fractals and disordered systems [5].

The seminal paper of M.I. Rabinovich and his collaborators [1], for the first time, predicted the possibility of a permanent *multifractal anomalous diffusion*, characterized by the law

$$\langle r^q(t)\rangle \sim t^{q\gamma(q)}/2, \quad \gamma(q) \neq 1, \tag{5.2}$$

M.A. Zaks (✉)
Humboldt University Berlin, Berlin, Germany
e-mail: zaks@physik.hu-berlin.de

A. Nepomnyashchy
Technion, Haifa, Israel
e-mail: nepom@math.technion.ac.il

I.S. Aranson et al. (eds.), *Advances in Dynamics, Patterns, Cognition*, Nonlinear Systems and Complexity 20, DOI 10.1007/978-3-319-53673-6_5

for the particle transport in a regular two-dimensional velocity field, periodic in space and time. For that kind of periodic flows, the transport anomaly is caused by the complexity of the underlying Poincaré map, which includes regions of mixing ("Lagrangian chaos") and stability islands ("traps") with no mixing. Disclosure of the dynamical mechanism standing behind the transport properties allowed the authors of [1] to call this phenomenon "anomalous dynamic diffusion." Later, anomalous diffusion was found also for random but correlated flows [9].

Is persistent anomalous transport possible in *steady* spatially periodic flows, where there is no Lagrangian chaos? A bit surprisingly, the answer is positive. In [23], the advection of particles was considered for a class of two-dimensional steady flows that were spatially periodic and contained isolated stagnation points. In its motion along the streamline of the flow, each tracer repeatedly visited the neighborhoods of stagnation points, coming arbitrarily close to them. The primary interest to that kind of flows was caused by the unusual, singular continuous, power spectrum of their Lagrangian observables. Due to interrelation between the spectral characteristics and the transport properties, evolution of ensembles of tracers, carried by such flows, turned out to be unconventional as well. In the deterministic context (absence of molecular diffusion) the particle dispersion at large time showed a subdiffusive asymptotics. Similar transport properties were disclosed in steady plane viscous flows past periodic arrays of solid obstacles [22]. In both cases, the origin of the anomalous transport is the unboundedness of the time required for a passage across a basic cell of the flow pattern ("return time"), caused by presence of stagnation points or zones [21]. Because of the low velocity of particles near the stagnation points, a direct numerical computation of the particle dispersion law is rather costly; in any case, extensive simulations of partial (or, at best, ordinary) differential equations over long time intervals are required. A significant acceleration of computations can be reached using the model of "special flow" [3] (for details, see Sect. 5.2 below).

In this paper, we present a systematic description of the anomalous transport in steady plane flows modeled by means of the special flow approach. We start in Sect. 5.2 with description of exemplary steady flow patterns with stagnation points that arise in incompressible fluids under the action of spatially periodic time-independent forcing. In Sect. 5.3, we explain the origin of logarithmic and power law-like singularities in the distributions of passage times for tracers carried along the streamlines of such flows; qualitatively, effects of those singularities upon transport are described in Sect. 5.4. In Sect. 5.5 we introduce the main tool of our research: the special flow, based on circle maps with irrational rotation numbers. In Sect. 5.6, results of numerical analysis are presented. We show that the monotonic (power-law or logarithmic) temporal evolution of dispersion, common for all rotation numbers, is "decorated" by a non-monotonic pattern determined by the particularities of the continued fraction representation of the given rotation number. In the case of power-law singularities, the numerically obtained estimates of the growth rate of dispersion match well the theoretical predictions. Furthermore, presence of singularities in the distributions of passage times makes the transport not only anomalous, but also multifractal.

5.2 Enforced Flow Patterns with Stagnation Points

We confine ourselves to the simple setup of steady two-dimensional flow patterns of incompressible fluids. Since in every point of the physical space the velocity is constant, the Eulerian observables are static. In contrast, the Lagrangian observables, like the velocity components of tracers carried along the streamlines, are, in general, time-dependent; epochs of slow drift can alternate with epochs of relatively fast motion. In a steady plane flow, incompressibility implies that the motion of a tracer obeys the Hamiltonian system with one degree of freedom in which the velocity components are canonical conjugate variables. The system is integrable: In the two-dimensional geometry there are no chaotic streamlines, and both Lyapunov exponents vanish identically.

To obtain nontrivial transport characteristics, we need the cumulative effect of repeated passages of tracers through the regions of fast and slow motion. The flow pattern should ensure infinitely many returns to those regions. A simple realization of this mechanism is delivered by a forced flow on the 2-torus with irrational rotation number, provided that the vector field on the torus turns into zero in certain points or along certain curves. In conventional hydrodynamical terms, this is a flow on a square with periodic boundary conditions and non-zero mean components. Due to irrational rotation number, the tracers repeatedly visit the stagnation region(s).

We start with the flow of the incompressible fluid with density ρ_0 and kinematic viscosity ν on the 2-torus ($0\leq x \leq 2\pi, 0\leq y \leq 2\pi$), governed by the forced Navier–Stokes equation

$$\frac{\partial}{\partial t}\mathbf{v} + (\mathbf{v}\cdot\nabla)\mathbf{v} = -\frac{\nabla P}{\rho_0} + \nu\,\nabla^2\mathbf{v} \;+\; \mathbf{F}, \qquad \nabla\cdot\mathbf{v} = 0 \tag{5.3}$$

with $\mathbf{v}$ and P being, respectively, the velocity and the pressure. The force $\mathbf{F} = (f\sin y, f\sin x, 0)$, parameterized by the amplitude f, is spatially periodic and time-independent; physically, such forces can be generated in thin layers of electroconducting fluids by regular arrays of electrodes positioned at the bottom [16]. The structure of the forcing term is reminiscent of the Kolmogorov flow [10]. In contrast to the latter, we impose periodic boundary conditions:

$$\mathbf{v}(x,y) = \mathbf{v}(x+2\pi, y) = \mathbf{v}(x, y+2\pi) \tag{5.4}$$

and introduce along both coordinates the fixed non-zero mean flow across the square domain, parameterizing it by the flow rates α and β, respectively:

$$\int_0^{2\pi} v_x dy\Bigg|_{x=0,2\pi} = 2\pi\alpha, \qquad \int_0^{2\pi} v_y dx\Bigg|_{y=0,2\pi} = 2\pi\beta \tag{5.5}$$

Introducing the stream function $\Psi(x, y) : \; v_x = \partial\Psi/\partial y, \;\; v_y = -\partial\Psi/\partial x$, we obtain the steady solution of the Navier–Stokes equation (5.3)

$$\Psi(x, y) = \alpha y - \beta x + \frac{f \; \sin(x - \phi_1)}{\sqrt{\alpha^2 + \nu^2}} - \frac{f \; \sin(y - \phi_2)}{\sqrt{\beta^2 + \nu^2}} \tag{5.6}$$

$$\phi_1 = \arctan\frac{\nu}{\alpha}, \quad \phi_2 = \arctan\frac{\nu}{\beta},$$

and, thereby, the components of time-independent velocity field:

$$\dot{x} = v_x = \alpha - \frac{f \; \cos(y - \phi_2)}{\sqrt{\beta^2 + \nu^2}}, \qquad \dot{y} = v_y = \beta - \frac{f \; \cos(x - \phi_1)}{\sqrt{\alpha^2 + \nu^2}} \tag{5.7}$$

In the absence of forcing ($f = 0$) this is the linear flow on the 2-torus with the rotation number α/β: the streamlines are straight, and the velocity is the same in every point. If α and β are commensurate, all streamlines are eventually closed. Below we focus on the generic case: irrational values of α/β; there, every streamline is dense on the torus, hence a passive tracer carried by the flow repeatedly passes arbitrarily close to any given position. Transport at $f = 0$ is the trivial translation: an ensemble of tracers, carried away from its original location, preserves its size and shape. Introduction of weak force deforms the streamlines (Fig. 5.1a) and makes the velocity field non-uniform. However, at sufficiently small $|f|$ the qualitative picture persists: due to local variations of velocity, the distance between the neighboring tracers mildly oscillates, but on the average there is no spreading of the ensemble.

The picture changes when the forcing amplitude reaches the critical value

$$|f_{\rm cr}| = \sqrt{\alpha^2\beta^2 + \nu^2 \max(\alpha^2, \beta^2)} :$$

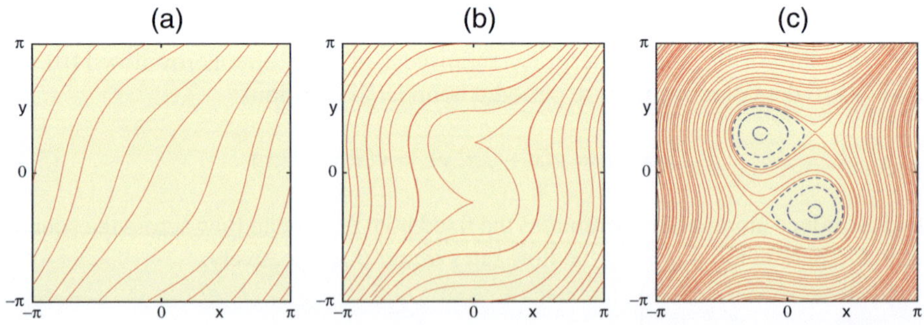

Fig. 5.1 Transformation of the flow pattern (5.6) under the increase of the forcing amplitude f. *Solid curves*: global component; *dashed curves*: nested closed streamlines inside the vortices. Displayed patterns correspond to (**a**) $|f| < f_{\rm cr}$, (**b**) $|f| = f_{\rm cr}$ and (**c**) $|f| > f_{\rm cr}$

cusp points appear upon two symmetric streamlines (Fig. 5.1b). In these points both components of velocity vanish. When $|f|$ is increased beyond f_{cr}, each cusp splits into a pair of stagnation points: the elliptic point and the saddle. Two counter-rotating vortices (eddies) appear on the torus. Each vortex is centered around one of the elliptic points, is filled by nested closed streamlines, and is encircled by the separatrix of one of the saddles. For $|f| > f_{cr}$ the phase space is mixed, with simple periodic motions inside the eddies, and the global component, populated by aperiodic motions with rotation number α/β.

Each trajectory outside the eddies is dense in the global component, hence a tracer repeatedly traverses the vicinities of the saddle points, where its velocity becomes arbitrarily small. Now, consider *two* initially close tracers that enter the region near the saddle point. The particle whose path lies closer to the saddle experiences a stronger slowdown; as a result, the former neighbors leave the stagnation region non-simultaneously and move apart. This decorrelating action is at the strongest when two tracers approach the saddle along opposite sides of the incoming separatrix: on passing the saddle, one of the particles departs from the slow region along the "non-closed" outgoing separatrix, whereas the other one is carried around the eddy and returns back to the stagnation zone where it slows down again. Although the tracers remain on the neighboring trajectories, the average distance between them grows in the course of many such events. Decorrelation, in its turn, is reflected in the spectral properties: the Fourier spectra of Lagrangian observables in Eq. (5.3) are neither discrete nor absolutely continuous with respect to the Lebesgue measure, but are supported by the fractal sets [23]. In a pattern like one in Fig. 5.1c with symmetric counter-rotating vortices, the relative accelerations and slowdowns roughly balance in the long run, and the autocorrelation of a Lagrangian observable (say, of a velocity component) does not ultimately decay. A more elaborate geometry of forcing allows to generate a flow with just one eddy inside a square-shaped cell; an example for the stream function

$$\Psi(x, y) = \alpha y - \beta x + \sin x - \cos y + \sin x \cos y \tag{5.8}$$

is shown in Fig. 5.2. This configuration lacks balance, and the autocorrelation of velocity displays algebraic decay [23].

5.3 Singularities of Passage Time

Spatial periodicity of the flow pattern prompts a reduction of dynamics in continuous time to the discrete Poincaré map, with boundaries of the square cell (e.g., any of the horizontal borders in Figs. 5.1 and 5.2) serving as natural secants. Remarkably, the one-dimensional Poincaré mapping induced by the flow is equivalent to the circle shift $\xi \to (\xi + \alpha/\beta) \bmod 1$ with rotation number α/β. Dynamics of such mappings is ordered, correlations do not decay along their orbits, and power spectra are discrete (pure point). In contrast, the spectra of the underlying flows include

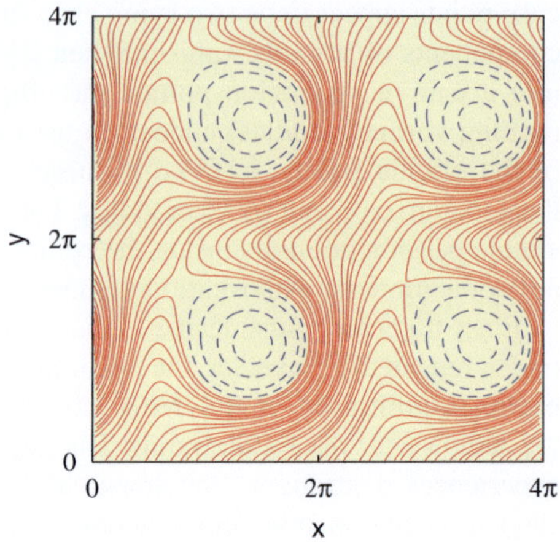

Fig. 5.2 Pattern with golden mean rotation number, generated by stream function (5.8) at $\alpha = 3/2$, $\beta = 3(\sqrt{5} - 1)/4$ (2×2 elementary cells). *Solid curves*: global component; *dashed curves*: nested closed streamlines inside the vortices

(singular) continuous component whose presence owes to singularities of temporal characteristics [12]. Indeed, in some sense an iteration of the map lasts "as long" as a trajectory of the flow requires for a passage from the secant onto the next secant—in other words, for a passage through the basic cell of the flow pattern. As long as stagnation points are absent, this time is obviously bounded, and there is no qualitative difference between the dynamics of the map and the dynamics of the underlying flow. Birth of vortices does not change the mapping but marks an important transition for the timescale: the return (passage) time $\tau_{\rm ret}(x)$ diverges when the initial position x of a tracer approaches the path which exactly hits the saddle stagnation point (the incoming separatrix of the saddle). Straightforward linearization shows that this divergence follows the law $\tau_{\rm ret}(x) \sim C_{l,r} \log |x - x_0|$, with x_0 being the coordinate of intersection of the separatrix with the Poincaré secant. The prefactors C_l and C_r refer to orbits which start on the Poincaré secant, respectively, to the left and to the right from x_0. Since trajectories on one side of x_0 make an excursion around the vortex, the tracers on them hover near the saddle twice before returning onto the secant, whereas the tracers from the opposite side of x_0 depart from the saddle and go directly to the secant. Therefore the prefactors C_l and C_r differ by the factor of 2. Since the flow pattern from Fig. 5.1 contains two vortices per basic cell, the passage time $\tau_{\rm ret}(x)$ diverges in two points; due to the symmetry the sum of both C_l equals the sum of both C_r. The passage time for the pattern from Fig. 5.2 has just one asymmetric ($C_r = 2C_l$) logarithmic singularity per basic cell.

Character of singularities of return time is reflected by dynamical properties of the flows. Recall a few mathematical results, relevant to this context. Flows on 2-tori without singular points (in our case, this is the flow (5.7) at $f < f_{\rm crit}$) have discrete spectrum and do not possess the property of mixing [8]. Flows on 2-tori with symmetric logarithmic singularities of return time (e.g., (5.7) at $f > f_{\rm crit}$) do not mix [7]. Flows on 2-tori with asymmetric logarithmic singularities of return time mix [15]. Finally, the flows on 2-tori with power-law singularities of return time mix [6].

In generic flows with vortices bounded by separatrices of hyperbolic stagnation points, the singularities of passage time are logarithmic. Power-law singularities are typical for non-hyperbolic equilibria, hence they occur at the criticality: at the bifurcation that marks the transition from the flow pattern without points of equilibrium to the pattern with equilibria. In this situation the departure from the newborn stagnation point is algebraic in time. In Eq. (5.7) at $f = f_{\rm crit}$, as well as in Eq. (5.8) and in similar flows without additional local degeneracies, the Hamiltonian saddle-center bifurcation takes place. Dynamics near the linearly neutral equilibrium can be brought to the form $\ddot{z} - z^2 = 0$, and the return time diverges near the separatrix as $|x - x_0|^{-1/6}$. If the velocity field near the neutral stagnation point has additional degeneracies, the quadratic term (as well as some of the higher-order terms) in the equation of motion can be missing. Altogether, for the appropriate normal form

$$\ddot{z} + z^n = 0, \qquad n \geq 2, \tag{5.9}$$

the time of passage near the neutral point diverges as $|x - x_0|^{-\kappa}$ with

$$\kappa = \frac{n-1}{2n+2} \tag{5.10}$$

For example, in the case of the local mirror symmetry $n = 3$, the "Hamiltonian pitchfork bifurcation" takes place, with $\kappa=1/4$.

For all these bifurcations of isolated equilibria in plane Hamiltonian flows the singularities of passage time, although stronger than logarithmic ones, are relatively weak: $\kappa < 1/2$. A stronger anomaly occurs in the hydrodynamically different situation, in which there is neither a forcing (except for the homogeneous pressure gradient which ensures the flow across the cells) nor a bifurcation: motion of a viscous flow past a regular array of solid obstacles. The flow pattern is shown in Fig. 5.3.

When viscosity is high enough, nonlinear terms in the equation of motion can be discarded; the resulting Stokes flow obeys, in terms of the stream function $\psi(x, y)$, the biharmonic equation

$$\Delta\Delta\psi = 0. \tag{5.11}$$

Due to no-slip boundary conditions, both components of velocity vanish not in isolated stagnation points, but along the entire border of each obstacle. From the point of view of dynamics, borders are continua of fixed points, and the motion along these continua is rather slow. In local coordinates, near the circular obstacle of radius R the solution of the biharmonic equation centered at the origin is

$$\psi(x, y) = \frac{(\alpha y - \beta x)(x^2 + y^2 - R^2)^2}{x^2 + y^2} + h.o.t. \tag{5.12}$$

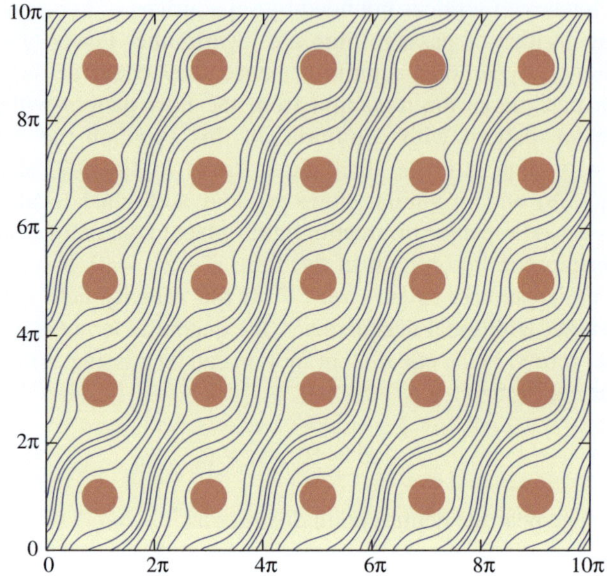

Fig. 5.3 Streamlines of the time-independent Stokes flow past the regular lattice of *solid circular cylinders*. For the description of the flow, see, e.g., [4, 14]

The time, required for a passage near the obstacle scales as $\tau_{\rm ret} \sim |x - x_0|^{-1/2}$ where x_0, as above, is the coordinate of the intersection of the Poincaré secant with the separatrix that ends on the cylinder. This singularity of return time is strong enough to ensure the power-law decay of the autocorrelation of Lagrangian observables [22].

5.4 Transport of Tracers Past Arrays of Obstacles

In all these examples of flow patterns, in spite of integrability of the underlying dynamical system, the fractal component is present in the spectra. Since transport properties are related to spectral characteristics, it is only natural that ensembles of tracers carried by such patterns display not quite conventional behavior. A compact cloud of passive tracers gets elongated and distorted with every new passage past a stagnation point or a solid obstacle. Figure 5.4 demonstrates the typical picture: after a sufficiently long time interval, the tracers from the initially small cloud can be found everywhere (outside the vortex) in the basic cell: an example of mixing in the integrable system.

The rate of the mixing process depends on the kind of singularity of passage time: mixing is slower in presence of logarithmic singularities and much faster for the power law-like ones. Vicinities of stagnation points/obstacles serve as a kind of

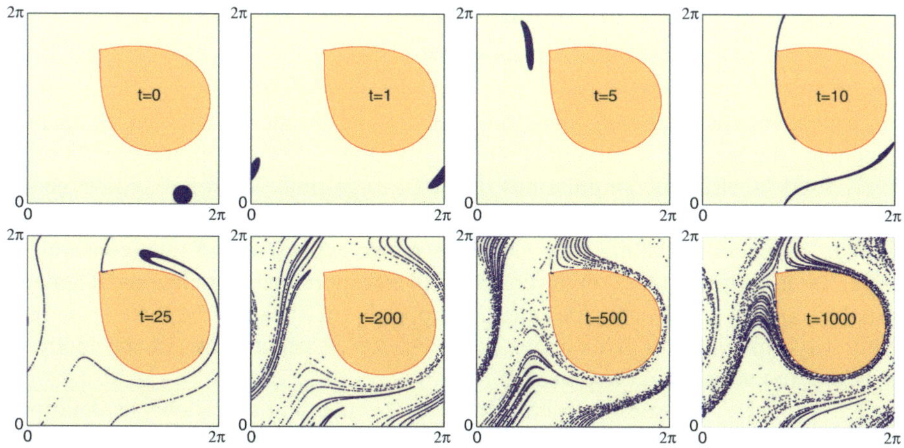

Fig. 5.4 Temporal evolution of the ensemble of 10^4 tracers transported by the pattern (5.8), projected upon the basic cell of the flow. *Dark region*: the eddy. Values of t indicate the time values at which the snapshots have been made

traps: when the ensemble moves past the trap, some of its elements get captured and spend a noticeable amount of time in the trap whilst the bulk of the ensemble is being carried further. One by one the tracers leave the trap, but in the meantime the new tracers get caught by further traps downstream, and the ensemble becomes stretched in the stream direction. As long as no molecular diffusion is allowed and the equations of motion stay completely deterministic, transport across the streamlines is absent. It is convenient to characterize stretching along the streamlines in terms of the time-dependent variance: we "unwrap" the torus onto the plane and consider the mean square elongation in the co-moving reference frame which drifts along with the center of the ensemble:

$$\xi^2(t) = \langle (x(\tau + t) - x(\tau) - \overline{v}_x t)^2 \rangle_\tau \tag{5.13}$$

where $\overline{v}_x$ is the mean velocity in the direction x and averaging is performed over the values of time τ.

Numerical estimates in [22, 23] have disclosed that the squared elongation is, on the average, growing in the course of time t, but the dependence on t was not a linear one: growth of $\xi^2(t)$ turned out to be strongly sublinear for flows past ensembles of vortices and mildly superlinear for the flow past a lattice of circular solid obstacles. This means that isolated stagnation points result in subdiffusion whereas repeated close passages along the no-slip boundaries produce superdiffusion. Since the effect is purely deterministic, we conclude that presence of either "hard" (solid bodies) or "soft" (vortices) obstacles in the flow pattern can be a source of "anomalous dynamic diffusion" in the sense of [1].

5.5 Special Flow Construction: Flow Over the Mapping

Direct numerical estimates of the transport characteristics through simulations of the hydrodynamical equations are cumbersome. In absence of positive Lyapunov exponents the process of mixing is remarkably slow, especially in the case of relatively weak logarithmic singularities of the passage time. Therefore, convergence of averages in Eq. (5.13) at non-small values of t requires billions of passages through basic cells, making such estimates rather expensive and time-consuming. Here, we draw conclusions on the character of transport from simulations of a much simpler model that preserves the essential properties of dynamics: repeated passages arbitrarily close to any given point and presence of singularities of the required character in the times of the passage.

This model is, in a sense, a modification of the mapping that allows its iterations to possess "duration." In expressions for average characteristics of conventional mappings, all iterations have the same weight. Account of duration allows us to assign higher weights to long slow passages from the secant to the next secant, and, respectively, lower weights to relatively short fast passages. This construction, suggested by von Neumann in [19], is known in ergodic theory as the "special flow" [3]; it is a flow built over the mapping. As sketched in Fig. 5.5, the phase space of the flow is the segment of the plane between the abscissa, parameterized by x, and the given function $T(x)$; the latter corresponds to the passage time $\tau_{\text{ret}}(x)$. The variable x (the variable of the circle mapping) changes its value in discrete events. The variable y mimics the passage time; it piecewise-continuously depends on time. Dynamics starts at some abscissa point and evolves as follows: The value of x remains fixed whereas the value of the coordinate y increases with unit speed ($\dot{y} = 1$) until reaching the value $T(x)$; from there, the point instantaneously jumps into the position $\big((x+\rho) \bmod 1, 0\big)$, starts the next epoch of vertical motion with unit speed, and so on.

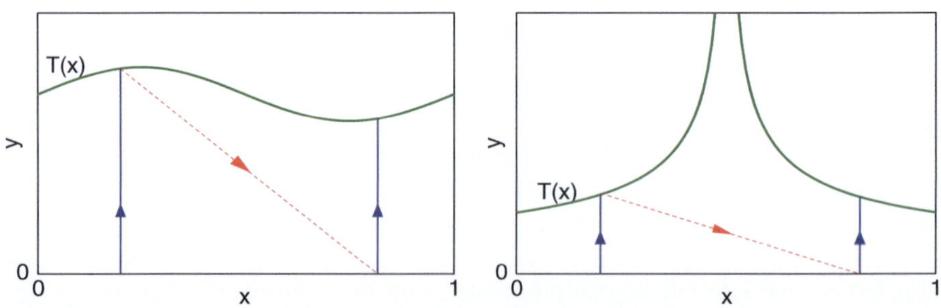

Fig. 5.5 Special flow over the circle map. Trajectory moves from the abscissa upwards with unit velocity, until hitting the curve $T(x)$. *Solid blue curve*: continuous evolution. *Dashed curve*: instantaneous iteration of the circle map. *Left panel*: passage time without singularities. *Right panel*: special flow with a singularity of passage time

Already von Neumann noted in [19] the role of singularities in $T(x)$: he showed that presence of a discontinuity disrupts the purely discrete spectrum and gives rise to the continuous spectral component. Later, special flows with different kinds of singularities were employed in studies of the mixing properties of flows on 2-tori with points of equilibrium [6, 15].

5.6 Results of Numerical Studies

To compute the transport characteristics, we took the special flow over the circle map $x \to (x + \rho) \bmod 1$ with golden mean rotation number $\rho = (\sqrt{5} - 1)/2$. To study the effects of logarithmic divergence of passage times, the functions

$$T(x) = 1 - \log|x - 1/2|$$

and

$$T(x) = \begin{cases} 1 - \log|1/2 - x|, & x < 1/2 \\ 1 - \log 2 - 2\log(x - 1/2), & x > 1/2 \end{cases}$$

were employed for, respectively, symmetric and asymmetric singularities. For power law singularities, the functions $T(x) = 2|x - 1/2|^{-\kappa}$ with $0 < \kappa < 1$ were taken. As an observable we have chosen the "phase" ϕ of the flow, defined for a point with coordinates $x(t), y(t)$ as the ratio $y/T(x)$; ϕ varies between 0 and 1. For the lift of ϕ onto the infinite line we calculated, in accordance with (5.13), the variance (mean squared elongation) $\xi^2(t)$. Trajectories of the special flow until $t = 10^{11}$ were computed numerically; this has allowed us to obtain good convergence of $\xi^2(t)$ in the range $0 < t < 10^5$.

5.6.1 *Logarithmic Singularities of Passage Time*

Figure 5.6 presents temporal evolution of variance $\xi^2(t)$ for two types of logarithmic singularities. Essential feature of both plots is remarkable non-monotonicity of $\xi^2(t)$ that owes to the properties of dynamics on the torus with the irrational rotation number. A trajectory on such torus never closes completely, but from time to time it returns arbitrarily close to any given point. Local minima of $\xi^2(t)$ correspond to the values of t at which the points on the average come especially close to their initial positions: the ensemble of tracers vaguely recalls its past. Rotation number ρ rigidly prescribes the ordering of returns; the closest returns occur at the values of t, proportional to denominators of the best rational approximations to ρ. For the golden mean σ the times of closest returns are proportional to the terms F_n of the Fibonacci sequence of integers; since $\lim_{n\to\infty} F_n/F_{n+1} = \sigma$, the positions of local minima of

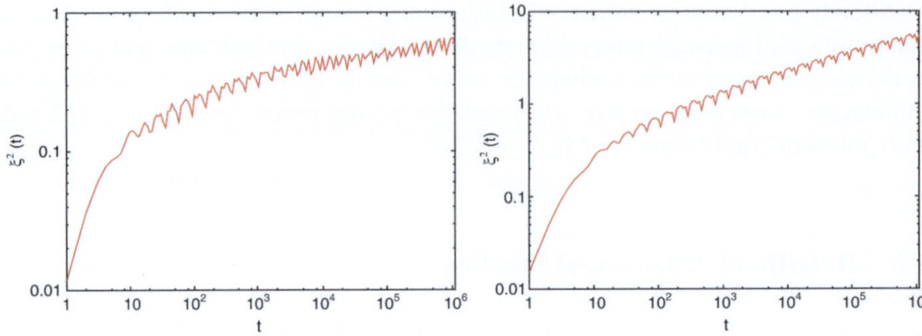

Fig. 5.6 Temporal evolution of variance in special flows with logarithmic singularities of passage time. *Left panel*: symmetric singularity. *Right panel*: asymmetric singularity

$\xi^2(t)$ form the log-periodic sequence, well recognizable in logarithmic coordinates. In the case of a different irrational rotation number ρ, the sequence of local minima is log-periodic whenever ρ is the quadratic irrational (i.e., its expansion into the continued fraction is periodic), otherwise that sequence looks disordered.

As seen in the plot, every next minimum lies higher than the previous one: the system gradually forgets its past, and the ensemble does not really contract to its initial size. Nevertheless, in the case of the symmetric singularity shown in the left panel, when the passages on different sides of the separatrix roughly compensate each other, dispersion along the streamline grows rather slowly: within the time, required for hundreds of thousands of turnovers on the torus, the mean deviation does not exceed 1: the size of just one basic cell. This is an example of the very slow subdiffusive process; on the average $\xi^2(t) \sim \log t$.

In the case of the asymmetric logarithmic singularity (right panel), the process goes noticeably faster. Again, we observe the general growing tendency, decorated by the log-periodic Fibonacci pattern. Fitting these results yields $\xi^2(t) \sim t^{0.1} \log t$

5.6.2 *Power-Law Singularities of Passage Time*

A faster subdiffusion is observed in special flows with singularities of the return time of the kind $T(x) \sim |x - x_0|^{-\kappa}$ with $0 < \kappa < 1/2$. In Sect. 5.3 we have shown that in the forced viscous flows the singularities of this kind correspond to critical situations: bifurcations of isolated points of equilibrium. In the left and right panels of Fig. 5.7 we present numerical results, respectively, for the cases $\kappa = 1/6$ (recall that this singularity occurs at the saddle-center bifurcation) and $\kappa = 1/4$ (pitchfork).

In both cases, a short initial segment with ballistic growth is followed by subdiffusion: the variance follows the power law with the exponent 2κ. The value of the exponent as well as the estimate for the corresponding global prefactor can be derived explicitly via the evaluation of the dispersion of the time that a special

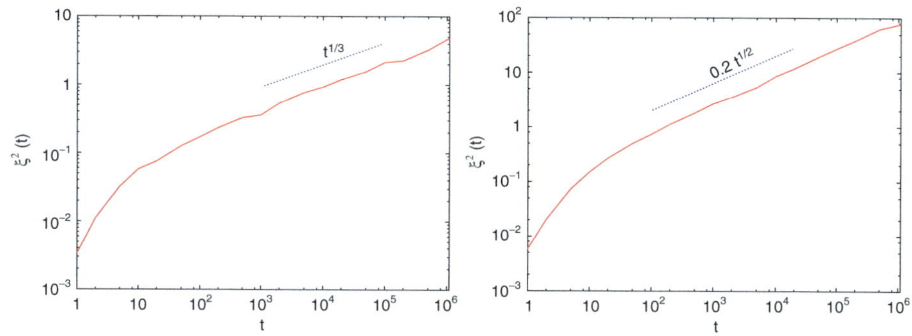

Fig. 5.7 Temporal growth of variance in special flows with power-law singularities of passage time. *Left panel*: $T(x) \sim |x - x_0|^{-1/6}$. *Right panel*: $T(x) \sim |x - x_0|^{-1/4}$. *Solid lines*: numerical data. *Dotted lines*: power laws

flow with this kind of power-law singularity requires for carrying out $n \gg 1$ iterations of the circle map; details of the derivation, based on the assumption of the *equidistant* distribution of the map iterations, will be published elsewhere [11]. The law of average growth $\xi^2(t) \sim t^{2\kappa}$ is common for irrational values of rotation numbers ρ; individual features of ρ prescribe the sequences of closest returns and invoke oscillations of the local prefactor before $t^{2\kappa}$. Since for the golden mean rotation number the return times form the geometric progression, these oscillations are log-periodic; unlike the case of logarithmic singularities, here they are masked by relatively fast growth of $\xi^2(t)$, but are well visible in plots of $\xi^2(t)/t^{2\kappa}$ vs. $\log t$.

As κ approaches 1/2 from below, the subdiffusive process gets closer to normal diffusion. Recall that the singularity $T(x) \sim |x - x_0|^{-1/2}$ does not occur in flows with isolated stagnation points, but is typical for flows past arrays of solid obstacles with no-slip borders. Remarkably, exactly at the value $\kappa = 1/2$ the transport turns to be slightly superdiffusive: numerical results in Fig. 5.8 indicate that the mean square elongation follows the law $\xi^2(t) \sim t \log t$. As a result, within the relatively short time the initially compact ensemble of tracers spreads in longitudinal direction along hundreds and thousand basic cells of the flow.

We performed analogous simulations for special flows with other values of κ in the range $1/2 \leq \kappa < 1$, invariably detecting superdiffusive transport. In this range the singularities are too strong for the estimates based on the equidistant distribution. Instead, we have considered the special flow based on the random circle map: on reaching $T(x)$, the point jumps to the position, randomly chosen from the uniform distribution on the unit interval. Describing this process as an unidirectional *continuous time random walk* (CTRW), we calculated the leading terms in the distribution of the number of jumps within the fixed time t (see [11] for details). For the marginal case $\kappa = 1/2$ this yields the estimate

$$\xi^2(t) = \frac{\sqrt{2}}{16}\, t \log t - \frac{5 + \log 2}{16\sqrt{2}}\, t + \cdots$$

which well matches the numerical results.

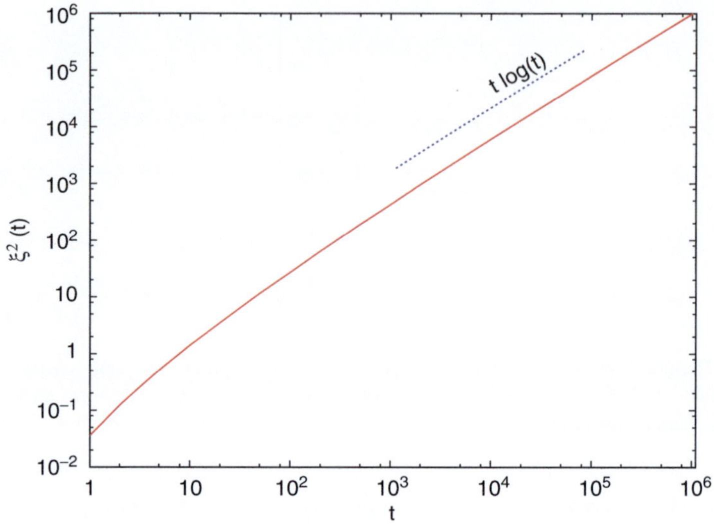

Fig. 5.8 Temporal growth of variance in the special flow with the singularity of passage time $T(x) \sim |x - x_0|^{-1/2}$. *Solid line*: numerical data. *Dotted line*: dependence $t \log t$

For $\kappa > 1/2$ the CTRW approach predicts superdiffusion with the leading term in the evolution of $\xi^2(t)$, proportional to $t^{3-1/\kappa}$; this prediction was corroborated by comparison with numerics.

Remarkably, transport in these special flows is not only superdiffusive but also multifractal. According to the CTRW calculation of the higher order central moments

$$m_n(t) = \langle (x(t) - \langle x(t) \rangle)^n \rangle,$$

the dominating term for $1/2 < \kappa < 1/$ is given by $m_n(t) \sim t^{n+1-1/\kappa}$ [11]. In Fig. 5.9 we demonstrate the effect of multifractality for the case of the special flow with $\kappa = 2/3$ by comparing the values of $r_n(t) = \big(m_n(t)\big)^{1/n}$.

5.7 Conclusions

Summarizing, we observe that plane steady laminar viscous flows past periodic arrays of liquid or solid obstacles can display anomalous transport properties: depending on the geometry of the flow pattern, motion of tracers can be subdiffusive or superdiffusive. Using the construction of special flow over the circle map with appropriate singularities of return time, we have obtained explicit estimates for the transport characteristics that well match the results of extensive numerical

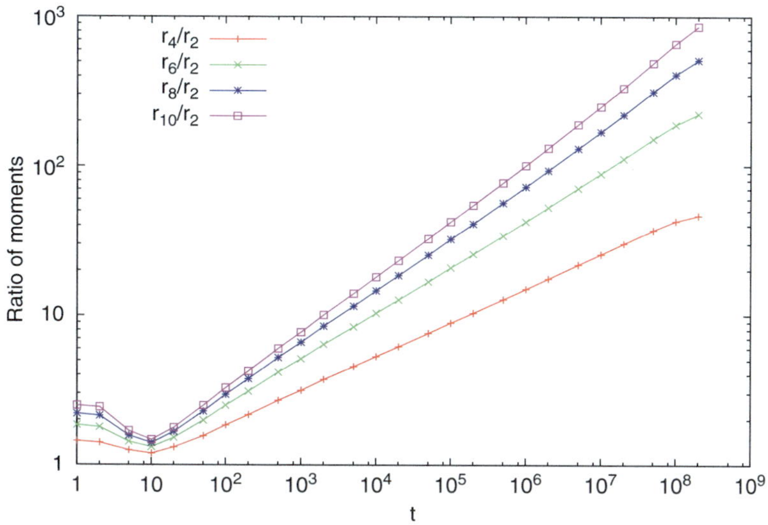

Fig. 5.9 Temporal evolution of ratios of central moments $r_n(t) = \langle (x(t) - \langle x(t) \rangle)^n \rangle^{1/n}$ for superdiffusive transport in the special flow with singularity $T(x) \sim |x - x_0|^{-2/3}$

simulations. Being deeply enrooted in deterministic dynamics of the flows, this kind of transport delivers yet another example of anomalous dynamic diffusion: the remarkable phenomenon, brought to our attention by the authors of [1].

Acknowledgements The authors feel deeply obliged to M.I. Rabinovich for his tireless proliferation of dynamical system approach to various fields of science. This research was supported by a Grant from the GIF, the German-Israeli Foundation for Scientific Research and Development, Grant I-1271-303.7/2014.

References

1. Aranson, I.S., Rabinovich, M.I., Tsimring, L.Sh.: Anomalous diffusion of particles in regular flows. Phys. Lett. A **151**, 523–528 (1990)
2. Bouchaud, P., George, A.: Anomalous diffusion in disordered media: statistical mechanisms, models and physical applications. Phys. Rep. **195**, 127–293 (1990)
3. Cornfeld, I.P., Fomin, S.V., Sinai, Ya.G.: Ergodic Theory. Springer, New York (1982)
4. Hasimoto, H.: On the periodic fundamental solution of the Stokes equations and their application to viscous flow past a cubic array of spheres. J. Fluid Mech. **5**, 317–328 (1959)
5. Havlin, S., Ben-Abraham, D.: Diffusion in disordered media. Adv. Phys. **36**, 695–798 (1987)
6. Kochergin, A.V.: On mixing in special flows over translation of intervals and in smooth flows on surfaces. Math. Sbornik **96**, 471–502 (1975)
7. Kochergin, A.V.: Nondegenerate saddle-points and absence of mixing. Math. Notes **19**, 453–468 (1976)
8. Kolmogorov, A.N.: On dynamical systems with an integral invariant on a torus. Dokl. Akad. Nauk. SSSR Ser. Mat. **93**, 763–766 (1953)

9. Majda, A.J., Kramer, P.R., Simplified models for turbulent diffusion: theory, numerical modelling, and physical phenomena. Phys. Rep. **314**, 237–254 (1999)
10. Meshalkin, L.D., Sinai, Ya,G.: Investigation of the stability of a stationary solution of a system of equations for the plane movement of an incompressible viscous liquid. J. Appl. Math. Mech. (Prikl. Mat. Mekh.) **25**, 1700–1705 (1961)
11. Nepomnyashchy, A., Zaks, M.A.: in preparation
12. Pikovsky, A.S., Zaks, M.A., Feudel, U., Kurths, J.: Singular continuous spectra in dissipative dynamical systems. Phys. Rev. E **52**, 285–296 (1995)
13. Pöschke, P., Sokolov, I.M., Nepomnyashchy, A.A., Zaks, M.A.: Anomalous transport in cellular flows: the role of initial conditions and aging. Phys. Rev. E **94**, 032128 (2016)
14. Sangani A.S., Yao, C.: Transport processes in random arrays of cylinders. II. Viscous flow. Phys. Fluids **31**, 2435–2444 (1988)
15. Sinai, Ya.G., Khanin, K.M.: Mixing for certain classes of special flows over the circle shift. Funct. Anal. Appl. **26**, 155–169 (1992)
16. Sommeria, J.: Experimental study of the two-dimensional inverse energy cascade in a square box. J. Fluid Mech. **170**, 139–168 (1986)
17. Taylor, G.I.: Diffusion by continuous movement. Proc. Lond. Math. Soc. **2**, 196–212 (1921)
18. Taylor, G.I.: Dispersion of soluble matter in solvent flowing slowly through a tube. Proc. R. Soc. A **219**, 186–203 (1953)
19. von Neumann, J.: Zur Operatorenmethode in der Klassichen Mechanik. Ann. Math. **33**, 587–642 (1932)
20. Young, W., Pumir, A., Pomeau, Y.: Anomalous diffusion of tracer in convection rolls. Phys. Fluids A **1**, 462–469 (1989)
21. Zaks, M.A.: Anomalous transport in steady plane flows of viscous fluids. In: Collet P., Courbage M., et al. (eds.) Chaotic Dynamics and Transport in Classical and Quantum Systems, pp. 401–412. Kluwer Academic Publishers, Dordrecht (2005)
22. Zaks, M.A., Straube, A.V.: Steady Stokes flow with long-range correlations, fractal Fourier spectrum and anomalous transport. Phys. Rev. Lett. **89**, 244101 (2002)
23. Zaks, M.A., Pikovsky, A.S., Kurths, J.: Steady viscous flow with fractal power spectrum. Phys. Rev. Lett. **77**, 4338–4341 (1996)

Part II
Synchronization and Networks

Chapter 6
Coherence–Incoherence Transition and Properties of Different Types of Chimeras in a Network of Nonlocally Coupled Chaotic Maps

Vadim S. Anishchenko, Tatiana E. Vadivasova, and Galina I. Strelkova

6.1 Introduction

Effects of synchronization, desynchronization, and different pattern formation in oscillatory ensembles, extended systems, and media are still the most important research directions in nonlinear dynamics and its applications, which has remained relevant (of significance) for many years. This is related to a large variety and complexity of the behavior of multicomponent systems, even in the case of a rather simple type of dynamics of individual elements [3, 11, 13, 15, 17, 18, 20, 21, 30, 32]. New effects are constantly discovered, new problems and challenges arise, and new models and methods are being developed. Recently a special type of cluster structures, which is observed in oscillatory networks and called chimera state [28], has attracted great interest. The distinctive feature of these patterns is that they consist of clearly identified spatial clusters of two types, i.e., parts with coherent behavior of oscillators (coherent clusters) and regions with inconsistent dynamics of elements (incoherent clusters). Chimera states appear in networks of identical oscillators with homogeneous interaction. This coupling is typically nonlocal, i.e., each oscillator is coupled with a group of others. Chimera states were first described in [1, 2, 14] for ensembles of phase oscillators, which are the simplest model of multicomponent oscillatory systems. Later on, similar structures were also found in a variety of one- and two-dimensional networks whose elements are represented by different oscillators, for example, Stuart-Landau harmonic self-sustained oscillators [35], oscillators with periodic unharmonic oscillations [24–26, 34]; discrete-time systems (maps) and differential systems in the regime of dynamical chaos [9, 22, 23, 33]. Several types of chimera states have been distinguished, e.g., phase and amplitude

V.S. Anishchenko (✉) • T.E. Vadivasova • G.I. Strelkova
Department of Physics, Saratov National Research State University, 83 Astrakhanskaya Street, 410012 Saratov, Russia
e-mail: wadim@info.sgu.ru; vadivasovate@yandex.ru; strelkovagi@info.sgu.ru

I.S. Aranson et al. (eds.), *Advances in Dynamics, Patterns, Cognition*, Nonlinear Systems and Complexity 20, DOI 10.1007/978-3-319-53673-6_6

chimeras and "chimera death" [9, 10, 24–26, 28, 35]. The peculiarities of chimera states and mechanisms of their formation with parameter variation depend largely on the type of network elements.

Chimeras which appear in ensembles of nonlocally coupled chaotic oscillators and discrete-time maps comprise a special type of chimera structures and their formation is closely related to the transition from complete chaotic synchronization to asynchronous dynamics (incoherent chaos) with decreasing coupling strength [9, 22, 23, 33]. Besides, chimera states can be observed when a partial element of a network exhibits a nonhyperbolic chaotic attractor [33]. More particularly, this must be a system in which a chaotic attractor results from a cascade of period doubling bifurcations of a periodic cycle. In such systems, the peculiarities of chimera formation are related to the fact that there are different combinations of phase shifts between interacting oscillators in the regime of 2-periodic (4-periodic, etc.) cycles as well as in the regime of multi-band chaotic attractors. This phenomenon is known as phase multistability [5, 7, 9, 10].

Chimera state is a nontrivial pattern with complex spatial and temporal behavior. They can be analyzed by using a number of different methods and characteristics. In this work we apply new methods and approach for studying spatial patterns which appear at the transition from complete chaotic synchronization to full incoherent chaos in an ensemble of chaotic oscillators with nonlocal coupling. Among these methods, the analysis of coupling characteristics of the ensemble elements and their cross-correlations can be distinguished. The use of these and other methods has significantly deepened the understanding of the bifurcation mechanisms of the transition, including the mechanisms of chimera structure formation, and has provided a more accurate classification of chimeras which appear in ensembles of chaotic oscillators.

We are very pleased to contribute to a collective monograph dedicated to the 75th birthday of the famous scientist, professor M.I. Rabinovich. We would like to note that 30 years ago one of the first (pioneering) papers on the study of dynamics in an ensemble of chaotic oscillators was published in co-authorship with Rabinovich [4].

6.2 Model and Problem Statement

We study a one-dimensional ring of nonlocally coupled identical chaotic oscillators, which is given by

$$x_i^{t+1} = f(x_i^t) + \frac{\sigma}{2P} \sum_{j=i-P}^{i+P} \left[f(x_j^t) - f(x_i^t) \right], \tag{6.1}$$

where $i = 1, 2, \ldots, N$ is the serial number of an individual oscillator, which can be considered to be a discrete spatial coordinate; N is the number of oscillators in the ensemble; t denotes the discrete time (iteration number); P defines the number of neighbors on each side of the ith oscillator, and σ is the coupling strength. The dynamics of all the oscillators in (6.1) is described by the well-known logistic map $f(x) = \alpha x(1 - x)$. The latter is one of the fundamental models of nonlinear dynamics [12, 31]. The bifurcation parameter α of the logistic map is fixed at the value $\alpha = 3.8$ which yields developed chaotic behavior in the individual element. We study numerically the model (6.1) for $N = 1000$ elements. The number of neighbors is set to be $P = 320$ that corresponds to the coupling radius $r = P/N = 0.32$.

The second term in (6.1) describes the overall effect of neighbors on the particular ith element of the ensemble

$$F(\sigma, r, x_i^t) = \frac{\sigma}{2P} \sum_{j=i-P}^{i+P} \left[f(x_j^t) - f(x_i^t) \right]. \qquad (6.2)$$

We will call $F(\sigma, r, x_i^t)$ the *coupling function.*

The dynamics of the system (6.1) was studied and described in many works, e.g., [9, 22, 23, 33], where the appearance of chimera-like states was mainly analyzed. Figure 6.1 shows a fragment of the numerically obtained bifurcation diagram for the model (6.1) in the (r, σ) plane. This diagram depicts regions of complete chaotic synchronization (1), periodic dynamics (2), and fully developed spatio-temporal chaos (fully incoherent oscillations) (3). Region 1 is bounded by the blowout bifurcation line [23]. When σ decreases inside region 2, the period-2, period-4, and period-8 oscillations emerge successively and exist in the regions separated by bifurcation lines. In the white regions shown in Fig. 6.1 complex oscillatory regimes and their bifurcations can be observed, including the appearance of chimera states (see also Fig. 6.2).

In terms of the notations introduced in [22, 23], regions 1 and 2 depicted in Fig. 6.1 correspond to coherence regions with wave numbers $k = 0$ and $k = 1$, respectively. The corresponding coherent states x_i in the system (6.1) are shown as snapshots (wave-like profiles) in the insets of Fig. 6.1. A wave-like profile or a snapshot is understood as a spatial distribution of the instantaneous values of dynamical variables x_i, which define the states of partial elements at a fixed time t. According to [22, 23], a network state is considered coherent if the following condition holds:

$$|x_i^t - x_{i+1}^t| < \delta, \;\; \delta \ll 1, \;\; t = \text{const} \gg 1, \\ i = 1, 2, \ldots, N. \qquad (6.3)$$

Thus, if the instantaneous distribution of dynamical variable values along the ring (instantaneous spatial profile or snapshot) can be described by a smooth, slowly changing function in space, then the network state is coherent. The regime of strong coherence corresponds to the condition $|x_i^t - x_{i+1}^t| = 0$ for any fixed t (see the upper right inset in Fig. 6.1). As can be seen from Fig. 6.1, when the coupling strength

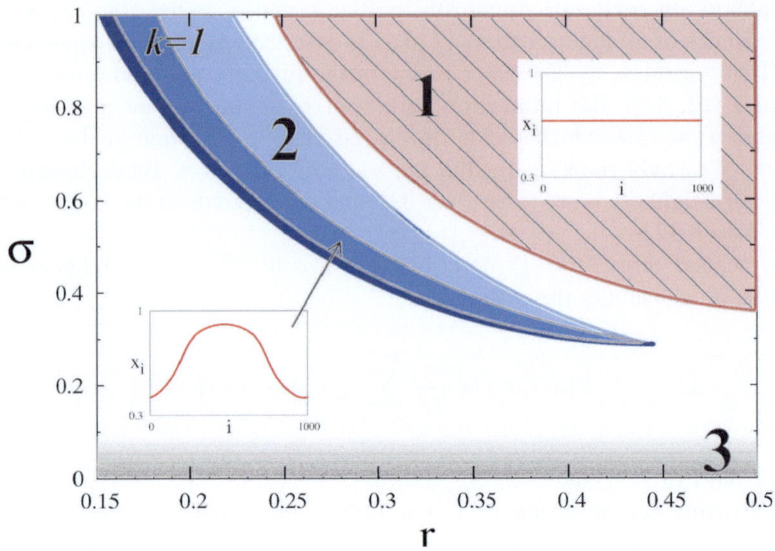

Fig. 6.1 Fragment of the numerically obtained bifurcation diagram for the system (6.1) in the (r, σ) parameter plane. Completely synchronized chaotic states exist in region 1 with wave number $k = 0$. Region 2 of coherent dynamics relates to 2^l periodic oscillations ($l = 1, 2, 3$) of the elements with $k = 1$. The corresponding snapshots are shown in the *insets*. Asynchronous (incoherent) dynamics is observed in region 3. The *white areas* correspond to different oscillatory regimes in the ring (6.1) and their bifurcations. Parameters: $a = 3.8$ and $N = 1000$

decreases from $\sigma = 1$ to $\sigma < 0.1$, a global transition from the fully coherent dynamics (complete chaotic synchronization) to the incoherence regime is realized in the system (6.1).

Let us consider in detail this transition using the different characteristic of the regimes. We will analyze numerically the behavior of the coupling function (6.2) and the dynamics of the network (6.1) with the parameter values given above and changing coupling strength σ. In contrast to the papers [22, 23, 33], where instantaneous values of the coordinate x_i^t (t = const) for all the network elements (snapshots) are mainly analyzed, we will use the so named space-time profiles [9]. For each element the accumulated 100 last iterations will be shown on the plots for x_i^t and $F(x_i^t)$. This method enables us to diagnose the type of oscillations (periodic or chaotic), to define the oscillation period and regimes of synchronization or desynchronization of the ensemble elements. If necessary, we will also display snapshots of the network states at a fixed iteration number t = const.

Besides these, a spatial coherence in the ensemble (6.1) will be analyzed using the cross-correlation coefficient (CCC) of the oscillations of different elements. We consider the first and the ith elements at the same time and the CCC for them is defined as follows:

$$\Psi_{1,i} = \frac{\langle \tilde{x}_1(t)\tilde{x}_i(t)\rangle}{\sqrt{\langle \tilde{x}_1^2(t)\rangle \langle \tilde{x}_i^2(t)\rangle}}, \tag{6.4}$$

where $\tilde{x}(t) = x(t) - \langle x(t)\rangle$ is a deviation from the mean value. The brackets $\langle \ldots \rangle$ denote the time averaging. Oscillations in the regime of chaotic dynamics are characterized by mixing and are ergodic. In this case the time averaging is equivalent to the ensemble averaging.

6.3 Evolution of the System Dynamics with Decreasing Coupling Strength

The evolution of the ensemble (6.1) dynamics was explored in [22, 23] when the coupling strength σ is varied and the coupling radius is fixed at $r = 0.32$. We reproduce those results for the ring of $N = 1000$ coupled logistic maps and analyze in more detail various spatio-temporal structures which appear in (6.1). For this purpose we consider the evolution of the coupling function and study spatial distributions of cross-correlations.

In order to obtain more detailed insight into the coherence–incoherence transition Fig. 6.2 illustrates a phase-parametric diagram for the coupling radius $r = 0.32$, i.e., a dependence of the variable x_i on the coupling strength σ. For each value of σ, the

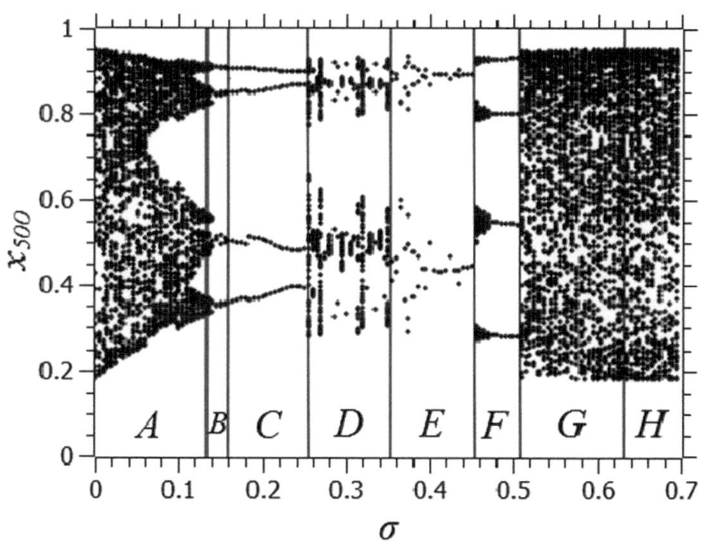

Fig. 6.2 Dependence of the x_i values on the coupling strength σ for $i = 500$ and $r = 0.32$. For each value of σ, the last 100 iterations of the system (6.1) are shown after discarding the transient of 30,000 iterations

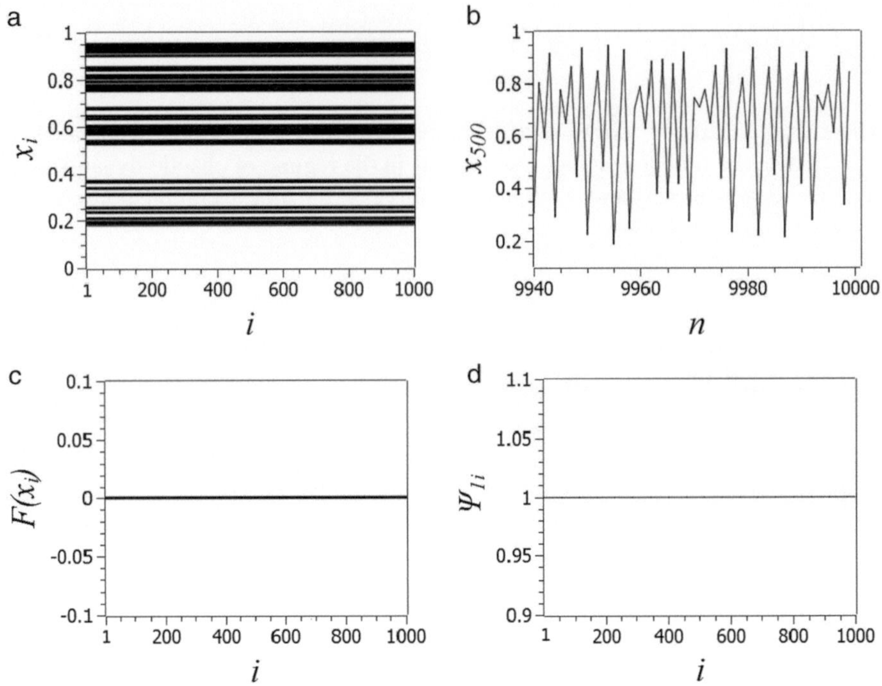

Fig. 6.3 Illustration of complete chaotic synchronization for $\sigma = 0.64$. (**a**) Space-time profile for the variables x_i, (**b**) temporal waveform of oscillations (time series) for the $i = 500$ oscillator of the system (6.1), (**c**) space-time profile for the coupling function $F(x_i)$, and (**d**) cross-correlation coefficient for the first and ith oscillators

last 100 iterations of the system (6.1) are shown in Fig. 6.2 for the chosen element $i = 500$. The initial conditions are chosen to be randomly distributed in the interval $0 < x_i^0 < 1$ and this realization is fixed for all numerical calculations.

Let us consider in detail the regimes which are typical for different regions marked in the diagram in Fig. 6.2. We start with analyzing region H which corresponds to complete chaotic synchronization ($\sigma \geq 0.63$). Numerical results for this regime are shown in Fig. 6.3.

As can be seen from Fig. 6.3, the regime of complete chaotic synchronization is observed in region H. Thus, although the oscillations of the network elements are chaotic (Fig. 6.3b), the instantaneous values of all the variables x_i coincide at any discrete time t (Fig. 6.3a). The values of the coupling function turn to zero (Fig. 6.3c), so that the dynamics of all the ensemble elements matches the oscillation mode of an individual element. The cross-correlation coefficient $\Psi_{1,i}$ is equal to unity for all $i = 2, 3, \ldots, N$ (Fig. 6.3d).

When the coupling coefficient σ decreases ($\sigma < 0.63$), the complete chaotic synchronization loses its stability in a blowout bifurcation. This bifurcation and related phenomena were studied for systems of two interacting identical oscillators

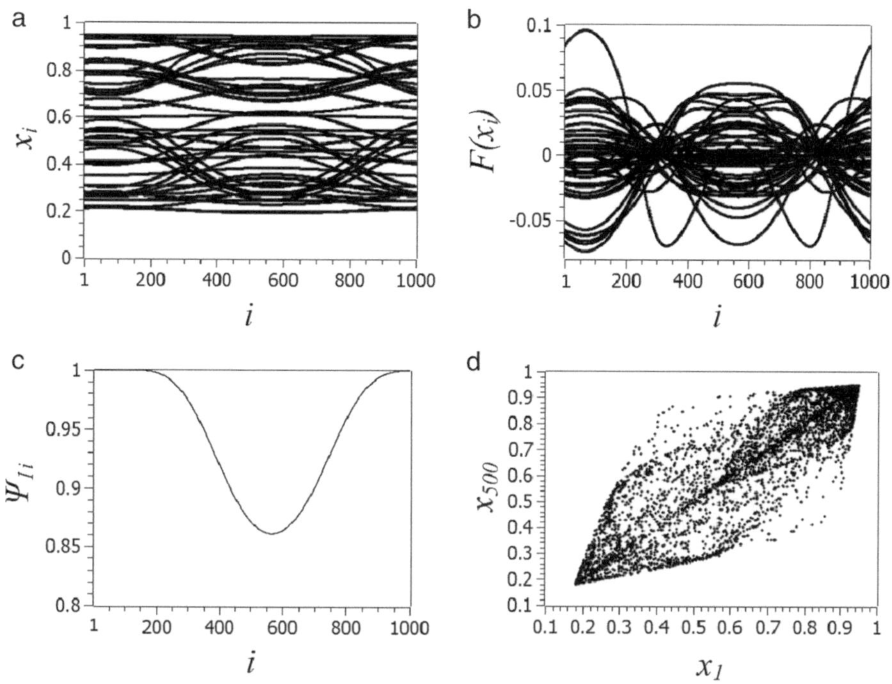

Fig. 6.4 Illustration of the regime of oscillating (partial) chaotic synchronization in the model (6.1) for $\sigma = 0.55$. (**a**) Space-time profile of the network dynamics, (**b**) space-time profile of the coupling function, (**c**) cross-correlation coefficient for the first and ith oscillators, and (**d**) projection of the phase trajectory on the (x_1, x_{500}) plane. System parameters: $r = 0.32$, $a = 3.8$, and $N = 1000$

in [6, 8, 16, 19, 27, 29] and for extended ensembles with nonlocal interaction only in [23]. It is interesting to note the following fact. When the coupling strength deviates slightly from the blowout bifurcation line, $\sigma < \sigma_{cr} \approx 0.63$, a regime is realized with snapshots which are different from straight horizontal lines. However, similarly to the regions with regular dynamics (for example, region 2 in Fig. 6.1), this wave-like profile can be still described by a smooth function of the spatial coordinate. Such a space-time profile for the variables x_i is exemplified in Fig. 6.4a. Following [22, 23], it corresponds to the coherence regime. The corresponding space-time profile for the coupling function is shown in Fig. 6.4b. As can be seen, it is similar to the x_i profile. As follows from Fig. 6.4c, the CCC now depends on the oscillator number i and decreases when the distance between the oscillators increases. The CCC achieves its minimal value ≈ 0.84 for $i = 500$. This means that there is no full correlation of the oscillations of the network elements. Phase trajectories no longer lie in the symmetric subspace U and this is illustrated by a projection on the plane of (x_1, x_{500}) variables in Fig. 6.4d.

The regime described above is observed in region G of Fig. 6.2. Up to now, nothing has been reported regarding this regime. Its peculiarities are as follows. Despite the fact that snapshots have smooth profiles, i.e., they correspond to the coherence regime, the cross-correlation coefficient along the ensemble (6.1) is not equal to 1. This implies that the complete chaotic synchronization no longer exists. This special regime of chaotic synchronization can be called *oscillating (or partial) chaotic synchronization*.

When $0.4 < \sigma < 0.5$ in regions E and F of Fig. 6.2 (region 2 in Fig. 6.1), the oscillations in the network (6.1) are periodic. As the coupling strength decreases, a cascade of period doubling bifurcations is realized. Quasiperiodic oscillations can also be observed for certain values of σ. Naturally, the CCC $\Psi_{1,i}$ for all $i = 2, 3, \ldots, N$ is equal to 1 in the regimes of regular (periodic and quasiperiodic) oscillations of the network elements.

It is established that for $\sigma \leq 0.4$, the wave front of the profile x_i becomes vertical, i.e., the spatial derivative tends to infinity. In [22, 23] it has been stated that it is the infinite derivative that causes the emergence of incoherent states. As follows from our calculations and the numerical results in [22, 23] chimera states are observed in numerical simulation when $\sigma \leq \sigma_{\rm cr} = 0.35$.

Figure 6.5 shows space-time profiles for $F(x_i^t)$ (left column) and snapshots of the states x_i^t (right column) for the system (6.1). It is seen that as σ decreases from 0.43 to 0.35, the coupling function behaves periodically and undergoes period-doubling bifurcations: period-2, period-4 cycles (Fig. 6.5a, b, respectively). When $\sigma = 0.35$, a weak chaos emerges in the system (Fig. 6.5d).

The snapshots displayed in Fig. 6.5a, b (right column) indicate the regime of coherence and the ensemble dynamics is periodic. At the critical point $\sigma \cong 0.35$ the situation changes dramatically (see Fig. 6.5c). The coupling function demonstrates the transition to the chaotic regime and the snapshot of the states x_i reveals the appearance of two regions of incoherent dynamics, i.e., a chimera-like state is born [22, 23]. Our calculations indicate that the temporal dynamics of the partial elements, as well as the coupling function, is characterized by a cascade of period-doubling bifurcation as σ decreases.

Therefore, the transition from coherent dynamics of the whole ensemble to the birth of incoherent regions (chimera-like states) results from chaos arising. The chimera-like states appear when σ reaches its critical value $\sigma_{\rm cr} \simeq 0.35$ at which the transition to the chaotic oscillation regime occurs for all the partial elements and the coupling function. As can be seen from Fig. 6.5b, the coupling function is periodic and the solution profiles x_i^t are characterized by an infinite derivative. However, no chimera-like states appear. They arise when the coupling function and the whole ensemble begin operating in the chaotic regime.

When one moves to region D in Fig. 6.2, two different kinds of chimeras can be distinguished, namely phase and amplitude chimera states [9, 10]. The regime of phase chimera state is related to the appearance of regions with spatially nonregular alternations of elements with in-phase and anti-phase oscillations. If a discrete map demonstrates 2-periodic oscillations (the oscillation period is equal to two iterations of the map), then the latter are considered to be anti-phase if oscillations in two

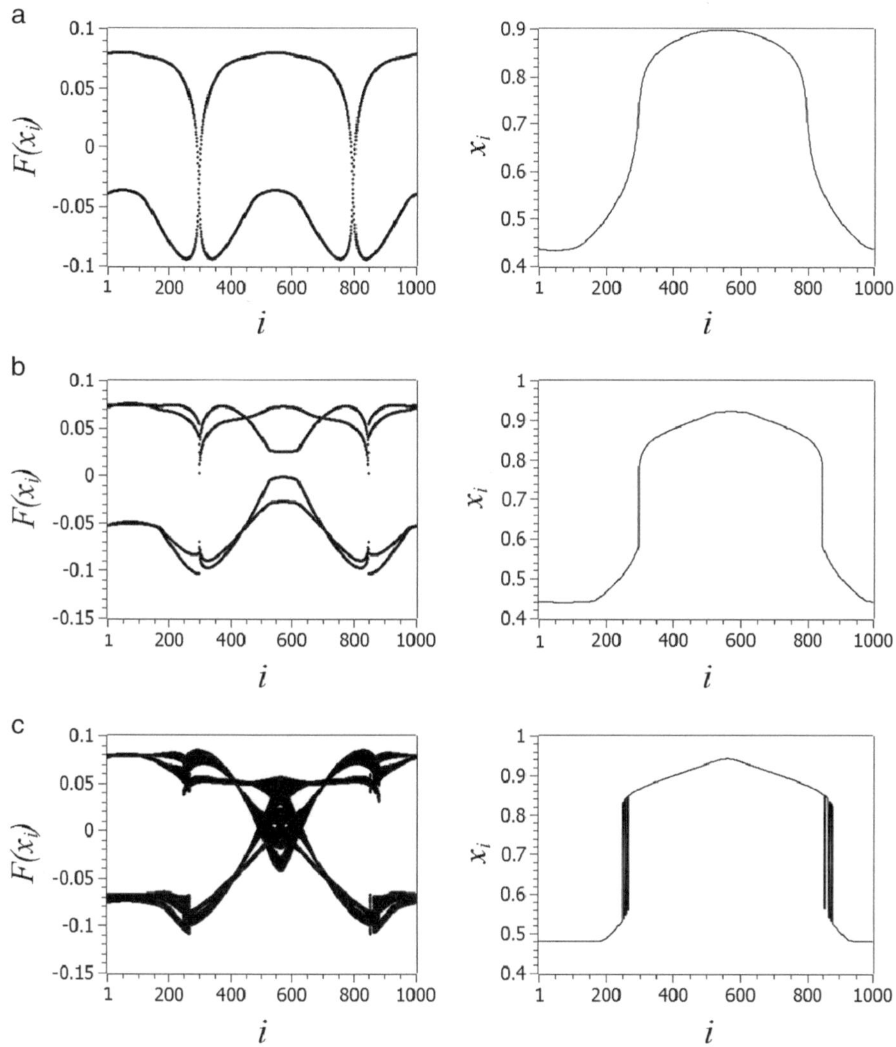

Fig. 6.5 Space-time profiles for the coupling function $F(x_i)$ (*left column*) and snapshots (*right column*) of the dynamics of the system (6.1) for different values of the coupling strength: (**a**) $\sigma = 0.43$, (**b**) $\sigma = 0.38$, and (**c**) $\sigma = 0.35$

neighboring oscillators are shifted by one iteration. Regions with alternations of phase shifts were called incoherence regions in [22, 23] and phase chimera states in [9, 10]. Inside these domains the phase shift between the oscillators occurs randomly in space, is defined by initial conditions, and is constant in time.

Figure 6.6 shows numerical results for the regime of phase chimera state obtained for $\sigma = 0.29$ and randomly chosen initial conditions. In coherence regions 1 and 2,

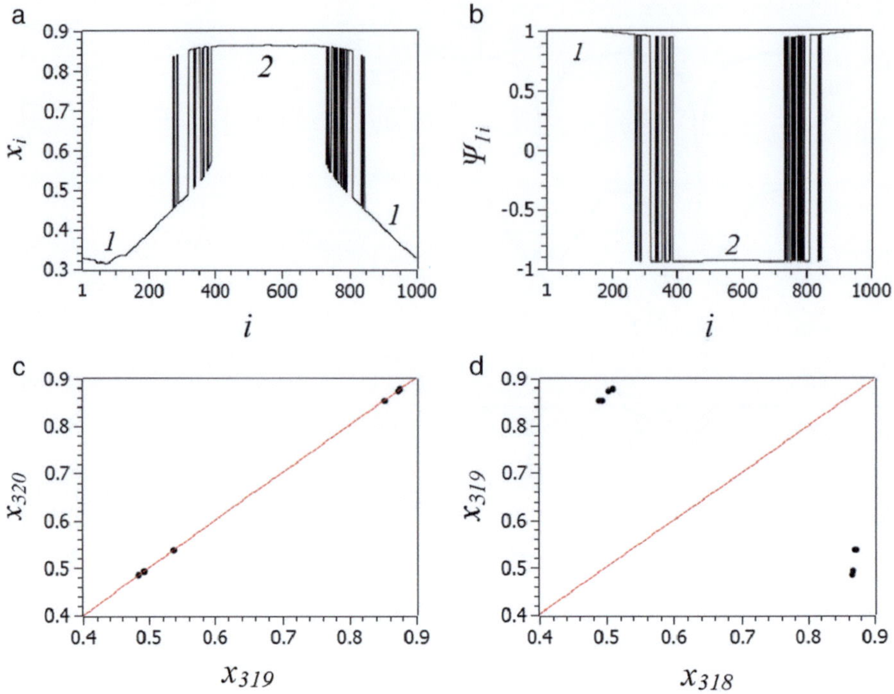

Fig. 6.6 Illustration of the regime of phase chimera state in the system (6.1) for $\sigma = 0.29$ and $r = 0.32$. (**a**) Snapshot of the network dynamics, (**b**) the cross-correlation coefficient $\Psi_{1,i}$, (**c**) projections of phase trajectories on the (x_{319}, x_{320}) plane (these oscillators belong to the phase chimera region and oscillate in-phase) and (**d**) on the (x_{318}, x_{319}) plane (these oscillators demonstrate anti-phase oscillations). The initial conditions are chosen randomly

which are indicated in Fig. 6.6a, the oscillations of the network elements differ only in phase and the CCC is close to 1 in absolute value, i.e., $\Psi_{1,i} = +1$ in region 1 and $\Psi_{1,i} = -1$ in region 2 (Fig. 6.6b). The difference in the $\Psi_{1,i}$ sign simply distinguishes between in-phase oscillations (region 1) and anti-phase oscillations (region 2) of the ith oscillator with respect to the first one. Figure 6.6c, d illustrates in-phase and anti-phase oscillations in the projections of relevant variables, respectively. In the spatial incoherence regions (phase chimeras) the cross-correlation coefficient $\Psi_{1,i}$ switches between the values ± 1 (Fig. 6.6b). This fact corroborates once again a random alternation between the in-phase and anti-phase oscillations. We note that although the CCC value is close to unity in its absolute value, there is no strong equality $|\Psi_{1,i}| = 1$. Actually, as our calculations have shown, $|\Psi_{1,i}| \leq 1$ for the phase chimera state. This finding can be explained by the fact that the transition to the regime of phase chimera is accompanied by the emergence of chaotic oscillations in the network elements in time. The degree of chaoticity is still low for $\sigma = 0.29$ but it is just the temporal chaotic dynamics that can lead to the situation when the absolute value of the CCC becomes slightly less than unity [9, 10].

The phase change occurs before the chimera states appear. As can be seen from Fig. 6.5a, the coupling function $F(x_i^t) \simeq 0$ for the oscillators $i \simeq 298$ and $i \simeq 798$. The period-2 oscillations in these elements degenerate into 1-period points and their values become equal to $x_{302}^t \simeq 0.73 = \text{const}$. As follows from Eq. (6.1), when $F(x_i) = 0$ we can obtain

$$x_i^{t+1} = ax_i^t(1 - x_i^t) = \text{const}, \quad i = 302. \tag{6.5}$$

Map (6.5) realizes the unstable equilibrium point $x^0 = ax^0(1 - x^0)$ with coordinate $x^0 = 0.7368$ and this agrees with the data shown in Fig. 6.5a. This fixed point x^0 is saddle and separates the basins of attraction of in-phase and anti-phase oscillations, that is described in detail in [10, 23].

When the coupling strength σ decreases further in interval D (Fig. 6.2), $0.25 < \sigma < 0.29$, the regions of phase shifts increase, as mentioned in [22, 23]. Moreover, a different type of chimera clusters, which have been referred to as amplitude chimera states [9, 10], appears in the network (6.1) simultaneously with phase chimera states. The incoherence region corresponding to the amplitude chimera is characterized by developed chaotic dynamics of all the elements in the incoherence cluster. The amplitudes of the oscillators in this incoherence region are highly chaotic and their behavior becomes completely asynchronous, while the neighboring oscillators oscillate almost periodically and are synchronized [9, 10]. It should be noted that in region D (Fig. 6.2), there can coexist a set of different chimera structures which are observed for different initial conditions. This effect is known as a huge multistability and is described in [22, 23].

An amplitude chimera regime is exemplified in Fig. 6.7 for $\sigma = 0.28$. As can be seen from the snapshot in Fig. 6.7a, the amplitude chimera cluster (region 2 in Fig. 6.7a) includes oscillators from $i = 120$ to $i = 290$. In addition, the whole spatial structure of the network dynamics also contains two phase chimera states (Fig. 6.7a, region 1). Within region 2, the partial units together with the coupling function display more developed chaotic dynamics (Fig. 6.7b, c, respectively). The partial elements in the amplitude incoherent cluster are characterized by completely asynchronous chaotic dynamics as shown in Fig. 6.7d.

The cross-correlation coefficient $\Psi_{1,i}$ for the regime at $\sigma = 0.28$ is still close to ± 1 for the phase chimera clusters (Fig. 6.8a) but decreases slightly in modulus for the amplitude chimera. As follows from Fig. 6.7b there are two domains of the chaotic set in the considered space region. The trajectory visits both of them regularly in one time moment (iteration). Thus, the time series $x_i(t)$ includes two components, 2-periodic and chaotic. It is the presence of periodic component that is the reason for a very slow decay of the CCC in the amplitude chimera domain ($\Psi_{1,i} \approx 0.9$, $120 < i < 290$). This regular (2-periodic) component can be excluded if every second point from the time series $x_i(t)$ is taken into account. In this case the resulting CCC $\Psi_{1,i}^{(2)}$ (Fig. 6.8b) is considerably reduced (up to the value of ≈ 0.68). The minimal correlation corresponds to the center of the amplitude chimera cluster where a more developed chaos is observed. The spatial correlation decay is

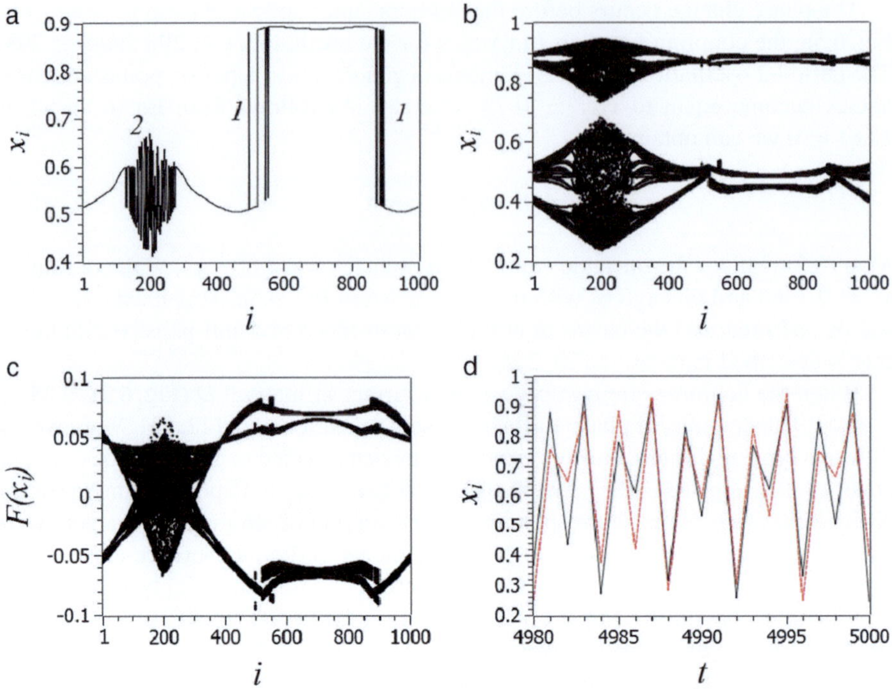

Fig. 6.7 Coexistence of phase (cluster 1) and amplitude (cluster 2) chimera states in the model (6.1) for $\sigma = 0.28$: (**a**) Snapshot of the network dynamics, (**b**) space-time profiles for variables x_i, (**c**) space-time profiles for the coupling function $F(x_i)$, and (**d**) time series x_i^t for $i = 194$ (*solid line*) and $i = 195$ (*dotted red line*) in the amplitude chimera cluster

related to asynchronous chaotic behavior of the oscillators in the amplitude chimera. Hence, in contrast to the phase chimera, the amplitude chimera state enables one to demonstrate "truly" incoherent oscillations which become incoherent both in space and in time.

If the coupling strength decreases further ($\sigma < 0.25$), 4-period oscillations can be observed in the ensemble (6.1) in time (region C in Fig. 6.2). They undergo a cascade of period doubling bifurcations in region B of Fig. 6.2. Despite the complete regularity in time, the spatial behavior in regions C and B is absolutely irregular. Similar regimes were observed in ensembles with local coupling [20]. The network dynamics are similar to those for the phase chimera regime described above. This means that the spatially incoherence regions, which appear in the phase chimera regime, expand and occupy the whole ring with decreasing σ. Due to the time periodicity the cross-correlation coefficient $\Psi_{1,i}$ is strictly equal to unity in absolute value for all i and its sign is changed randomly in space following the phases of the partial oscillators. A similar behavior also persists in the region of longer periodic oscillations (region B in Fig. 6.2) up to the point of emergence of spatio-temporal chaos.

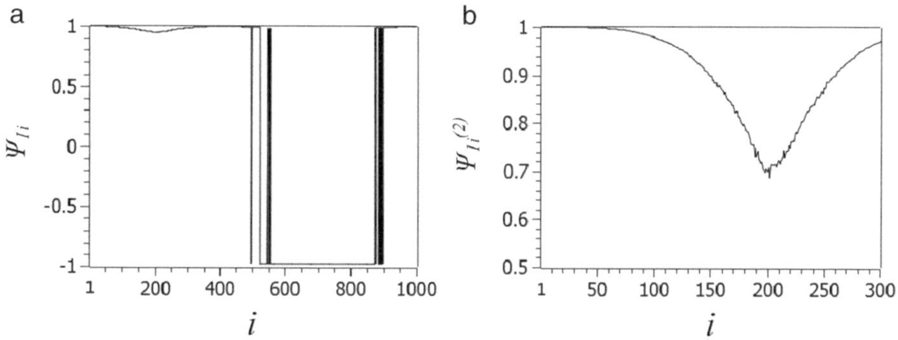

Fig. 6.8 (**a**) Cross-correlation coefficient $\Psi_{1,i}$ and (**b**) cross-correlation coefficient $\Psi_{1,i}^{(2)}$ for the amplitude chimera when the time series contains every second point

When $\sigma < 0.13$, a transition to temporal chaos is observed (region 3 in Fig. 6.1 and region A in Fig. 6.2). Oscillations x_i^t in this region correspond to the regime of spatio-temporal chaos, i.e., the partial oscillators behave chaotically in time and are completely desynchronized. The spatial behavior is fully irregular (incoherent). With decreasing coupling strength the chaotic dynamics develops as a result of merging bifurcations. They consist in merging parts of a chaotic set, which occurs in all the ensemble oscillators. The points corresponding to the instantaneous states of the oscillators form horizontal stripes of different width in the snapshots of the network dynamics. So, a two-band chaotic set can be seen for $\sigma = 0.1$ (Fig. 6.9a) and a developed chaos with a combined region of variable values is observed at $\sigma = 0.05$ (Fig. 6.9c). Since the temporal dynamics is chaotic and there is mixing in the system, the cross-correlation coefficient for various oscillators becomes significantly less than 1 in absolute value. Moreover, in contrast to the regime of oscillating chaotic synchronization, $\Psi_{1,i}$ decays instantaneously in space with increasing i (Fig. 6.8d). In the case of the two-band chaotic set, when the phase trajectory switches between the bands regularly in time, the CCC does not decrease to zero and oscillates at the level ± 0.87 (Fig. 6.9b). When a single-band (developed) chaotic set is realized, the CCC almost vanishes (Fig. 6.9d).

6.4 Mechanism of the Coherence–Incoherence Transition

As follows from the results presented above, the transition from completely coherent dynamics to the regime of complete incoherence occurs as the coupling strength decreases, $0.64 > \sigma > 0.05$. First, when $0.35 < \sigma < 0.45$, both the partial oscillators and the coupling function undergo a cascade of period-doubling bifurcations. At the critical point $\sigma_{cr} \simeq 0.35$ they all demonstrate weak chaotic oscillations and, finally, a phase chimera-like state appears. Further on, when $\sigma < 0.35$, amplitude chimera states arise and the number of incoherent clusters increases. As a result, the

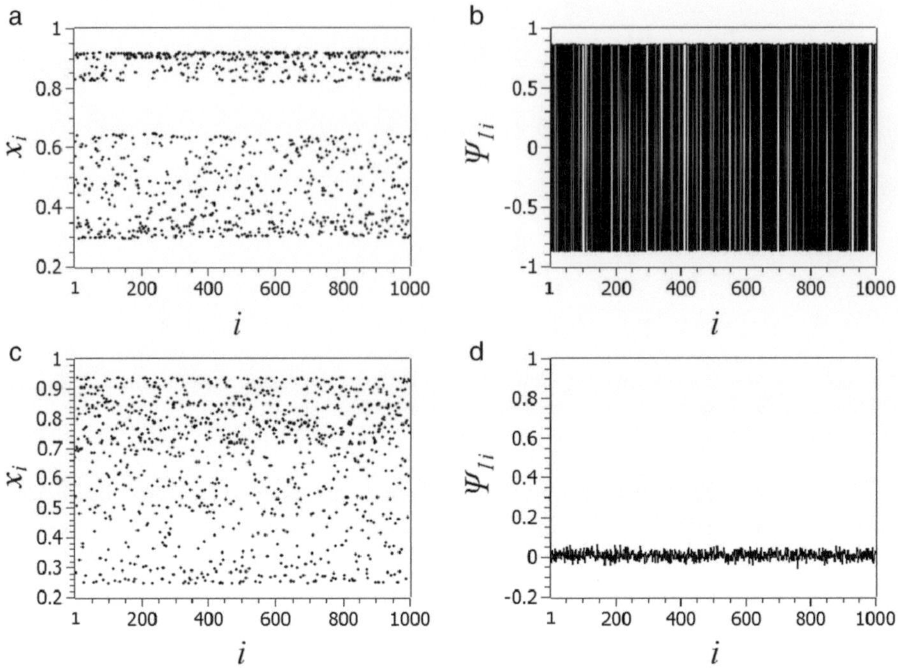

Fig. 6.9 Regime of incoherent chaos. (**a**) Snapshot of the network dynamics and (**b**) cross-correlation coefficient $\Psi_{1,i}$ for a two-band chaotic set at $\sigma = 0.1$; (**c**) snapshot and (**d**) $\Psi_{1,i}$ for a single-band chaotic set at $\sigma = 0.05$. The points in (**a**) and (**c**) plots are not connected by lines so that the structure of the chaotic sets can be better visualized and distinguished

whole ensemble displays asynchronous chaotic dynamics (region 3 in Fig. 6.1). To order to establish the mechanism of coherence–incoherence transition we turn to the original equation (6.1) which can be rewritten in the following form:

$$x_i^{t+1} = a(1-\sigma)x_i^t(1-x_i^t) + \frac{\sigma}{2P}\sum_{j=i-P,\, j\neq i}^{i+P} a x_j^t(1-x_j^t). \tag{6.6}$$

Equation (6.6) includes two bifurcation parameters, the coupling strength σ and the number of neighbors P (the coupling radius). In our paper, the coupling strength σ can be considered as most important because its values govern the dynamics of the first and second terms in (6.6) and enables one to understand the mechanism of the bifurcation transitions. The first term characterizes the ensemble of oscillators without coupling between neighbors. The coupling coefficient σ, as seen from Eq. (6.6), reduces the effective parameter of the logistic map $f(x_i^t)$ as the nonlinearity parameter is now set to be $a(1-\sigma)$. The second term takes into account the effect of nonlocal coupling of P neighbors and here σ also changes the effective parameter a that is now $\sigma \cdot a$. Two limit values of σ can be distinguished. In the first case, when

$\sigma \to 1$, the first term in Eq. (6.6) vanishes and the second term is characterized by a larger parameter σa, that can result in the transition to chaotic dynamics in all the elements. Increasing the coupling strength ($\sigma \to 1$) gives rise to the regime of completely synchronous chaotic oscillations in the ring (region 1 in Fig. 6.1).

When $\sigma \to 0$, that is another limit case for σ, the first term in Eq. (6.6) describes chaotic oscillations in each partial element ($a = 3.8$) while the second term can be neglected due to its smallness. The ring demonstrates the regime of incoherent chaotic oscillations, that corresponds to region 3 in Fig. 6.1. Bifurcation effects in the ring can occur only in a certain parameter range $\sigma_1 < \sigma < \sigma_2$ where both terms in Eq. (6.6) play a significant role. As the numerical results above have shown, the bifurcations can be observed in the range $0.05 < \sigma < 0.63$.

The influence of coupling leads to a decrease of the control parameter in the logistic map and the partial elements of the ensemble oscillate periodically. The periodic oscillations in each element experience the impact of P neighboring oscillators, which is given by the second term in Eq. (6.6). Thus, the coupling causes two effects, namely, (1) a decrease of the nonlinearity parameter of the partial oscillators [the first term in Eq. (6.6)] and (2) an effective increase of the parameter due to nonlocal coupling [the second term in Eq. (6.6)].

With decreasing coupling strength σ the effective nonlinearity parameter of the logistic map grows and thus, causes a cascade of period-doubling bifurcations in the ensemble elements. When the critical point of chaos birth in the partial elements is reached, clusters of incoherent oscillations appear in certain parts of the ring. With further decrease of σ (with increasing subcriticality in the partial oscillators) the whole ring demonstrates incoherent chaotic oscillations. Thus, the appearance of a finite number of incoherence clusters (a chimera-like state) is nothing more than an intermediate state of the ensemble in the global transition from the coherence regime, when all the elements in the ring demonstrate in-phase chaotic oscillations ($\sigma \to 1$), to the completely incoherent dynamics of the whole ring ($\sigma \to 0$). The appearance of a finite number of incoherent intervals depends on different initial conditions randomly distributed over the network. Our studies have shown that the described transition does not change qualitatively when the initial conditions x_i^0 vary randomly in the interval [0; 1].

6.5 Temporally Intermittent Chimera Structure

We have considered the evolution of spatio-temporal dynamics of the network (6.1) in a selected cross-section of the bifurcation diagram (Fig. 6.1) and have found that it is sufficiently complicated. The especially complex behavior which manifests itself in the formation of various clusters and high multistability is typical for the region of chimera state existence. Chimera states can be observed in a wide range of the coupling radius r variation. For certain values of $r < 0.32$, more specific regimes of the ensemble (6.1) dynamics can be realized, for example, a regime of temporally intermittent chimera states. In this case the phase shift between the

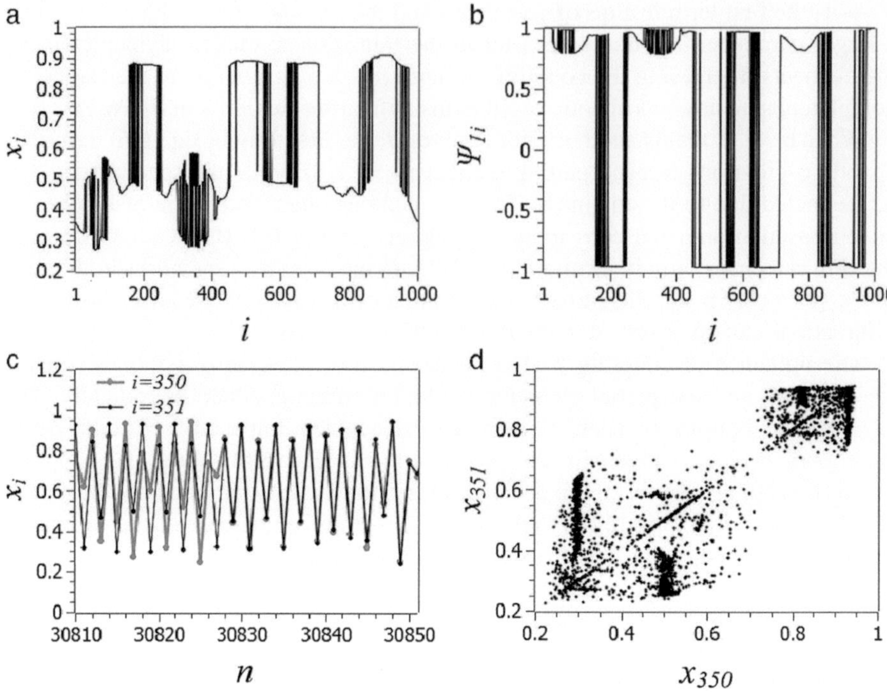

Fig. 6.10 Example of an intermittent chimera for $r = 0.08$ and $\sigma = 0.25$. (**a**) Snapshot of the variables x_i, (**b**) cross-correlation coefficient $\Psi_{1,i}$, (**c**) part of the time series for the oscillators $i = 350$ and $i = 351$, and (**d**) projection of the phase trajectory on the (x_{350}, x_{351}) plane

oscillators belonging to the incoherence cluster may vary over time. A temporally intermittent chimera state is exemplified in Fig. 6.10 for $r = 0.08$ and $\sigma = 0.25$. As can be seen from the snapshot (Fig. 6.10a), this chimera structure includes several incoherence clusters which are typical for the phase chimera and one cluster ($300 < i < 390$) with special intermittent behavior. The cross-correlation coefficient (Fig. 6.10b) is close to 1 in modulus for some oscillators, while at the same time, there are oscillators for which $0 < \Psi_{1,i} < 1$ that indicates the presence of weak chaos. Figure 6.10c shows a fragment of the time series for the oscillators $i = 350$ and $i = 351$. It is seen that these elements demonstrate almost anti-phase oscillations on the time interval $30{,}810 \leq t \leq 30{,}828$ (this is typical for the phase chimera) and in-phase but different in amplitude oscillations on another time interval $30{,}828 \leq t \leq 30{,}850$ (amplitude chimera). The intermittent behavior can also be observed on the projection of the phase trajectory on the plane of (x_{350}, x_{351}) variables, which is shown in Fig. 6.10d. A part of the points is located on the bisectrix (in-phase behavior) and another part is found in the region away from the bisectrix. We note that the lifetime of the intermittent chimera states strongly depends on the initial conditions and may be finite. In this case, as our calculations have shown, when the time of observation (number of iterations) of the system dynamics $t \geq 40{,}000$, the regime of intermittency can change (switch) to the regime of phase or amplitude chimera.

6.6 Discussion and Conclusion

In the present paper we have described numerical results for the bifurcation phenomena which are observed at the transition from complete chaotic synchronization (complete coherence) to fully asynchronous spatio-temporal chaos (complete incoherence) in the one-dimensional ensemble of nonlocally coupled chaotic maps (6.1). A complex set of bifurcations which are realized in the system (6.1) by varying the parameters σ and r (Fig. 6.2) has been known in the scientific literature [22, 23]. However, many of the details of these effects are not sufficiently studied. In our opinion, this is due to the fact that the indicated transitions have been basically analyzed using qualitative characteristics without applying any quantitative criteria. In our research we have used different characteristics of the dynamical regimes, both qualitative and quantitative. We have considered both instantaneous spatial profiles (snapshots) of dynamical variables and space-time profiles which combine a set of snapshots at different time moments. We have also analyzed space-time profiles for the coupling function in order to understand the mechanisms of mutual influence of the ensemble elements inside various clusters and to explain the reasons for the bifurcation phenomena. Besides this, as one of the important quantitative characteristics of dynamical regimes, we have used the cross-correlation coefficient (CCC) of different elements of the ensemble. Such an approach has enabled us to reveal a number of important details and clarify the considered mechanism. The following main results can be formulated.

We have described in detail the transition from complete chaotic synchronization (region 1 in Fig. 6.1) to oscillating (partial) chaotic synchronization (transition from region H to region G in Fig. 6.2). It has been shown that whereas the CCC is strictly equal to 1 in region H, it is less than 1 in region G. This indicates the destruction of complete chaotic synchronization (Figs. 6.3b and 6.4c). However, the regimes which are characterized by the snapshots like in Fig. 6.4a can be treated to be coherent in the sense of the definition (6.2).

We have shown that the vertical front formation in the wave-like solution profile x_i^t is indeed an important factor which precedes the appearance of a chimera-like state. In terms of the ensemble (6.1) dynamics, the infinite spatial derivative is caused by the fact that the phase in neighboring oscillators is shifted by one iteration. This shift is realized when the difference between the maximal and minimal oscillation amplitudes of a 2-period cycle becomes the largest. This condition is necessary but not sufficient for the appearance of chimera states. When the solution profile x_i^t becomes discontinuous, chimera states emerge only in the case when the coupling strength σ corresponds to the chaos birth in both the ensemble elements and the coupling function. It is the chaotic dynamics of the network that leads to the effect of phase synchronization loss and induces the intermittency between the in-phase and anti-phase oscillations of a localized cluster of oscillators. A chimera state appears which can be naturally called a phase chimera state. Thus, the vertical front formation and slightly chaotic dynamics in all the elements of the ensemble are the necessary and sufficient conditions for realizing phase chimera states.

It has been found that in the phase chimera mode (Fig. 6.6), the CCC is close to 1 in modulus and randomly changes its sign from $+$ to $-$ for the elements belonging to this incoherence cluster, where the oscillations x_i^t are shifted in time by one iteration. From the standpoint of the correlation analysis, the phase chimera is not strictly an incoherence regime although it is characterized by a random spatial distribution of the phase shifts of oscillations in the partial elements in the incoherence cluster.

Development of chaotic temporal behavior in the ensemble leads to the appearance of new chimera states (see Fig. 6.7) which are characterized by a random distribution of the instantaneous oscillation amplitudes of the elements within the incoherent interval. This type of chimera states can be naturally called an amplitude chimera state. As the calculations have shown, amplitude and phase chimera states may coexist in the ensemble of coupled chaotic maps (Fig. 6.7). The appearance of amplitude chimera states is related to the loss of stability of chaotic synchronization in the network elements included in the coherent cluster. This is manifested by the fact that the intensity of chaotic oscillations of the elements in this cluster increases significantly.

We have revealed the differences in cross-correlations for phase and amplitude chimera states. The CCC is less than 1 in modulus (Fig. 6.8) for all the elements belonging to the amplitude chimera (cluster 2 in Fig. 6.7a). The decrease of the CCC in this case shows that the oscillations of different elements in the amplitude chimera are truly incoherent (non-synchronous) and fully comply to the definition of chimera state [1, 14].

The effect of intermittency between phase and amplitude chimera states has been established. In this case the time series of the ith oscillator x_i^t demonstrates a random alternation (switching) between phase and amplitude chimeras (Fig. 6.10).

We have explored the transition to the region of fully incoherent chaos which is observed for weak coupling (transition from C to A in Fig. 6.2). It has been shown that the appearance of fully asynchronous spatio-temporal chaos is preceded by temporal periodic oscillations of oscillators whose phase differences are completely random in space. When moving to region A in Fig. 6.2, a completely incoherent spatiotemporal chaotic regime is realized in all the network elements (Fig. 6.9).

Acknowledgements The authors acknowledge support from SFB910. Vadim S. Anishchenko, acknowledges support from the Russian Science Foundation (grant No. 16-12-10175). Tatiana E. Vadivasova and Galina I. Strelkova acknowledge support from RFBR (grants No. 14-52-12002 and No. 15-02-02288).

References

1. Abrams, D.M., Strogatz, S.H.: Chimera States for Coupled Oscillators. Phys. Rev. Lett. **93**, 174102 (2004)
2. Abrams, D.M., Strogatz, S.H.: Chimera states in a ring of nonlocally coupled oscillators. Int. J. Bifurcation Chaos **16**, 21–37 (2006)
3. Afraimovich, V.S., Nekorkin, V.I., Osipov, G.V., Shalfeev, V.D.: Structures and Chaos in Nonlinear Synchronization Networks. World Scientific, Singapore (1995)
4. Anishchenko, V.S., Aranson, I.S., Postnov, D.E., Rabinovich, M.I.: Spatial synchronization and chaos development bifurcations in a chain of coupled oscillators. Dokl. USSR Acad. Sci. **286**(5), 1120–1124 (1986) (in Russian)
5. Anishchenko, V.S., Astakhov, V.V., Vadivasova, T.E., Neiman, A.B., Schimansky-Geier, L.: Nonlinear Dynamics of Chaotic and Stochastic Systems. Tutorial and Modern Development. Springer, Berlin (2007)
6. Ashwin, P., Buescu, J., Stewart, I.: From attractor to chaotic saddle: tale of transverse instability. Phys. Lett. A **193**, 126–139 (1994)
7. Astakhov, V.V., Bezruchko, B.P., Ponomarenko, V.I.: Multistability formation, isomer classification and their evolution in coupled Feigenbaum's systems. Izv. Vuz. Radiophys. **34**, 35–39 (1991) (in Russian)
8. Astakhov, V., Shabunin, A., Kapitaniak, T., Anishchenko, V.: Loss of chaos synchronization through the sequence of bifurcations of saddle periodic orbits. Phys. Rev. Lett. **79**, 1014–1017 (1997)
9. Bogomolov, S.A., Strelkova, G.I., Schöll, E., Anishchenko, V.S.: Amplitude and Phase Chimeras in an Ensemble of Chaotic Oscillators. Tech. Phys. Lett. **42**(7), 763–766 (2016)
10. Bogomolov, S.A., Slepnev, A.V., Strelkova, G.I., Schöll, E., Anishchenko, V.S.: Mechanisms of Appearance of Amplitude and Phase Chimera States in Ensembles of Nonlocally Coupled Chaotic Systems. Commun. Nonlinear Sci. Numer. Simul. **43**, 25–36 (2017)
11. Epstein, I.R., Pojman, J.A.: An Introduction to Nonlinear Chemical Dynamics: Oscillations, Waves, Patterns, and Chaos. Oxford University Press, New York (1998)
12. Feigenbaum, M.J.: Quantitative universality for a class of nonlinear transformations. J. Stat. Phys. **19**, 25–52 (1978)
13. Kuramoto, Y.: Chemical Oscillations, Waves, and Turbulence. Springer Series in Synergetics. Springer, Berlin (1984)
14. Kuramoto, Y., Battogtokh, D.: Coexistence of Coherence and Incoherence in Nonlocally Coupled Phase Oscillators. Nonlinear Phenom. Complex Syst. **4**, 380–385 (2002)
15. Landa, P.S.: Nonlinear Oscillations and Waves in Dynamical Systems. Springer, Berlin (1996)
16. Maistrenko, Yu., Kapitaniak, T.: Different types of chaos synchronization in two coupled piecewise linear maps. Phys. Rev. E **54**, 3285–3292 (1996)
17. Malchow, H., Petrovskii, S.V., Venturino, E.: Spatiotemporal Patterns in Ecology and Epidemiology: Theory, Models, and Simulation. Chapman and Hall/CRC, London/Boca Raton, FL (2007)
18. Mikhailov, A.S., Loskutov, A.: Foundation of Synergetics. Complex Patterns. Springer, Berlin (1995)
19. Nagai, Y., Lai, Y.Ch.: Characterization of blowout bifurcation by unstable periodic orbits. Phys. Rev. E **55**, R1251–R1254 (1997)
20. Nekorkin, V.I., Velarde, M.G.: Synergetic Phenomena in Active Lattices. Springer, Berlin (2002)
21. Oispov, G.V., Kurths, J., Zhou, Ch.: Synchronization in Oscillatory Networks. Springer, Berlin (2007)
22. Omelchenko, I., Maistrenko, Y., Hövel, P., Schöll, E.: Loss of coherence in dynamical networks: spatial chaos and chimera states. Phys. Rev. Lett. **106**, 234102 (2011)
23. Omelchenko, I., Riemenschneider, B., Hövel, P., Maistrenko, Y., Schöll, E.: Transition from spatial coherence to incoherence in coupled chaotic systems. Phys. Rev. E **85**, 026212 (2012)

24. Omelchenko, I., Omelchenko, O.E., Hövel, P., Schöll, E.: When nonlocal coupling between oscillators becomes stronger: patched synchrony or multichimera states. Phys. Rev. Lett. **110**, 224101 (2013)
25. Omelchenko, I., Zakharova, A., Hövel, P., Siebert, J., Schöll, E.: Nonlinearity of local dynamics promotes multi-chimeras. Chaos **25**, 083104 (2015)
26. Omelchenko, I., Provata, A., Hizanidis, J., Schöll, E., Hövel, P.: Robustness of chimera states for coupled FitzHugh-Nagumo oscillators. Phys. Rev. E **91**, 022917 (2015)
27. Ott, E., Sommerer, J.C.: Blowout bifurcation in chaotic dynamical systems. Phys. Lett. A **188**, 39–47 (1994)
28. Panaggio, M.J., Abrams, D.M.: Chimera states: Coexistence of coherence and incoherence in networks of coupled oscillators. Nonlinearity **28**, R67(1–18) (2015)
29. Pikovsky, A.S., Grassberger, P.: Symmetry breaking bifurcation for coupled chaotic attractors. J. Phys. A **24**, 4587–4597 (1991)
30. Pikovsky, A.S., Rosenblum, M.G., Kurths, J.: Synchronization – A Universal Concept in Nonlinear Sciences. Cambridge University Press, Cambridge (2001)
31. Rabinovich, M.I.: Stochastic self-oscillations and turbulence. Physics–Uspekhi **21**, 443–469 (1978)
32. Rabinovich, M.I., Trubetskov, D.I.: Introduction to the Theory of Oscillations and Waves, 1st edn. Nauka, Moscow (1984); Regular and Chaotic Dynamics, 2nd edn. Moscow–Izhevsk (2000) (in Russian)
33. Semenova, N., Zakharova, A., Schöll, E., Anishchenko, V.: Europhys. Lett. **112**, 40002 (2015)
34. Vüllings, A., Hizanidis, J., Omelchenko, I., Hövel, P.: New J. Phys. **16**, 123039 (2016)
35. Zakharova, A., Kapeller, M., Schöll, E.: Phys. Rev. Lett. **112**, 154101 (2014)

Chapter 7
Regular and Chaotic Transition to Synchrony in a Star Configuration of Phase Oscillators

Vladimir N. Belykh, Maxim I. Bolotov, and Grigory V. Osipov

M.I. Rabinovich is one of the pioneers in the field of synchronization in complex networks of oscillators [1]. We dedicate the present piece of work to Misha's 75th birthday.

7.1 Introduction

Synchronous behavior is one of the ubiquitous collective phenomena in ensembles of oscillatory systems [2–5]. Among different phase oscillators the Kuramoto model is arguably the most studied model that describes synchronization [6–8]. This model captures essential features of synchronization, observed in science and applications. Examples are arrays of coupled Josephson junctions [9], semiconductor laser arrays [10], the ensembles of the cells in the heart [11], Hodgkin–Huxley neurons [12], central pattern generator for animal locomotion [13], rhythmic applause [14], pedestrian crowd synchrony on London's Millennium bridge [15], microwave oscillator arrays [16], etc. For other examples, see [17, 18] and [19]. Another popular example of phase oscillators is the networks of coupled phase-locked loops (PLL) [20–22]. In this paper, an analytical study of a star motif of phase oscillators is presented. The system consists of a central node, the hub, connected to an arbitrary number of peripheral nodes. The star configuration of ensemble of oscillators for the different models has been studied. In [23], starting from the analysis of the topological properties of the star configuration, some analytical considerations have been applied to derive the bifurcation diagram of the system with respect to the parameter mismatch between peripheral oscillators and hub and to the coupling strength. The analysis revealed that the system may become fully synchronized (more precisely, the peripheral oscillators are completely synchronized among each other and phase synchronized with the hub). In the case of star-coupled ensemble of

V.N. Belykh (✉)
Volga State University of Water Transport, Nesterova str., 5, Nizhny Novgorod 603005, Russia
e-mail: belykh@vgavt-nn.ru

M.I. Bolotov • G.V. Osipov
Nizhny Novgorod University, Gagarin Ave., 23, Nizhny Novgorod 603022, Russia
e-mail: maxim.i.bolotov@gmail.com; osipov@vmk.unn.ru

I.S. Aranson et al. (eds.), *Advances in Dynamics, Patterns, Cognition*,
Nonlinear Systems and Complexity 20, DOI 10.1007/978-3-319-53673-6_7

phase oscillators the analytical description of the parameter regions of existence of different synchronous regimes has been obtained in [24]. It was shown that peripheral oscillators compete for the synchronization with the hub and only a given number of peripheral oscillators can win this competition.

7.2 Models Under Study

In this paper we consider two examples of phase oscillators connected in a star configuration, when one leading oscillator controls a network of noncoupled slave oscillators.

7.2.1 *Phase-Locked Loops (PLLs) with a Parallel Coupling*

Consider a network of coupled PLLs with block diagram depicted in Fig. 7.1. The leading oscillator (LO) drives the identical voltage controlled oscillators (VCO) via the sum of outputs of similar phase detectors (PD_i) passing through the similar low-pass filters (LPF).

The equations of this system can be written in the form [22]

$$p\phi_i + \frac{1}{\beta p + 1}\sum_{j=1}^{n} a_{ij}F(\phi_j) = \Delta_i, \tag{7.1}$$

$p \equiv \frac{d}{dt}$, $F(\phi_i) = \sin\phi_i$, $i = 1, 2, \ldots, n$, where ϕ_i is the phase difference of VCO_i and LO phases, Δ_i is the frequency difference between the VCO_i and LO frequencies, a_i are the coefficients of the weighted sum.

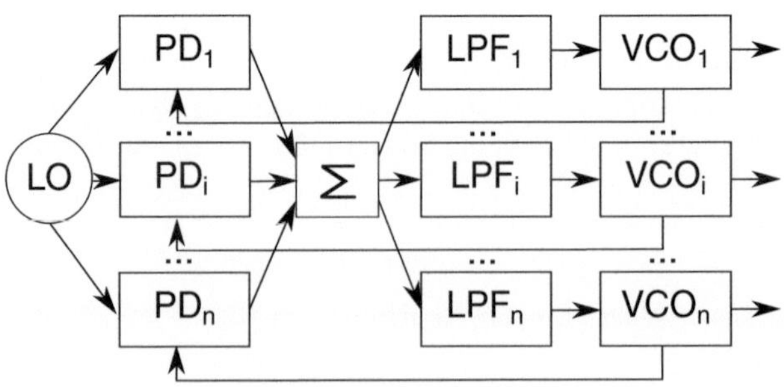

Fig. 7.1 The block diagram of the network of coupled PLLs

The system (7.1) is equivalent to the following system

$$\beta\ddot{\phi}_i + \dot{\phi}_i = -\sum_{j=1}^{n} a_{ij}\sin\phi_j + \Delta_i, \tag{7.2}$$

$i = 1, 2, \ldots, n.$

The same system can be obtained for the star-like topology of the Kuramoto model with inertia.

7.2.2 *Kuramoto Phase Model with Inertia*

Consider the Kuramoto phase model with inertia of N coupled phase oscillators [25, 26]

$$\beta\ddot{\theta}_i + \dot{\theta}_i = \omega_i + \frac{1}{N}\sum_{j=1}^{N} K_{ij}\sin(\theta_j - \theta_i),$$
$$i = 1, 2, \ldots, N, \tag{7.3}$$

where θ_i is the instantaneous phase, ω_i is the natural frequency of the ith oscillator, $K_{i,j}$ are the entries of a coupling symmetric matrix $K = \{K_{ij}\}_N^N$, β is a positive parameter representing an inertia of oscillators. In the case $\beta = 0$ and $K_{ij} = K$, $i, j = 1, 2, \ldots, N$ the system (7.3) becomes the original paradigmatic Kuramoto model [6–8].

We consider the star configuration of coupling [27–29] when the matrix K has the form

$$K = \begin{pmatrix} 0 & K_{12} & \cdots & K_{1N} \\ K_{21} & 0 & \cdots & 0 \\ \vdots & \vdots & \ddots & \vdots \\ K_{N1} & 0 & \cdots & 0 \end{pmatrix},$$

i.e., the first element is the hub of the configuration, and the system (7.3) reads

$$\begin{cases} \beta\ddot{\theta}_1 + \dot{\theta}_1 = \omega_1 + \dfrac{1}{N}\displaystyle\sum_{j=2}^{N} K_{1j}\sin(\theta_j - \theta_1), \\ \beta\ddot{\theta}_2 + \dot{\theta}_2 = \omega_2 + \dfrac{1}{N}\Big(K_{21}\sin(\theta_1 - \theta_2)\Big), \\ \ldots, \\ \beta\ddot{\theta}_N + \dot{\theta}_N = \omega_N + \dfrac{1}{N}\Big(K_{N1}\sin(\theta_1 - \theta_N)\Big). \end{cases} \tag{7.4}$$

We introduce new variables and new parameters

$$\phi_i = \theta_{i+1} - \theta_1, \Delta_i = \omega_{i+1} - \omega_1, \frac{K_{i+1,1}}{N} = a_i, \frac{K_{1,i+1}}{N} = b_i, \tag{7.5}$$
$$i = 1, 2, \ldots, n = N - 1,$$

where ϕ_i is the phase difference between the hub and each another peripheral oscillator, Δ_i is the frequency mismatch of $(i+1)$'s peripheral oscillator and the first hub oscillator. Using (7.5) we rewrite the system (7.4) in the form

$$\begin{cases} \beta\ddot{\phi}_1 + \dot{\phi}_1 = \Delta_1 - ((a_1 + b_1)\sin\phi_1 + b_2 \sin\phi_2 + \cdots + b_n \sin\phi_n), \\ \beta\ddot{\phi}_2 + \dot{\phi}_2 = \Delta_2 - (b_1 \sin\phi_1 + (a_2 + b_2)\sin\phi_2 + \cdots + b_n \sin\phi_n), \\ \ldots, \\ \beta\ddot{\phi}_n + \dot{\phi}_n = \Delta_n - (b_1 \sin\phi_1 + b_2 \sin\phi_2 + \cdots + (a_n + b_b)\sin\phi_n). \end{cases} \tag{7.6}$$

The phase synchronization of oscillators in the model (7.3) is defined as an attractor of the system (7.6) which trajectories $(\phi_i^s(t), \dot{\phi}_i^s(t))$ satisfy the conditions

$$|\phi_i^s(t)| < \varepsilon, \langle \dot{\phi}_i^s(t) \rangle = 0, i = 1, 2, \ldots, n, \tag{7.7}$$

where $\langle\cdot\rangle$ denotes the mean value, and parameter $\varepsilon < \pi/2$ is a measure of synchronization.

Respectively, the steady trajectories $(\phi_i^a(t), \dot{\phi}_i^a(t))$ of the system (7.6) defining asynchronous mode of the oscillators in (7.3) satisfy the condition

$$\langle \dot{\phi}_i^a(t) \rangle \neq 0, i = 1, 2, \ldots, n. \tag{7.8}$$

The mutual synchronization of the peripheral oscillators is characterized by the rotation numbers

$$r_{ij} = \lim_{t\to\infty} \frac{\phi_i^a(t)}{\phi_j^a(t)}, i, j = 1, 2, \ldots, n. \tag{7.9}$$

Below we consider both the PLL and Kuramoto similar models (7.2), (7.3) in the form of the system (7.6).

7.2.3 *Regular Transition to Synchrony*

In this section for simplicity we put $a_i = b_i$. The main candidate for synchronous state is the stable equilibrium point of the system (7.6).

The phase space of the system (7.6) written in Cauchy form with coordinates $(\phi_i, y_i = \dot{\phi}_i)$, $i = 1, 2, \ldots, n$, is the cylinder $G = \{\mathbb{R}^n \times \mathbb{T}^n\}$.

From the boundedness of the right parts in (7.6) follows the next.

Proposition 2.1 *The solid torus* $G_0 = \{\phi_i \in \mathbb{S}^1, |y_i| < \Delta_i + N\bar{a}, i = 1, 2, \ldots, n\}$, $G_0 \subset G$, *where* $\bar{a} = \max_{i \in \{1,2,\ldots,n\}} a_i$, *is the absorbing domain of the system (7.6).*

Equilibria of the system (7.6) are the solutions of the system

$$MS = \Gamma, \tag{7.10}$$

where

$$M = \begin{pmatrix} 2 & 1 & \cdots & 1 \\ 1 & 2 & \cdots & 1 \\ \vdots & \vdots & \ddots & \vdots \\ 1 & 1 & \cdots & 2 \end{pmatrix},$$

$S = \text{column}(a_1 \sin\phi_1, a_2 \sin\phi_2, \ldots, a_n \sin\phi_n)$, $\Gamma = \text{column}(\Delta_1, \Delta_2, \ldots, \Delta_n)$.

Since $\det M = n + 1 = N$ the system (7.10) has a unique solution

$$a_i \sin\phi_i = \tilde{\Delta}_i, i = 1, 2, \ldots, n, \tag{7.11}$$

where $\tilde{\Delta}_i$ are the entries of the column $M^{-1}\Gamma$, and (7.11) reads

$$\sin\phi_i = \frac{1}{a_i} \sum_{j=1}^{n} \left(\frac{\Delta_i}{n} - \frac{\Delta_j}{N} \right). \tag{7.12}$$

The system (7.11) for $|\tilde{\Delta}_i| < a_i$ has 2^n equilibria in G. The principal equilibrium corresponding to the synchronous mode is the point $O(\phi_1^+, \phi_2^+, \ldots, \phi_n^+)$, $\phi_i^+ = \arcsin \tilde{\Delta}_i a_i^{-1}$. All the rest equilibria coordinates are obtained from O by changing $\phi_{i_k}^+$ to $\phi_{i_k}^- = \pi - \phi_{i_k}^+$, $k = 1, 2, \ldots, n$.

The stability of the equilibria is defined by the variational linear system of ODE ($\phi_i = \phi_i^+ + x_i$)

$$\begin{cases} \beta\ddot{x}_1 + \dot{x}_1 + 2\alpha_1 x_1 + \alpha_2 x_2 + \cdots + \alpha_n x_n = 0, \\ \beta\ddot{x}_2 + \dot{x}_2 + \alpha_1 x_1 + 2\alpha_2 x_2 + \cdots + \alpha_n x_n = 0, \\ \ldots, \\ \beta\ddot{x}_n + \dot{x}_n + \alpha_1 x_1 + \alpha_2 x_2 + \cdots + 2\alpha_n x_n = 0, \end{cases} \tag{7.13}$$

where $\alpha_i = a_i \cos\phi_i^+ = \sqrt{a_i^2 - \tilde{\Delta}_i^2}$ (or $\alpha_i = a_i \cos\phi_i^- = -\sqrt{a_i^2 - \tilde{\Delta}_i^2}$).

We seek a solution of the system (7.13) in the form $x_i = c_i e^{pt}$ and obtain the next characteristic equation for the system (7.13)

$$\det \begin{pmatrix} \sigma_1 & 1 & \cdots & 1 \\ 1 & \sigma_2 & \cdots & 1 \\ \vdots & \vdots & \ddots & \vdots \\ 1 & 1 & \cdots & \sigma_n \end{pmatrix} = 0, \tag{7.14}$$

where $\sigma_i = (\beta p^2 + p + 2\alpha_i)\alpha_i^{-1}$.

Hence, the equilibrium O is stable, i.e. the complete synchronization regime is stable, when the real parts of $2n$ roots of Eq. (7.14) for $\alpha_i = \sqrt{a_i^2 - \tilde{\Delta}_i^2}$ are negative.

Let us consider the particular case of the system (7.6)

$$\alpha_i = \sqrt{a_i^2 - \tilde{\Delta}_i^2} = \alpha = \text{const}, \quad i = 1, 2, \dots, n. \tag{7.15}$$

We study the case for $\alpha_i = \sqrt{a_i^2 - (\Delta/N)^2} = \alpha$, and $\sigma_i = \sigma$. In this case Eq. (7.14) takes the form

$$(\sigma - 1)^{n-1}(\sigma + n - 1) = 0. \tag{7.16}$$

Then $\sigma - 1 = 0$ gives the equation

$$\beta p^2 + p + \alpha = 0, \tag{7.17}$$

and $\sigma + n - 1 = 0$ leads to the equation

$$\beta p^2 + p + N\alpha = 0. \tag{7.18}$$

From (7.17) repeated $(n-1)$ times and (7.18) we conclude that the real parts of all the roots of Eq. (7.14) are negative and therefore the equilibrium O is asymptotically stable. In this case from Eq. (7.4) we obtain the following expression for frequency of complete synchronization

$$\omega_s^G = N^{-1} \sum_{i=1}^{N} \omega_i. \tag{7.19}$$

It's easy to verify that in homogeneous case (7.15) all the rest equilibria are saddles with different dimensions of unstable manifolds.

Hence, we proved the next statement

Theorem 2.1 *Let for $\Delta_i = \Delta$, $a_i = a$ the following condition holds*

$$|\Delta| < Na. \tag{7.20}$$

Then the system (7.6) has the stable equilibrium point O, corresponding to the synchronous mode of the system (7.3) when the hub oscillator synchronizes the enclosing ones.

Corollary 2.1 *The stability of the equilibrium point O is preserved for small mismatch $|\alpha_i - \alpha| < \mu$ due to its structural stability. For large mismatch the stability conditions one can derive using (7.14).*

Corollary 2.2 *The system (7.6) has no equilibrium points if* $|\tilde{\Delta}_i| > a_i$ *at least for one* $i = i_1 \in \{1, 2, \ldots, n\}$.

Indeed, in this case the system (7.11) has no solution, and the synchronization loss occurs due to disappearance of the stable equilibrium O via saddle-node bifurcation.

Corollary 2.3 *The emergence of synchrony in the system (7.6) is defined by the* ***regular transition*** *via saddle-node bifurcation.*

This statement follows from Corollaries 2.1, 2.2.

7.3 The Uniform Coupling in Star Configuration

Consider the case of uniform coupling when $a_i = a = \text{const}$, $\Delta_i = \Delta = \text{const}$, but rewrite the system (7.6) in another form

$$\beta\ddot{\phi}_i + \dot{\phi}_i = \Delta - a\sin\phi_i - b\sum_{j=1}^{n}\sin\phi_j, i = 1, 2, \ldots, n, \tag{7.21}$$

where new parameter $b > 0$ is not necessarily equal to a.

Lemma 3.1 *The system (7.21) has the invariant manifold* $M : \{u_1 = u_2 = \cdots = u_n = u\}$ *where* $u_i = (\phi_i, \dot{\phi}_i = y_i)$, $\dim M = 2$. *The dynamics in the manifold M is determined by the pendulum equation*

$$\beta\ddot{\phi} + \dot{\phi} = \Delta - \alpha\sin\phi, \tag{7.22}$$

where $\alpha = a + bn$.

Indeed, each equation (7.21) after substitution $u_i = u$ becomes one and the same equation (7.22), and any trajectory of the system (7.21) with uniform initial condition $u_i = u^0 \in M$ does not leave M.

The local stability of the manifold M is defined by the variational equation

$$\beta\ddot{\xi}_i + \dot{\xi}_i = \cos\phi\left[-a\xi_i - b\sum_{j=1}^{n}\xi_j\right], \tag{7.23}$$

where $\xi_i = \phi_i - \phi$ and ϕ is driven by the system (7.22).

First we present the well-known [30–32] bifurcational diagram and qualitative phase pictures of the pendulum equation (7.22) which for new time $t' = \sqrt{\dfrac{\alpha}{\beta}}t$ and new parameters $\lambda = (\alpha\beta)^{-1/2}$, $\gamma = \Delta/\alpha$ takes the form

$$\ddot{\phi} + \lambda\dot{\phi} + \sin\phi = \gamma. \tag{7.24}$$

For $\lambda > 0$ the bifurcations in this equation are saddle-node for $|\gamma| = 1$ and the homoclinic loop encircling the cylinder $(\phi, y = \dot{\phi})$ at $|\gamma| = \gamma_h(\lambda)$, where $\gamma_h(\lambda)$ is the Tricomi curve satisfying the conditions

$$\gamma_h(0) = 0; \gamma_h'(0) = \frac{4}{\pi}; \gamma_h'(\lambda) > 0, 0 < \lambda < \lambda_{\text{sn}}; \tag{7.25}$$

$$\gamma_h(\lambda) = 0, \lambda \geq \lambda_{\text{sn}}, \tag{7.26}$$

where the value $\lambda_{\text{sn}} \approx 1.2$ corresponds to the homoclinic loop of the saddle-node. The condition $\gamma_h'(0) = \frac{4}{\pi}$ one can obtain using averaging method for small parameters λ and γ in (7.24). The property $\gamma_h'(\lambda) > 0$ follows from clockwise rotation of the vector field $(\dot{\phi} = y, \dot{y})$ given by Eq. (7.24).

For the parameters β, α, Δ of Eq. (7.22) the bifurcations read

$$|\Delta| = \alpha\gamma_h((\beta\alpha)^{-1/2}) \tag{7.27}$$

is the homoclinic bifurcation, and

$$|\Delta| = \alpha \tag{7.28}$$

is the saddle-node bifurcation.

Lemma 3.2 *The system given by Eq. (7.22)*

$$\dot{\phi} = y, \beta\dot{y} + y + \alpha\sin\phi = \Delta \tag{7.29}$$

in the phase cylinder $G = \{\phi \in \mathbb{S}^1, y \in \mathbb{R}^1\}$ has the following phase portraits:

1) In the parameter domain

$$d_1 : \{|\Delta| < \alpha\gamma_h((\beta\alpha)^{-1/2})\} \tag{7.30}$$

the system (7.29) is globally asymptotically stable (Fig. 7.2, S) such that the stable focus (or node) $O_f(\phi_f = \arcsin\frac{\Delta}{\alpha}, y_f = 0)$ attracts the whole cylinder besides the stable separatrices of the saddle $O_s(\phi_s = \pi - \phi_f, y_s = 0)$.

2) In the parameter domain

$$d_2 : \{\alpha\gamma_h((\beta\alpha)^{-1/2}) < |\Delta| < \alpha, (\beta\alpha)^{-1/2} < \lambda_{\text{sn}}\} \tag{7.31}$$

the system (7.29) is bistable: it has the stable focus (node) and the unique stable limit cycle $l_c(\phi = \phi_c(t), y = y_c(t))$ encircling the cylinder; the basins of the focus and the cycle are separated by the stable separatrices of the saddle (Fig. 7.2, B).

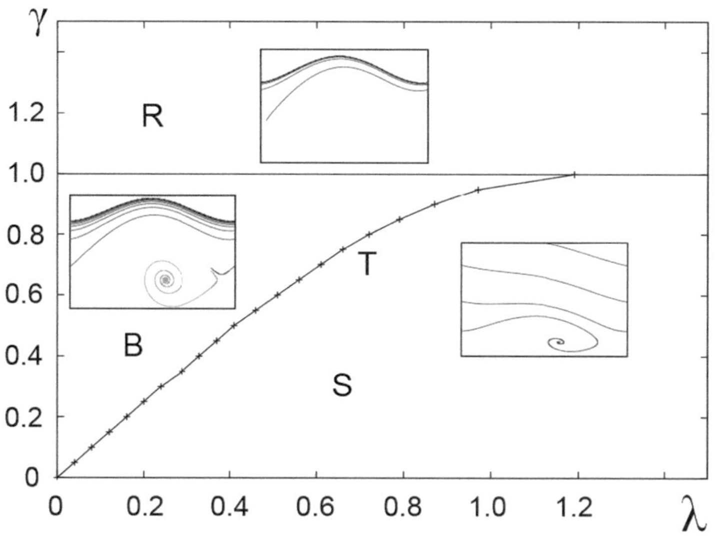

Fig. 7.2 Phase portraits of the system (7.22) for different value of parameters

3) In the parameter domain

$$d_3 : \{|\Delta| > \alpha\} \tag{7.32}$$

the system (7.29) has the unique limit cycle attracting the whole cylinder (Fig. 7.2, R).

Figure 7.2 corresponds to $\Delta > 0$. For $\Delta < 0$ the phase portraits are the same for the reverse coordinates $(\phi, y) \to (-\phi, -y)$.

Consider the local stability of the trajectories $l^*(\phi^*(t), y^*(t))$ in the invariant manifolds M, especially of the limiting set, which consist of the stable focus (node) $O_f(\phi, 0)$, the saddle $O_s(\phi_s, 0)$, and limit cycle $l_c(\phi_c, y_c)$.

Obviously, the stability along the manifold M is defined by the variational equation of Eq. (7.22) ($\phi = \phi^* + \xi$)

$$\beta\ddot{\xi} + \dot{\xi} + \alpha\cos\phi^*(t)\xi = 0, \alpha = a + nb, \tag{7.33}$$

which obviously determines, type of the equilibria stability and for the limit cycle, $\phi^* = \phi_c$, gives one zero and one negative ($\mathrm{div}(\dot{\phi}, \dot{y}) = -(\beta\alpha)^{-1/2} < 0$) Lyapunov exponents.

The variational equation (7.23) for original system (7.21) along the manifold M, i.e. along the vector $(1, 1, \ldots, 1)$ gives the same equation (7.33).

In the transverse direction to the manifold (transversally to the vector $(1, 1, \ldots, 1)$) the system (7.23) takes the form

$$\beta\ddot{\eta}_i + \dot{\eta}_i + a\cos\phi^*(t)\eta_i = 0, \tag{7.34}$$

where $\eta_i = \xi_i - \xi_{i+1}$, $i = 1, 2, \ldots, n-1$, and $\phi^*(t)$ is driven by Eq. (7.22). Equation (7.34) similarly to (7.14) determines the stability of the focus (node) O_f in the transverse direction and unstability of the saddle O_s in the transverse direction. Equation (7.34) for the trajectories $\phi^*(t)$ from the basin of node O_f lying in the band $|\phi| < \frac{\pi}{2}$, providing $\cos\phi^*(t) > 0$, at least for $(\beta a)^{-1/2} \geq \lambda_{\text{sn}}$ guarantees the local stability of this part of the manifold M. The transverse stability of the limit cycle is defined by Eq. (7.34) for $\phi^* = \phi_c(t)$. Since $\dot{\phi}_c(t) = y_c(t) > 0$ the phase $\phi_c(t)$ rotates and the term $\cos\phi_c(t)$ in (7.34) changes the sign thereby creating a problem of the cycle transverse stability. We solve it in the case when $|\Delta| = (\beta a)^{-1/2} + \varepsilon$, for small enough $\varepsilon > 0$. In this case the cycle just appearing from the homoclinic loop of the system (7.29) passes a small neighborhood of the saddle O_s and therefore spends the most time (of order $1/\varepsilon$) in the neighborhood $|\phi - \phi_s| < \varepsilon$. Since $\cos\phi_s < 0$, due to (7.34) the limit cycle l_c is born being unstable.

From the above reasoning we conclude

Proposition 3.1 *1) If the Lyapunov–Floquet exponents from (7.34) for $\phi^*(t) = \phi_c(t)$ are negative, then the asynchronous mode is such that the peripheral oscillators are synchronized with rotation numbers equaled* 1.

2) The homoclinic bifurcation of the system (7.21) leads to an asynchronous mode of the peripheral oscillators.

For sufficiently large inertia such that $(\beta a)^{-1/2} < \lambda_{\text{sn}}$, the transition from coherence to incoherence of oscillators is hysteretic. When the frequency difference Δ increases the transition from the stable equilibrium O_f (coherence) to the rotation mode in the solid torus G_0 (incoherence) occurs via the saddle-node bifurcation $|\Delta| = a$. Obviously, this rotation mode can be the stable cycle $l_c = (\phi_c(t), \dot{\phi}_c(t))$ in the manifold M. In this case one observes the transition from complete phase synchronization to the synchronous state of the peripheral oscillators being asynchronous to the hub oscillator with mean frequency difference $\langle\dot{\phi}_c(t)\rangle$. When the frequency difference Δ decreases from large values corresponding to the rotation mode at the bifurcation of homoclinic orbit of the saddle $|\Delta| = a\gamma_h((\beta a)^{-1/2})$ the reverse transition to the complete synchronization due to Proposition 3.1 occurs only from the asynchronous mode of the peripheral oscillators. Note that this hysteretic behavior being similar to the transitions in the Josephson junction model [33] was discussed in the recent paper [34].

7.4 Nonsymmetric Coupling

We consider the general case of nonsymmetric coupling but, as an example, for three oscillators in the star configuration. Similarly to symmetrical case (7.4), (7.6) we obtain the system

$$\begin{cases} \beta_1\ddot{\phi}_1 + \dot{\phi}_1 + a_1\sin\phi_1 = \Delta_1 - b_1\sin\phi_2, \\ \beta_2\ddot{\phi}_2 + \dot{\phi}_2 + a_2\sin\phi_2 = \Delta_2 - b_2\sin\phi_1, \end{cases} \tag{7.35}$$

where $\beta_{1,2}$ are the different inertias of peripheral oscillators, $a_{1,2}$, $b_{1,2}$ are the coupling matrix entries, and $\Delta_{1,2}$ are the frequency differences. Our goal is to introduce several parameter domains exhibiting different simple and complicated dynamics of the system (7.35).

7.4.1 Equilibria

The system (7.35) has four equilibria in the region

$$\tilde{\Delta}_{1,2} < \delta, \tag{7.36}$$

where $\tilde{\Delta}_{1,2} = a_{2,1}\Delta_{1,2} - b_{1,2}\Delta_{2,1}$, $\delta = a_1a_2 - b_1b_2$, defined by the equation

$$\sin\phi_{1,2} = \tilde{\Delta}_{1,2}\delta^{-1}, \tag{7.37}$$

giving solutions $\phi_{1,2}^{+(-)}$ similarly to (7.11). For $\beta_1 = \beta_2 = \beta$ the stability of the equilibria in this case is defined by the equation

$$\sigma^2 + r\sigma + \alpha_1\alpha_2\delta = 0, \tag{7.38}$$

where $r = a_1\alpha_1 + a_2\alpha_2$, $\alpha_{1,2} = \cos\phi_{1,2}^{+(-)}$, $\sigma = \beta p^2 + p$. Due to (7.38) the principal equilibrium $O_f(\phi_1^+, \phi_2^+)$ is stable and three other equilibria are saddles.

7.4.2 Comparison Systems

We rewrite Eq. (7.35) in the form of the systems

$$\begin{cases} \dot{\phi}_{1,2} = y_{1,2}, \\ \beta_{1,2}\dot{y}_{1,2} = \Delta_{1,2} - y_{1,2} - a_{1,2}\sin\phi_{1,2} - b_{1,2}\sin\phi_{2,1}. \end{cases} \tag{7.39}$$

Introduce two comparison $2D$ systems (see [35] and reference within) for each subsystem in (7.39) $A_i^{+(-)}$:

$$\begin{cases} \dot{\phi}_i = y_i, \\ \beta_i\dot{y}_i = \Delta_i \pm b_i - y_i - a_i\sin\phi_i, \end{cases} \tag{7.40}$$

$i = 1, 2$, acting in $2D$-cylinder $G = \mathbb{R}^1 \times \mathbb{S}^1$. The systems $A_i^{+(-)}$ are the system (7.29) with $\tilde{\Delta}_i = \Delta_i \pm b_i$, $i = 1, 2$, standing for Δ.

The vector projection of the system (7.40) on the cylinder (ϕ, y) is rotated clockwise (counterclockwise) relatively to the vector of A_i^+ (A_i^-, respectively) in

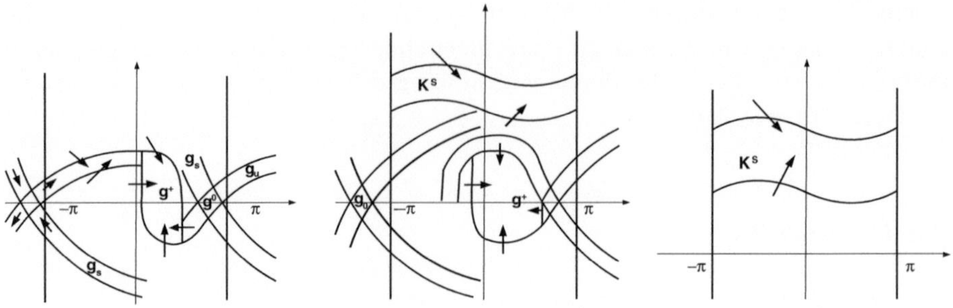

(a) (b) (c)

Fig. 7.3 Illustration of the comparison system (7.40)

the half-cylinder ($\phi, y > 0$) and vice versa in the half-cylinder ($\phi, y < 0$). Now we depict the separatrices of the saddles and cycles simultaneously for the systems A_i^+ and A_i^-. The unstable (stable) separatrices form the strips between them—separatrix channels g_i^u (g_i^s, respectively). Introduce the intersection $g_i^o = g_i^u \cap g_i^s$, called the saddle cell [35], the annulus K_i^s bounded by the stable cycles of the system A_i^+ and A_i^-, and the absorbing domain g_i^+ bounded by unstable separatrices of the systems A_i^+ and A_i^-, and segments $\phi = \text{const}$ (see Fig. 7.3), $i = 1, 2$. We select three pairs of parameter domains $d_{ki} = d_{ki}^+ \cup d_{ki}^-$, where $d_{ki}^{+(-)}$, $k = 1, 2, 3$, $i = 1, 2$, are domains d_k, $k = 1, 2, 3$ from Lemma 3.2 with $\tilde{\Delta}_i = \Delta_i \pm b_i$, a_i and β_i, $i = 1, 2$, standing for Δ, a and β, respectively. Both systems A_i^+ and A_i^- have the same qualitative phase portraits in each of these domains d_{ki}, $k = 1, 2, 3$, forming mutual arrangement of the saddle channels, annulus and absorbing domain depicted in Fig. 7.3. Using the above geometric structures we obtain the following findings.

Theorem 4.1 *Let the parameters of the system (7.39) be in the domain* $\overline{d}_1 = d_{11} \cup d_{12}$ *when the comparison systems* $A_1^{+(-)}$ *and* $A_2^{+(-)}$ *form the same structure of Fig. 7.3a. Then the equilibrium point* $O_f(\phi_1^+, \phi_2^+)$ *is globally asymptotically stable. Herewith three oscillators are globally synchronized.*

Proof The system (7.39) has no entire trajectories besides the saddles in the domains $g_1^o \times g_2^o$, $g_{1,2}^o \times g_{1,2}^o$. It follows from Lyapunov–Chetaev function for monotone functions $\sin\phi_{1,2}$ at the intervals of $\phi_{1,2}$ for $g_{1,2}^{o(+)}$ in the system (7.39). The domain $\overline{g}^+ = g_1^+ \cup g_2^+$ is the attracting domain of the trajectories of the system (7.39) (besides the stable manifolds of the saddles) for the parameter region $d_{11} \cup d_{12}$ due to the directing property of the comparison systems (see Fig. 7.3a). The stability of the locally stable point O_f in the globally attracting domain $\overline{g}^+$ can be derived with the Lyapunov function using the monotonicity of $\sin\phi_{1,2}$ in the square $|\phi_{1,2} - \phi_{1,2}^+| < \varepsilon$ corresponding to $\overline{g}^+$.

Theorem 4.2 *Let the parameters of the system (7.39) be in the domain* $\overline{d}_3 = d_{31} \cup d_{32}$, *when the comparison systems* $A_1^{+(-)}$ *and* $A_2^{+(-)}$ *form the same structures of*

Fig. 7.3c. Then the solid torus $\overline{K}^+ = K_1^s \times K_2^s$ *attracts all the trajectories of the system (7.39). The nonwandering set of trajectories in* $\overline{K}^+$ *is rotating and defines the asynchronous mode of the oscillators.*

This statement immediately follows from the simple structure of Fig. 7.3c forcing all trajectories of the system (7.39) to enter $\overline{K}^+$.

Theorem 4.3 *Let the parameters of the system (7.39) be in the domain* $\overline{d}_2 = d_{21} \cup d_{22}$ *with the same structures of comparison systems (Fig. 7.3b). Then the system (7.39) is fourfold-stable, that is, it has four separate components of limiting set in four absorbing domains* $\overline{g}^+$, $\overline{K}^+$, $M_1^+ = g_1^+ \times K_2^s$, $M_2^+ = g_2^+ \times K_1^s$.

Proof In the parameter domain $\overline{d}_2$ a trajectory of the system (7.39) given by a solution $\phi_i = \tilde{\phi}_i(t, \phi_1^o, y_1^o, \phi_2^o, y_2^o)$, $y_i = \tilde{y}_i(t, \phi_1^o, y_1^o, \phi_2^o, y_2^o)$, $i = 1, 2$, with initial conditions $(\phi_i^o, y_i^o) \in g_i^+(K_i^s)$, $i = 1, 2$, is such that due to the comparison principal the coordinates of the first (second, respectively) subsystem remain in the first (second, respectively) absorbing domain, $(\tilde{\phi}_i, \tilde{y}_i) \in g_i^+(K_i^s)$, $i = 1, 2$, for any $t > 0$. This implies that in $\mathbb{R}^2 \times \mathbb{T}^2$ the trajectory of the system (7.39) with any initial point in the domain $\overline{g}^+$ ($\overline{K}^+$, $\overline{M}_1^+$, $\overline{M}_2^+$, respectively), stays in these domains for any $t > 0$.

7.4.3 Chaotic Transitions to Synchrony

First we note that in all cases of system the bifurcations of equilibria are simple and occur via the saddle-node when the frequency differences increase. In order to exhibit the complicated bifurcations leading to emergence of chaos we consider the reduced system (7.35) for $b_2 = 0$ corresponding to the unidirectional coupling of one of the peripheral oscillators. The second ("master") equation in (7.35) has the pendulum dynamics and in the simple case $|\Delta_2| > a_2$ which we consider has the unique rotating limit cycle $\phi_c(t) = \phi_c(t + T)$ (see Lemma 3.2). The first ("slave") equation is the pendulum one as well but it is driven by periodic force $-b_1 \sin \phi_c(t)$. Using the results from [33, 35] we obtain the next

Theorem 4.4 *1) In the parameter region* $b_2 = 0$, $|\Delta_2| > a_2$, $|\Delta_1| < a_1 - b_1$ *the system (7.35) in the solid torus* $g_1^o \times K_2^s$ *has a unique saddle cycle* l_s *which stable* W_1^s *and unstable* W_1^u *manifolds lie in the channels* $W_1^s \subset g_1^s \times K_2^s$, $W_1^u \subset g_1^u \times K_2^s$ *and have mutual arrangement corresponding for* d_{11} *to Fig. 7.3a and for* d_{12} *to Fig. 7.3b.*

2) There exists an interval $|\Delta_1 - a_1\gamma_h((\beta_1 a_1)^{-1/2})| < \varepsilon$ *corresponding to a structurally stable homoclinic orbit to the cycle* l_s *providing a chaotic component of the system (7.35) limiting set containing infinite numbers of saddle cycles.*

3) There exists an interval linking to the bifurcational point $\Delta_1 = a_1\gamma_h((\beta_1 a_1)^{-1/2}) + \varepsilon$ *for which the system (7.35) has a quasi-strange rotating attractor which defines the chaotic transition to synchrony.*

The proof of the theorem is based [33, 35] on the fact that when the parameter Δ_1 increases from the values from the domain d_{11} (corresponding to Fig. 7.3a) up to the values from the domain d_{12} (corresponding to Fig. 7.3b) the channels g_1^s and g_1^u as well as the manifolds W_1^s and W_1^u change their mutual arrangement causing the birth, existence and death of the homoclinic orbits $H_1 = W_1^s \cap W_1^u$. Omitting the details we note the important property of the hysteretic bifurcational transition in the system (7.35). When the parameter Δ_1 decreases *the transition from asynchronous rotation to the synchrony occurs at random values of the parameter* Δ_1 from the interval corresponding to the quasi-attractor existence. This complexity of the system (7.35) dynamics is similar to that of the shunted Josephson junction [35].

Hence, the example of three oscillators exhibits the complexity of the system dynamics which obviously is typical for the general system (7.3).

Finally we emphasize that the main reason of the complicated dynamics is relatively large inertia of oscillators. Indeed, in the limiting case $\beta_{1,2} \to \infty$ ($\lambda_{1,2} \to \infty$ for rescaled system (7.24)) the system (7.35) reduces to the divergence-free nonintegrable system

$$\begin{cases} \ddot{\phi}_1 + a_1 \sin\phi_1 + b_1 \sin\phi_2 = \Delta_1, \\ \ddot{\phi}_2 + a_2 \sin\phi_2 + b_2 \sin\phi_1 = \Delta_2. \end{cases} \tag{7.41}$$

In this system due to KAM theory the invariant tori breakdowns with increase of $b_{1,2}$ from zero.

Acknowledgements This work was supported by the RSF (Project No. 14-12-00811) (Sections 1, 2) and by the RFBR (project 15-01-08776) (Section 3).

References

1. Arfaimovich, V.S., Verichev, N.N., Rabinovich, M.I.: Stochastic synchronization of oscillations in dissipative systems. Radiophys. Quantum Electron. **29**(9), 795–803 (1986)
2. Pikovsky, A.S., Rosenblum, M.G., Kurths, J.: Synchronization - A Universal Concept in Nonlinear Science. Cambridge University Press, Cambridge (2001)
3. Osipov, G.V., Kurths, J., Zhou, Ch.: Synchronization in Oscillatory Networks. Springer, Berlin, Heidelberg (2007)
4. Mosekilde, E., Maistrenko, Yu., Postnov, D.: Chaotic Synchronization. Applications to Living Systems. World Scientific, Singapore (2002)
5. Manrubia, S.C., Mikhailov, A.S., Zanette, D.H.: Emergence of Dynamical Order: Synchronization Phenomena in Complex Systems. World Scientific, Singapore (2004)
6. Winfree, A.T.: Biological rhythms and the behavior of coupled oscillators. J. Theor. Biol. **16**, 15–42 (1967)
7. Kuramoto, Y.: In: Araki, H. (ed.) Proceedings of International Symposium on Mathematical Problems in Theoretical Physics. Lecture Notes in Physics, vol. 39. Springer, New York (1975)
8. Kuramoto, Y.: Chemical Oscillations, Waves and Turbulence. Springer, Berlin/Düsseldorf (1984)
9. Wiesenfeldt, K., Colet, P., Strogatz, S.: Frequency locking in Josephson junction arrays: connection with the Kuramoto model. Phys. Rev. E **57**, 1563–1567 (1998)

10. Kozyrev, G., Vladimirov, A.G., Mandel, P.: Global coupling with the time delay in an array of semiconductor lasers. Phys. Rev. Lett. **85**(18), 3809–3812 (2000)
11. Michaels, D.C., Matyas, E.P., Jalife, J.: Mechanisms of sinoatrial pacemaker synchronization a new hypothesis. Circ. Res. **61**(5), 704–714 (1987)
12. Brown, E., Holmes, P., Moehlis, J.: Globally coupled oscillator networks. In: Kaplan, E., Marsden, J.E., Sreenivasan, K.R. (eds.) Perspectives and Problems in Nonlinear Science: A Celebratory Volume in Honor of Larry Sirovich, pp. 183–215. Springer, Berlin (2003)
13. Kopell, N., Ermentrout, G.B.: Coupled oscillators and the design of central pattern generators. Math. Biosci. **90**, 87–109 (1988)
14. Neda, Z., Ravasz, E., Vicsek, T., Brecht, Y., Barabasi, A.-L.: Physics of the rhythmic applause. Phys. Rev. E **61**, 6987–6992 (2000)
15. Strogatz, S.H., Abrams, D.M., McRobie, A., Eckhardt, B., Ott, E.: Theoretical mechanics: crowd synchrony on the Millenium Bridge. Nature **438**(70640), 43–44 (2005)
16. York, R.A., Compton, R.C.: Quasi-optical power combining using mutually synchronized oscillator arrays. IEEE Trans. Autom. Control **57**(4), 920–935 (2012)
17. Dorfler, F., Bullo, F.: Synchronization in complex networks of phase oscillators: a survey. Automatica **50**(6), 1539–1564 (2014)
18. Acebron, J.A., Bonilla, L.L., Vicente, C.J.P., Ritort, F., Spigler, R.: The Kuramoto model: a simple paradigm for synchronization phenomena. Rev. Modern Phys. **77**(1), 137–185 (2005)
19. Belykh, V.N., Petrov, V.S., Osipov, G.V.: Dynamics of the finite-dimensional Kuramoto model: global and cluster synchronization. Regul. Chaotic Dyn. **20**(1), 37–48 (2015)
20. Lindsey, W.C.: Synchronization Systems in Communication and Control. Pearson Education, Upper Saddle River, NJ (1972)
21. Shalfeev, V.D., Matrosov, V.V.: Nonlinear Dynamics of the Phase Synchronization Systems. Publishing House of the Nizhny Novgorod State University, N. Novgorod (2013) (in Russian)
22. Arfaimovich, V.S., Nekorkin, V.I., Osipov, G.V., Shalfeev, V.D.: Stability, Structures and Chaos in Nonlinear Synchronization Networks. World Scientific, Singapore (1994)
23. Frasca, M., Bergner, A., Kurths, J., Fortuna, L.: Bifurcations in a star-like network of Stuart-Landau oscillators. Int. J. Bifurcation Chaos **22**(7), 1250173 (2012)
24. Kazanovich, Y., Burylko, O., Borisyuk, R.: Competition for synchronization in a phase oscillator system. Physica D **261**, 114–124 (2013)
25. Tanaka, H.A., Lichtenberg, A.J., Oishi, S.: First order phase transition resulting from finite inertia in coupled oscillator systems. Phys. Rev. Lett. **78**, 2104 (1997)
26. Tanaka, H.A., Lichtenberg, A.J., Oishi, S.: Self-synchronization of coupled oscillators with hysteretic responses. Physica D **100**, 279 (1997)
27. Pecora, L., Carrol, T.: Synchronization in chaotic systems. Phys. Rev. Lett. **64**, 821 (1990)
28. Belykh, V.N., Belykh, I.V., Hasler, M.: Connection graph stability method for synchronized coupled chaotic systems. Physica D **195**, 159–187 (2004)
29. Belykh, V.N., Osipov, G.V., Petrov, V.S., Suykens, J., Vandewalle J.: Cluster synchronization in oscillatory networks. Chaos **13**, 037106 (2008)
30. Tricomi, F.: Integrazione di un' equazione differenziale presentatasi in elettrotecnica. Annali della R. Scuola Normale Superiore di Pisa Ser. 11 **2**, 1 (1933)
31. Urabe, M.: The least upper bound of a damping coefficient ensuring the existence of a periodic motion of a pendulum under constant torque. J. Sci. Hiroshima Univ. Ser. A **18**, 379–389 (1955)
32. Belykh, V.N., Pedersen, N., Soerenses, O.: Shunted-Josephson-junction model. I. The autonomous case. Phys. Rev. B **16**, 4853 (1977)
33. Belykh, V.N., Pedersen, N., Soerenses, O.: Shunted-Josephson-junction model. II. The nonautonomous case. Phys. Rev. B **16**, 4860 (1977)
34. Olmi, S., Navas, A., Boccaletti, S., Torcini, A.: Hysteretic transitions in the Kuramoto model with inertia. Phys. Rev. E **90**, 042905 (2014)
35. Belykh, V.N.: Homoclinic and heteroclinic linkages in concrete systems: nonlocal analysis and model maps. Adv. Math. Sci., Amer. Math. Soc. Transl. **2**(200), 51–62 (2000)

Chapter 8
Lotka–Volterra Like Dynamics in Phase Oscillator Networks

Christian Bick

8.1 Introduction

The human brain as a complex network consists of many individual neural oscillators whose activities are coordinated on multiple scales. Such coordination is, for example, given by metastable dynamics that show transient switching between different patterns of activity [1, 12, 39]. Misha Rabinovich and coworkers pioneered the idea that to obtain in-depth understanding of brain dynamics one has to focus on understanding the dynamical principles that underly the generation of reproducible sequential activity of metastable states [33, 35, 36]. This applies to both small circuits [26, 34] and cognitive functions involving many neurons [10, 36] that involve many neurons. In particular in the latter case, generalized Lotka–Volterra equations have provided suitable mathematical models to generate sequential switching behavior, exhibited by trajectories close to heteroclinic structures involving saddle equilibria (or more general sets). However, as coarse-grained models for higher order brain functions these Lotka–Volterra type equations neglect the microscopic dynamics of individual neural oscillators.

At the same time, phase oscillator networks provide paradigmatic models to describe the collective dynamics of weakly interacting oscillators [3, 15]. Consider a network of M populations with N oscillator each where the phase $\theta_k^\sigma \in \mathbf{T} := \mathbb{R}/2\pi\mathbb{Z}$ of the kth oscillator in population σ evolves according to

$$\dot{\theta}_k^\sigma = \omega_k^\sigma + \sum_{\tau=1}^{M} \frac{K_{\sigma\tau}}{N} \sum_{j=1}^{N} \sin\left(\theta_j^\tau - \theta_k^\sigma - \alpha_{\sigma\tau}\right). \tag{8.1}$$

C. Bick (✉)
Centre for Systems, Dynamics and Control and Department of Mathematics, University of Exeter, Exeter EX4 4QF, UK
e-mail: C.Bick@exeter.ac.uk

I.S. Aranson et al. (eds.), *Advances in Dynamics, Patterns, Cognition*, Nonlinear Systems and Complexity 20, DOI 10.1007/978-3-319-53673-6_8

where ω_k^σ is the intrinsic frequency of the oscillator and the coupling is given by phase lags $\alpha_{\sigma\tau}$ and strengths $K_{\sigma\tau}$. For globally and identically coupled oscillators, that is, $K_{\sigma\tau} = K$, $\alpha_{\sigma\tau} = \alpha$ for all σ, τ, we recover the original Kuramoto–Sakaguchi equations [37, 38]. In the continuum limit of $N \to \infty$ oscillators, there is an invariant manifold on which the dynamics are fully determined by a closed set of equations for the local Kuramoto order parameter of each population [28]. This reduction has been particularly useful to analyze networks with identical frequencies, (8.1) with ω_k^σ sampled from a single distribution, that supports particular solutions where one population is desynchronized while the rest is fully phase synchronized. These solutions are commonly known as chimeras [2, 16, 31].

In this chapter, we point out possible links between the dynamics of the mean field equations of (8.1) in the continuum limit and generalized Lotka–Volterra systems. Such similarities have been observed before for oscillators with nonresonant interaction [14], but here we are primarily interested in populations with identical frequency distributions that support chimeras. While the global dynamics for two populations [25] of identical oscillators—where chimeras were originally studied [2]—are now fairly well understood [21, 22], for three populations the attention has been limited to certain invariant subspaces [20] and global results are lacking. Here we observe that the dynamics of (8.1) in the continuum limit bears similarity to a generalized Lotka–Volterra system with a restricted set of parameters. This provides a first attempt to characterize the global dynamics of a network of an arbitrary number of identical phase oscillator populations bridging both microscopic dynamics on the level of a single oscillator and macroscopic mean field dynamics.

The remainder of this chapter is organized as follows. In Sect. 8.2 we discuss sequential switching dynamics in Lotka–Volterra type equations: we review results how Lotka–Volterra dynamics can be used to model cognitive functions and derive limit on cognitive processing capabilities. In Sect. 8.3 we derive general mean field equations for coupled populations of sinusoidally coupled phase oscillators in the continuum limit explicitly. Finally, in Sect. 8.4, we look at the dynamics of the oscillator populations in the continuum limit which resemble a Lotka–Volterra type system close to a cluster state before giving some concluding remarks.

8.2 Lotka–Volterra Type Equations in Neural Dynamics

Generalized Lotka–Volterra equations describe the dynamics of a network of $M \in \mathbb{N}$ competitively interacting agents where the state $a_\sigma \geq 0$ of node σ evolves according to

$$\dot{a}_\sigma = a_\sigma \left(s_\sigma - p_\sigma a_\sigma - \sum_{\tau \neq \sigma} r_{\sigma\tau} a_\tau \right). \tag{8.2}$$

These equations provide a classical model for the interaction of different species [23] where a_k corresponds to the population size of population σ which has some growth s_σ, a nonlinear term leading to saturation, and interaction given by $r_{\sigma\tau}$. More recently, generalized Lotka–Volterra have been employed to model sequential neural information processing in the brain [33, 35]: *winnerless competition dynamics* yield switching dynamics between subsequent temporary winners induced, for example, through heteroclinic structures in phase space. In this section assume that the self-inhibition is normalized, i.e., $p_\sigma = 1$.

Robust heteroclinic sequences can arise in Lotka–Volterra type equations (8.2) in the invariant subspaces given by $\mathcal{A}_\sigma := \{(a_1, \ldots, a_M) \mid a_\sigma = 0\}$. More precisely, a heteroclinic sequence is a sequence of distinct saddle equilibria with a one-dimensional unstable manifold together with heteroclinic orbits that connect consecutive saddles. For (8.2) we have equilibria $A_\sigma := (0, \ldots, 0, s_\sigma, 0, \ldots, 0)$ where s_σ is the σth entry of A_σ. The saddle A_σ is dissipative if $\nu(A_\sigma) = -\operatorname{Re}(\lambda_{2\sigma})/\operatorname{Re}(\lambda_{1\sigma}) > 1$ where $\lambda_{j\sigma}$ are the eigenvalues of the linearization around A_σ ordered monotonically decreasing by real part. Let $\iota = (\sigma_1, \ldots, \sigma_q)$, $q \leq M$, be a sequence of distinct indices and for $n \in \mathbb{N}$ write $[n] = \{1, \ldots, n\}$.

As shown in [5], for given ι there are parameters that give rise to a heteroclinic sequence with saddles A_{σ_k}, $k \in [M]$, where the heteroclinic connections between saddles A_{σ_k} and $A_{\sigma_{k+1}}$ lie in $\bigcap_{\tau \notin \{\sigma_k, \sigma_{k+1}\}} A_\tau$. In fact, parameters can be chosen such that the heteroclinic sequence is stable, that is all saddles are dissipative. Stable heteroclinic sequences give rise to a *stable heteroclinic channel*, which is a suitable ε-neighborhood $V(\iota, \varepsilon)$ of the stable heteroclinic sequence so trajectories that have their initial point in the vicinity of A_1 stay in $V(\iota, \varepsilon)$ for some finite time as they traverse neighborhoods of A_{σ_k} sequentially according to ι; cf. [11]. In the context below, it makes sense to consider networks where the parameters are sampled randomly. In joint work with Misha Rabinovich, we build upon [5] to prove the following theorem.

Theorem 2.1 (Bick and Rabinovich [11]) *Let ι be a sequence of equilibria of length $q \in \mathbb{N}$ and suppose that the entries of s corresponding to indices in ι form a Fibonacci sequence, i.e., $s_{\sigma_1} = a$, $s_{\sigma_2} = ac$, $s_{\sigma_{k+1}} = s_{\sigma_k} + s_{\sigma_{k-1}}$ with $s_{\sigma_k} < 2s_{\sigma_{k+1}}$, $k = 1, \ldots, q$ and the remaining entries are sufficiently small.*

Then there are intervals J, J' on the positive real axis and $\mu > 0$ large enough such that for any $c \in J, a > 0$ and sufficiently small $\varepsilon > 0$ the system (8.2) *with a coupling matrix r with coefficients $r_{\sigma_{k-1}\sigma_k}, r_{\sigma_{k+1}\sigma_k}$ sampled randomly from J' and $r_{\tau\sigma_k}, \tau \notin \{\sigma_{k-1}, \sigma_k, \sigma_{k+1}\}$ sampled randomly from (μ, ∞) has a robust stable heteroclinic channel in $V(\iota, \varepsilon)$.*

We subsequently applied this result to a model of sequential working memory based on winnerless competition dynamics [10]. Let $a_\sigma(t) \geq 0$ denote the activity of a neural cluster or pattern of an informational item σ whose dynamics are given by (8.2). Here $r_{\sigma\tau} \geq 0$ describe the inhibitory connections between the τth and σth clusters and s_σ represent the level of self-excitation. In a neural setting, these parameters depend on additional effects, are determined by the sequence to be stored and modulated by attention and other cognitive processing capabilities. Working memory recall is now given by winnerless competition dynamics between informational items induced by a heteroclinic sequence of length q.

Assuming that the model parameters are bounded, Theorem 2.1 now gives a bound on the length q of the memory sequence—the number of items that can be recalled in sequence. Both μ and the interval J' depend on the mutual inhibition $r_{\sigma\tau}$. This relationship can be expressed by a "scaling function" $\Xi(q)$ depending only on the length of the sequence q—see [10] for a more detailed definition—which satisfies

$$\Xi(q) \leq \frac{r_{\tau\sigma_k}}{r_{\sigma_{k-1}\sigma_k}} \tag{8.3}$$

with $\tau \notin \{\sigma_{k-1}, \sigma_k, \sigma_{k+1}\}$. The scaling function provides a lower bound on the strength of the inhibitory connections relative to lateral connections that impose order on the heteroclinic sequence. This function grows like a geometric sequence with increasing q. Given an upper bound on the relative connection strengths, this implies that there is an upper bound on the length of the sequence. Experimental neurobiological data suggests that an upper bound for Ξ of about 10–20 is realistic. The resulting limit of the sequence length adequately matches the "magical number seven" [24] for working memory which provides a purely dynamical explanation for the limitation of sequential working memory.

8.3 Mean Field Equations for Populations of Phase Oscillators

Now consider the Kuramoto–Sakaguchi equations (8.1) with non-local coupling between M populations of N oscillators where the intrinsic frequencies ω_k^σ of the oscillators are sampled from a Lorentzian distribution with density g_σ. Let $i = \sqrt{-1}$ and $\bar{z}$ denote the complex conjugate of $z \in \mathbb{C}$. In the continuum limit of $N \to \infty$ the dynamics can be described by the probability density $f_\sigma(\omega^\sigma, \theta^\sigma, t)$ of the oscillators in population $\sigma = 1, \ldots, M$ [38] which must satisfy the continuity equation

$$\frac{\partial f_\sigma}{\partial t} + \frac{\partial}{\partial \theta}(f_\sigma v_\sigma) = 0 \tag{8.4}$$

where $v_\sigma(\omega^\sigma, \theta^\sigma, t)$ is their velocity

$$v_\sigma = \omega^\sigma + \sum_{\tau=1}^{M} K_{\sigma\tau} \int_{-\infty}^{\infty} \int_0^{2\pi} f_\tau(\omega^\tau, \theta^\tau, t) \sin(\theta^\tau - \theta^\sigma - \alpha_{\sigma\tau})\, \mathrm{d}\theta^\tau\, \mathrm{d}\omega^\tau \tag{8.5}$$

$$= \omega^\sigma + \sum_{\tau=1}^{M} \frac{K_{\sigma\tau}}{2i} \left(z_\tau \exp(-i(\theta^\sigma + \alpha_{\sigma\tau})) - \bar{z}_\tau \exp(i(\theta^\sigma + \alpha_{\sigma\tau}))\right). \tag{8.6}$$

Here we allow for general interaction between populations given by a coupling strength $K_{\sigma\tau}$ and a phase lag $\alpha_{\sigma\tau}$. The dynamics for $M = 2$ populations were studied in detail in [22].

The reduction by Ott and Antonsen [28, 29] to mean field dynamics has become standard: there is an invariant manifold of probability densities on which the dynamics can be expressed by dynamical equations for the order parameters

$$z_\sigma(t) = \int_{-\infty}^{\infty}\int_0^{2\pi} \exp(i\theta^\sigma) f_\sigma(\omega^\sigma, \theta^\sigma, t)\, \mathrm{d}\theta^\sigma\, \mathrm{d}\omega^\sigma \tag{8.7}$$

of population $\sigma = 1, \ldots, M$. More specifically, considering probability densities

$$f_\sigma(\omega^\sigma, \theta^\sigma, t) = \frac{g_\sigma(\omega^\sigma)}{2\pi}\left(1 + \sum_{n=1}^{\infty} (a_\sigma(\omega^\sigma, t)\exp(i\theta^\sigma))^n + \text{c.c.}\right), \tag{8.8}$$

where c.c. denotes the complex conjugate, we find that

$$0 = \frac{\partial a_\sigma}{\partial t} + i\omega^\sigma a_\sigma - \frac{1}{2}\sum_{\tau=1}^{2} K_{\sigma\tau}\big(\exp(i\alpha_{\sigma\tau})\bar{z}_\tau - \exp(-i\alpha_{\sigma\tau}) z_\tau a_\sigma^2\big) \tag{8.9}$$

where

$$z_\sigma(t) = \int_{-\infty}^{\infty} \bar{a}_\sigma(\omega^\sigma, t) g_\sigma(\omega^\sigma)\, \mathrm{d}\omega^\sigma. \tag{8.10}$$

With density

$$g_\sigma(\omega^\sigma) = \frac{\Delta_\sigma/\pi}{(\omega^\sigma - \Omega_\sigma) + \Delta_\sigma^2}, \tag{8.11}$$

where Ω_σ gives the center and Δ_σ the width (half width at half maximum) of the Lorentzian, the last integral can be solved. We have $z_\sigma(t) = \bar{a}_\sigma(\Omega_\sigma - i\Delta_\sigma, t)$ and evaluating (8.9) and (8.10) at the poles $\omega^\sigma = \Omega_\sigma - i\Delta_\sigma$, we obtain

$$\frac{\partial \bar{z}_\sigma}{\partial t} = -(\Delta_\sigma + i\Omega_\sigma)\bar{z}_\sigma + \frac{1}{2}\sum_{\tau=1}^{M} K_{\sigma\tau}\left(\exp(i\alpha_{\sigma\tau})\bar{z}_\tau - \exp(-i\alpha_{\sigma\tau}) z_\tau \bar{z}_\sigma^2\right) \tag{8.12}$$

$$= -(\Delta_\sigma + i\Omega_\sigma)\bar{z}_\sigma + \frac{1}{2}\sum_{\tau=1}^{M} \left(\bar{c}_{\sigma\tau}\bar{z}_\tau - c_{\sigma\tau} z_\tau \bar{z}_\sigma^2\right) \tag{8.13}$$

where $c_{\sigma\tau} = K_{\sigma\tau}\exp(i\alpha_{\sigma\tau})$ describe the interaction between the populations.

Equation (8.13) describes the mean field dynamics on the invariant manifold—the Ott–Antonsen (OA) manifold—given by the family of probability densities f whose Fourier coefficients $f_n(t)$ satisfy $f_n(t) = a(t)^n$. We will consider identical frequency distributions $\Omega_\sigma = \Omega$, $\Delta_\sigma = \Delta$ in the limit $\Delta \to 0$; while the manifold is globally attracting for distributions with nonzero Δ_σ [29, 30] the limit has been discussed, for example, in [17, 19, 22, 32]. Thus, $z_\sigma(t) = \bar{a}_\sigma(t)$, and the equations for the mean field dynamics are given by

$$\frac{\partial \bar{z}_\sigma}{\partial t} = -i\Omega\bar{z}_\sigma + \frac{1}{2}\sum_{\tau=1}^{M}\left(\bar{c}_{\sigma\tau}\bar{z}_\tau - c_{\sigma\tau}z_\tau\bar{z}_\sigma^2\right), \tag{8.14}$$

on $\mathbb{C}^M$. Writing $z_k = \rho_k \exp(i\phi_k)$ with $\rho_k \in \mathbb{R}_+, \phi_k \in \mathbf{T}$ for the complex order parameter yields polar coordinates for (8.14) on $(\mathbb{C} \setminus \{0\})^M$: angles and amplitudes $(\rho, \phi) = (\rho_1, \dots, \rho_M, \phi_1, \dots, \phi_M) \in \mathbb{R}_+^M \times \mathbf{T}^M$ evolve according to

$$\dot{\rho}_\sigma = \frac{1}{2}(1-\rho_\sigma^2)\sum_{\tau=1}^{M} K_{\sigma\tau}\rho_\tau \sin(\phi_\tau - \phi_\sigma + \alpha_{\sigma\tau}), \tag{8.15a}$$

$$\dot{\phi}_\sigma = \Omega - \frac{1+\rho_\sigma^2}{2\rho_\sigma}\sum_{\tau=1}^{M} K_{\sigma\tau}\rho_\tau \cos(\phi_\tau - \phi_\sigma + \alpha_{\sigma\tau}). \tag{8.15b}$$

Note that the "amplitude" $\rho_\sigma = |z_k|$ contains information about the synchronization of populations σ: if $\rho_\sigma = 1$, then the population is phase synchronized.

Equations (8.15) have a phase shift symmetry $\mathbf{T}$ which acts by shifting all phases by a constant angle. Thus, by going into a suitable co-rotating frame we may assume $\Omega = 0$. Reducing the phase shift symmetry—for example by introducing coordinates $\psi_k = \phi_k - \phi_M$—yields a dynamical system on $\mathbb{R}_+^M \times \mathbf{T}^{M-1}$.

Additional permutational symmetries of the coupling between the populations induce dynamically invariant subspaces [13]. Let $\mathbf{S}_M$ denote the group of permutations of M symbols. Suppose that $\Gamma \subset \mathbf{S}_M$ acts on the indices of the populations and the system (8.15) is Γ-equivariant—if Γ acts transitively, the populations are identical; cf. [6–8]. Then the fixed point subspaces of Γ are dynamically invariant. For $M = 2$ symmetrically coupled populations ($\Gamma = \mathbb{Z}/2\mathbb{Z}$) we have $K_{11} = K_{22}, K_{12} = K_{21}$ and $\alpha_{11} = \alpha_{22}, \alpha_{12} = \alpha_{21}$ [22]. Reduce the continuous $\mathbf{T}$ symmetry by writing a single equation for $\psi = \phi_1 - \phi_2$. The nontrivial element $\mathfrak{t} \in \mathbb{Z}/2\mathbb{Z}$ acts by $\mathfrak{t}(\rho_1, \rho_2, \psi) = (\rho_2, \rho_1, -\psi)$. Thus we have

$$\mathrm{Fix}(\mathfrak{t}) = \{(\rho, \rho, \psi) \mid \rho \in \mathbf{T}, \psi \in \{0, \pi\}\} \tag{8.16}$$

which corresponds to the invariant ray described in [21, 22]. Note that for $M > 2$ there will be more general invariant subspaces which depend on $\Gamma \subset \mathbf{S}_M$.

8.4 Lotka–Volterra Like Dynamics Close to Cluster States

Observe that (8.15) has dynamically invariant subspaces that are not induced by symmetry. Let $\mathcal{S}^\rho_\sigma = \{\rho \in \mathbb{R}^M_+ \mid \rho_\sigma = 1\}$, $\mathcal{D}^\rho_\sigma = \{\rho \in \mathbb{R}^M_+ \mid \rho_\tau = 1 \ \forall \tau \neq \sigma\}$ and $\mathcal{S}_\sigma = \mathcal{S}^\rho_\sigma \times \mathbf{T}^M$, $\mathcal{D}_\sigma = \mathcal{D}^\rho_\sigma \times \mathbf{T}^M$. Now for any set $I \subset [M]$ the sets $\mathcal{S}_I = \bigcap_{\tau \in I} \mathcal{S}_\tau$ and $\mathcal{D}_I = \bigcap_{\tau \in [M] \setminus I} \mathcal{S}_\tau$ are dynamically invariant. On $\mathcal{S} := \mathcal{S}_{[M]}$ all populations are phase synchronized, that is, $\rho_k = 1$ for all $k = 1, \ldots, M$, and the dynamics of the phases are given by

$$\dot{\phi}_\sigma = \Omega - \sum_{\tau=1}^{M} K_{\sigma\tau} \cos(\phi_\tau - \phi_\sigma + \alpha_{\sigma\tau}). \tag{8.17}$$

Definition 4.1 A *cluster state* $(\Phi_{\sigma\tau})$ is an equilibrium of (8.15) on $\mathcal{S}_{[M]}$ relative to the phase shift symmetry, that is, there are constants $\Phi_{\sigma\tau}$, Ω_{eq} such that

$$\Omega_{\text{eq}} = \sum_{\tau=1}^{M} K_{\sigma\tau} \cos(\Phi_{\sigma\tau} + \alpha_{\sigma\tau}). \tag{8.18}$$

for all $\sigma = 1, \ldots, M$. The cluster state is *hyperbolic* if it is a hyperbolic equilibrium in the reduced system on $\mathbb{R}^M_+ \times \mathbf{T}^{M-1}$.

For given $K_{\sigma\tau}, \alpha_{\sigma\tau}$ cluster states can be found by solving a linear equation. Existence and stability of cluster states in a system with full permutational symmetry has been considered in [27] in a more generalized setting; linearizing around these points in the full system (8.15) yields stability in the transversal directions. The number of cluster states depends on the number of populations. For $M = 2$ symmetrically coupled populations, there are either two isolated cluster states with phase differences $\Phi_{12} = 0$ or $\Phi_{12} = \pi$ [22] or $\mathcal{S}$ is a continuum of cluster states. By contrast, for $M = 3$ symmetrically coupled populations there are additional cluster states; cf. Fig. 2 in [20].

Considering the dynamics of (8.15a) while keeping $\phi_\tau - \phi_\sigma$ fixed suggests that the order parameter dynamics relate to a generalized Lotka–Volterra system. Set $Y_{\sigma\tau} = K_{\sigma\tau} \sin(\phi_\tau - \phi_\sigma + \alpha_{\sigma\tau})$ and write $R_\sigma = 1 - \rho_\sigma$ where R_σ describes the deviation from phase synchrony. We have

$$\dot{R}_\sigma = R_\sigma \left(\sum_{\tau=1}^{M} \left(-1 + R_\tau + \frac{1}{2} R_\sigma - \frac{1}{2} R_\tau R_\sigma \right) Y_{\sigma\tau} \right) \tag{8.19}$$

for $\sigma \in [M]$. With $\widehat{Y}'_\sigma = \sum_{\tau \neq \sigma} Y_{\sigma\tau}$ we obtain the system

$$\dot{R}_\sigma = R_\sigma \left(-\left(\widehat{Y}'_\sigma + Y_{\sigma\sigma}\right) + \frac{1}{2}\left(\widehat{Y}'_\sigma + 3Y_{\sigma\sigma}\right) R_\sigma + \sum_{\tau \neq \sigma} R_\tau Y_{\sigma\tau} \right) - \frac{1}{2} \sum_{\tau=1}^{M} R_\tau R_\sigma^2 Y_{\sigma\tau} \tag{8.20}$$

Note that—up to a higher order correction term which respects the invariant subspace structure—this is a Lotka–Volterra type equation (8.2) with $s_\sigma = -\left(\widehat{Y}'_\sigma + Y_{\sigma\sigma}\right)$, $p_\sigma = -\frac{1}{2}\left(\widehat{Y}'_\sigma + 3Y_{\sigma\sigma}\right)$, $r_{\sigma\tau} = -Y_{\sigma\tau}$.

In the full system (8.15) the phase dynamics close to a cluster state $(\Phi_{\sigma\tau})$ will be influenced by the dynamics of R_σ. Consider the dynamics on the invariant subspace $\mathcal{D}_\kappa$, $\kappa \in [M]$, where each population $\sigma \neq \tau$ is fully synchronized. For the cluster state $(\Phi_{\sigma\tau})$ define $X_{\sigma\tau} := K_{\sigma\tau}\cos(\Phi_{\sigma\tau}+\alpha_{\sigma\tau}) = \mathrm{Re}(c_{\sigma\tau})$ and $Y_{\sigma\tau} := K_{\sigma\tau}\sin(\Phi_{\sigma\tau}+\alpha_{\sigma\tau}) = \mathrm{Im}(c_{\sigma\tau})$. Let $\Delta\Phi_{\sigma\tau} = (\phi_\tau - \phi_\sigma) - \Phi_{\sigma\tau}$ denote the deviation from the cluster state $(\Phi_{\sigma\tau})$. The dynamics of (8.15) on $\mathcal{D}_\kappa$ are given by

$$\dot{\phi}_\kappa = \Omega - \sum_{\tau=1}^{M} K_{\kappa\tau}\cos(\phi_\tau - \phi_\kappa + \alpha_{\kappa\tau}) - R_\kappa X_{\kappa\kappa} - \frac{R_\kappa^2}{2}\sum_{\tau=1}^{M} X_{\kappa\tau} + O\left(\|(R_\kappa, \Delta\Phi_{\sigma\tau})\|^3\right), \tag{8.21a}$$

$$\dot{\phi}_\sigma = \Omega - \sum_{\tau=1}^{M} K_{\sigma\tau}\cos(\phi_\tau - \phi_\sigma + \alpha_{\kappa\tau}) + R_\kappa X_{\sigma\kappa} - R_\kappa \Delta\Phi_{\sigma\kappa} Y_{\sigma\kappa} + O\left(\|(R_\kappa, \Delta\Phi_{\sigma\tau})\|^3\right), \tag{8.21b}$$

$$\dot{R}_\kappa = R_\kappa\left(-\left(\widehat{Y}'_\kappa + Y_{\kappa\kappa} + \sum_{\tau=1}^{M}\Delta\Phi_{\kappa\tau}X_{\kappa\tau}\right) + \frac{1}{2}\left(\widehat{Y}'_\kappa + 3Y_{\kappa\kappa}\right)R_\kappa\right) + O\left(\|(R_\kappa, \Delta\Phi_{\sigma\tau})\|^3\right), \tag{8.21c}$$

$$\dot{R}_\sigma = 1, \tag{8.21d}$$

with the Landau O notation.

Note that up to third order, the influence of R_κ to the phase dynamics is solely by $X_{\kappa\tau}, X_{\tau\kappa}$, $\tau \in [M]$. A relation of the phase shift parameters $\alpha_{\sigma\tau}$ and the phase differences $\Phi_{\sigma\tau}$ of $\alpha_{\sigma\tau} + \Phi_{\sigma\tau} = \frac{\pi}{2} \mod \pi$ imply that close to the cluster state the interactions are effectively small. In fact, these nontrivial chimera equilibria have been found in $M = 2, 3$ populations close to the cluster state with phase differences $\Phi_{\sigma\tau} = 0$ for phase shift parameters $\alpha_{\sigma\tau} \approx \frac{\pi}{2}$ [2, 20, 22]. We will make this observation more rigorous elsewhere.

8.5 Discussion

While the invariant subspace structure close to a cluster state in the continuum limit mean field equations of populations of phase oscillators (8.15) resembles that of a generalized Lotka–Volterra system (8.2) there are some important differences. The appearance of higher order terms suggests a more generic bifurcation behavior.

In fact, chimera equilibria for (8.15) and $M = 2, 3$ populations arise as a pair in a saddle node bifurcation [2, 20] whereas (8.2) only has a single nontrivial equilibrium A_σ on the invariant subspace $\mathcal{A}_\sigma$. Moreover, there is typically bistability between the cluster state and chimera equilibria [21]. Note that the invariant subspace structure in (8.15) only exists if the intrinsic frequencies of all oscillators are identical. If the distribution of the intrinsic frequencies has nonzero width [18], the cylinder $[0, 1]^M \times \mathbf{T}^M$ is still dynamically invariant.

At the same time, the similarity between generalized Lotka–Volterra equations and the dynamics of the absolute value of the local order parameter (8.15) suggests that there may be parameter values that give rise to heteroclinic connections which join a sequence of saddle equilibria. Such heteroclinic connections would yield transient synchronization and desynchronization of individual oscillator populations—an effect observed in interacting populations of oscillators whose mean frequencies are well separated [14]—in networks of identical oscillators which support chimera states. While an attempt to directly use the same argument as in [4] for (8.15) is likely to fail if additional equilibria are present in the invariant subspaces, a suitable approximation result (given elsewhere) may be a way to prove the existence of heteroclinic connections.

Thus, understanding the global dynamics of $M \geq 3$ provides exciting new directions. In future research we anticipate to find parameter values such that there are heteroclinic connections between chimera equilibria. The resulting dynamics will exhibit transitions between localized synchronization that are organized by the network structure rather than by control [9]. Such dynamical transitions between chimeras would also bridge synchronization the microscopic level of individual oscillators and macroscopic transitions between different neural activity patterns induced by winnerless competition.

Acknowledgements The author is indebted to M.I. Rabinovich for his guidance and support over the years. Moreover, he would like to thank P. Ashwin and E.A. Martens for many helpful discussions. This work has received funding from the People Programme (Marie Curie Actions) of the European Union's Seventh Framework Programme (FP7/2007–2013) under REA grant agreement no. 626111.

References

1. Abeles, M., Bergman, H., Gat, I., Meilijson, I., Seidemann, E., Tishby, N., Vaadia, E.: Cortical activity flips among quasi-stationary states. Proc. Natl. Acad. Sci. USA **92**(19), 8616–8620 (1995)
2. Abrams, D.M., Mirollo, R.E., Strogatz, S.H., Wiley, D.A.: Solvable model for chimera states of coupled oscillators. Phys. Rev. Lett. **101**(8), 084103 (2008)
3. Acebrón, J., Bonilla, L., Pérez Vicente C., Ritort, F., Spigler, R.: The Kuramoto model: a simple paradigm for synchronization phenomena. Rev. Mod. Phys. **77**(1), 137–185 (2005)
4. Afraimovich, V.S., Rabinovich, M.I., Varona, P.: Heteroclinic contours in neural ensembles and the winnerless competition principle. Int. J. Bifurcation Chaos **14**(4), 1195–1208 (2004)

5. Afraimovich, V.S., Zhigulin, V.P., Rabinovich, M.I.: On the origin of reproducible sequential activity in neural circuits. Chaos **14**(4), 1123–1129 (2004)
6. Ashwin, P., Burylko, O.: Weak chimeras in minimal networks of coupled phase oscillators. Chaos **25**, 013106 (2015)
7. Bick, C.: Isotropy of angular frequencies and weak chimeras with broken symmetry. J. Nonlinear Sci. **27**(2), 605–626 (2017). http://doi.org/10.1007/s00332-016-9345-2
8. Bick, C., Ashwin, P.: Chaotic weak chimeras and their persistence in coupled populations of phase oscillators. Nonlinearity **29**(5), 1468–1486 (2016)
9. Bick, C., Martens, E.A.: Controlling chimeras. New J. Phys. **17**(3), 033030 (2015)
10. Bick, C., Rabinovich, M.I.: Dynamical origin of the effective storage capacity in the brain's working memory. Phys. Rev. Lett. **103**(21), 218101 (2009)
11. Bick, C., Rabinovich, M.I.: On the occurrence of stable heteroclinic channels in Lotka-Volterra models. Dyn. Syst. **25**(1), 97–110 (2010)
12. Friston, K.J.: Transients, metastability, and neuronal dynamics. NeuroImage **5**(2), 164–171 (1997)
13. Golubitsky, M., Stewart, I.: The Symmetry Perspective. Progress in Mathematics, vol. 200. Birkhäuser, Basel (2002)
14. Komarov, M., Pikovsky, A.: Effects of nonresonant interaction in ensembles of phase oscillators. Phys. Rev. E **84**(1), 16210 (2011)
15. Kuramoto, Y.: Chemical Oscillations, Waves, and Turbulence. Springer Series in Synergetics, vol. 19. Springer, Berlin (1984)
16. Kuramoto, Y., Battogtokh, D.: Coexistence of coherence and incoherence in nonlocally coupled phase oscillators. Nonlinear Phenom. Complex Syst. **5**(4), 380–385 (2002)
17. Laing, C.R.: Chimera states in heterogeneous networks. Chaos **19**(1), 013113 (2009)
18. Laing, C.R.: The dynamics of chimera states in heterogeneous Kuramoto networks. Physica D **238**(16), 1569–1588 (2009)
19. Laing, C.R.: Disorder-induced dynamics in a pair of coupled heterogeneous phase oscillator networks. Chaos **22**(4), 043104 (2012)
20. Martens, E.A.: Bistable chimera attractors on a triangular network of oscillator populations. Phys. Rev. E **82**(1), 016216 (2010)
21. Martens, E.A., Panaggio, M.J., Abrams, D.M.: Basins of attraction for chimera states. New J. Phys. **18**(2), 022002 (2016)
22. Martens, E.A., Bick, C., Panaggio, M.J.: Chimera states in two populations with heterogeneous phase-lag. Chaos **26**(9), 094819 (2016)
23. May, R.M., Leonard, W.J.: Nonlinear aspects of competition between three species. SIAM J. Appl. Math. **29**(2), 243–253 (1975)
24. Miller, G.: The magical number seven, plus or minus two: some limits on our capacity for processing information. Psychol. Rev. **63**, 81–97 (1956)
25. Montbrió, E., Kurths, J., Blasius, B.: Synchronization of two interacting populations of oscillators. Phys. Rev. E **70**(5), 056125 (2004)
26. Nowotny, T., Rabinovich, M.I.: Dynamical origin of independent spiking and bursting activity in neural microcircuits. Phys. Rev. Lett. **98**(12), 1–4 (2007)
27. Orosz, G., Moehlis, J., Ashwin, P.: Designing the dynamics of globally coupled oscillators. Prog. Theor. Phys. **122**(3), 611–630 (2009)
28. Ott, E., Antonsen, T.M.: Low dimensional behavior of large systems of globally coupled oscillators. Chaos **18**(3), 037113 (2008)
29. Ott, E., Antonsen, T.M.: Long time evolution of phase oscillator systems. Chaos **19**(2), 023117 (2009)
30. Ott, E., Hunt, B.R., Antonsen, T.M.: Comment on "Long time evolution of phase oscillator systems" [Chaos 19, 023117 (2009)]. Chaos **21**(2), 025112 (2011)
31. Panaggio, M.J., Abrams, D.M.: Chimera states: coexistence of coherence and incoherence in networks of coupled oscillators. Nonlinearity **28**(3), R67–R87 (2015)
32. Pikovsky, A., Rosenblum, M.: Partially integrable dynamics of hierarchical populations of coupled oscillators. Phys. Rev. Lett. **101**(26), 1–4 (2008)

33. Rabinovich, M.I., Varona, P., Selverston, A., Abarbanel, H.D.I.: Dynamical principles in neuroscience. Rev. Mod. Phys. **78**(4), 1213–1265 (2006)
34. Rabinovich, M.I., Huerta, R., Laurent, G.: Transient dynamics for neural processing. Science **321**(5885), 48–50 (2008)
35. Rabinovich, M.I., Afraimovich, V.S., Bick, C., Varona, P.: Information flow dynamics in the brain. Phys. Life Rev. **9**(1), 51–73 (2012)
36. Rabinovich, M.I., Simmons, A.N., Varona, P.: Dynamical bridge between brain and mind. Trends Cogn. Sci. **19**(8), 453–461 (2015)
37. Sakaguchi, H., Kuramoto, Y.: A soluble active rotater model showing phase transitions via mutual entertainment. Prog. Theor. Phys. **76**(3), 576–581 (1986)
38. Strogatz, S.H.: From Kuramoto to Crawford: exploring the onset of synchronization in populations of coupled oscillators. Physica D **143**(1–4), 1–20 (2000)
39. Tognoli, E., Scott Kelso, J.A.: The metastable brain. Neuron **81**(1), 35–48 (2014)

Chapter 9
Intrinsic Stability, Time Delays and Transformations of Dynamical Networks

Leonid Bunimovich and Benjamin Webb

9.1 Introduction

The great majority of real-world networks, i.e. multicomponent systems, are time delayed. Time delays are caused by the finite transmission speed of some quantity such a signal, energy, etc. moving from one network element to another as well as the time it takes to process this quantity once it arrives. These time delays make analyzing the dynamics of a network much more complicated, which is the reason why the theory of time-delayed systems is much less developed than the theory of undelayed systems. An important fact that demonstrates the difference between the dynamics of delayed and undelayed systems is that it is possible to destabilize an undelayed system by introducing delays into its dynamics. This suggests that if we are to understand the dynamics of a time-delayed system this cannot be done by analyzing the system without its time-delays. However, we show that this is not always the case.

Here we introduce the notion of intrinsic stability (see Sect. 9.2), which is a type of stability that can be exhibited by dynamical networks and other multidimensional systems. This notion of intrinsic stability is stronger than the standard notion of global stability of dynamical systems and networks. An intrinsically stable system has the remarkable property it remains stable even if time-delays are added or removed from the system provided that these delays are *nondistributed*, i.e. the

L. Bunimovich
School of Mathematics, Georgia Institute of Technology, 686 Cherry Street, Atlanta, GA 30332, USA
e-mail: leonid.bunimovich@math.gatech.edu

B. Webb (✉)
Department of Mathematics, Brigham Young University, Provo, UT 84602, USA
e-mail: bwebb@mathematics.byu.edu

I.S. Aranson et al. (eds.), *Advances in Dynamics, Patterns, Cognition*, Nonlinear Systems and Complexity 20, DOI 10.1007/978-3-319-53673-6_9

variable x_i depends on at most one of the previous states of the variable x_j (see Theorems 2.3 and 2.4). In general, x_i can depend on any number of the previous states of x_j.

The condition of intrinsic stability is fairly easy to verify. Therefore, by making use of this notion it is possible to determine whether a time-delayed network is globally stable by performing a much simpler calculation on the corresponding undelayed network. In fact, this procedure can be simplified even further by removing delays that are not formally time delays but are what we refer to as *implicit time-delays* (see Sect. 9.3). That is, even in a system in which there are no explicit time-delays it still takes time to propagate through the network as it must pass from one network element to another. By removing these implicit delays we both reduce the network in size making it easier to analyze and preserve certain spectral properties of the network. Because a reduced network is a collapsed version of the original network it can be used to determine whether the original network is stable when other methods fail. That is, a network is stable if any one of its reductions is intrinsically stable (see Theorem 3.1).

Not only can a network be reduced in size but it can also be expanded while certain spectral and structural properties of the network are preserved (see Sect. 9.4). This is important since networks are not only dynamic in terms of the behavior of their elements but also in terms of the structure of their interactions. In this regard expansions can be used as a tool for modeling network growth. Moreover, similar to time delays network growth can have a destabilizing effect on the network. We show that if a network is intrinsically stable then it will remain intrinsically stable as it is expanded (see Theorem 4.2). Hence, an intrinsically stable network will remain stable as its structure evolves via any sequence of expansions. These results are illustrated using examples of Cohen–Grossberg neural networks. Section 9.5 contains some concluding remarks.

9.2 Stability of Dynamical Networks

A network is composed of a set of *elements*, which are the individual units that make up the network, and a collection of interactions between these elements. An *interaction* between two network elements can be thought of as an element's ability to directly influence the state, function, or behavior of the other network element. More generally, there is a *directed interaction* between the *i*th and *j*th elements of a network if the *i*th network element can influence the state of the *j*th network element (where there may be no influence of the *j*th network element on the *i*th). The dynamics of a network can be defined as follows.

Definition 2.1 (Dynamical Network) Let $F : X \to X$ be a continuous map on the product space $X = \oplus_{i=1}^{n} X_i$ where each (X_i, d) is a complete metric space. For each $i = 1, \ldots, n$ the *i*th component function is defined as

$$F_i : \bigoplus_{j \in I_i} X_j \to X_i, \quad \text{for some } I_i \subseteq \{1, \ldots, n\}.$$

The dynamical system (F, X) generated by iterating the function F is called a *dynamical network.*

The component function $F_i : \bigoplus_{j \in I_i} X_j \to X_i$ describes the dynamics of the ith network element, where the set $I_i \subseteq \{1, \ldots, n\}$ indexes those elements that directly interact with the ith network element. Thus, we refer to the function $F : X \to X$ as the network's set of interactions, or simply its *interaction.*

To give a concrete example of a network and also to illustrate the results found in this and other sections we will use Cohen–Grossberg neural (CGN) networks.

Example 2.1 (Cohen–Grossberg Neural Networks) For $W \in \mathbb{R}^{n \times n}$, $\varphi : \mathbb{R} \to \mathbb{R}$, and $c_i, \epsilon \in \mathbb{R}$ let $(C, \mathbb{R}^n)$ be the dynamical network with components

$$C_i(\mathbf{x}) = (1 - \epsilon)x_i + \sum_{j=1}^{n} W_{ij}\varphi(x_j) + c_i, \quad 1 \leq j \leq n, \tag{9.1}$$

which is a special case of a Cohen–Grossberg neural network in discrete-time [9]. The function φ is assumed to be a bounded, have a bounded derivative, and be monotonically increasing.

In a CGN network the variable x_i represents the *activation* of the ith neuron. The function φ is a bounded monotonically increasing function, which describes the ith neuron's response to inputs. The matrix W gives the interaction strengths between each pair of neurons and describes how the neurons are connected within the network. The constants c_i indicate constant inputs from outside the network.

One standard question regarding the dynamics of a CGN network is whether the network is stable. In a *stable* network (F, X) the state of the network tends towards an equilibrium irrespective of its present state. That is, there is a *globally attracting fixed point* $y \in X$ such that for any $x \in X$, $F^k(x) \to y$ as $k \to \infty$.

Global stability is observed in a number of important systems including neural networks [6–9, 11], in epidemic models [12], and is also important in the study of congestion in computer networks [1]. In such systems the globally attracting equilibrium is typically a state in which the network can carry out a specific task. Whether or not this equilibrium stays stable depends on a number of factors including external influences but also internal processes such as the network's own growth both of which can destabilize a network.

To determine whether a given network (F, X) is stable we associate it with the following matrix.

Definition 2.2 (Stability Matrix) For $F : X \to X$ suppose there are constants $(M_F)_{ij} \geq 0$ such that

$$d_{\max}(F_i(\mathbf{x}), F_i(\mathbf{y})) \leq \sum_{j \in I_i} (M_F)_{ij} d(x_j, y_j) \quad \text{for all } \mathbf{x}, \mathbf{y} \in X,$$

where $d_{\max}(\mathbf{x}, \mathbf{y}) = \max_i d(x_i, y_i)$. Then $M_F \in \mathbb{R}^{n \times n}$ is called a *stability matrix* of (F, X).

For a matrix $A \in \mathbb{R}^{n\times n}$ we let $\sigma(A)$ denote the *eigenvalues* of A including multiplicities and let $\rho(A) = \max_i\{|\lambda| : \lambda \in \sigma(A)\}$ denote the *spectral radius* of A. This allows us to state the following result regarding the stability of a general dynamical network (F, X) (see [2, 4]).

Theorem 2.1 (Network Stability) *If* $\rho(M_F) < 1$, *then the dynamical network* (F, X) *is stable.*

It is worth noting that if the function $F : X \to X$ is differentiable and each $X_i \subseteq \mathbb{R}$ then the matrix $M_F \in \mathbb{R}^{n\times n}$ is given by

$$(M_F)_{ij} = \sup_{\mathbf{x}\in X}\left|\frac{\partial F_i}{\partial x_j}(\mathbf{x})\right|. \tag{9.2}$$

From a computational point of view, the stability matrix M_F of (F, X) can be quite straightforward to find by use of (9.2). For example, one can quickly compute that the stability matrix M_C of the CGN network given by (9.1) is $M_C = |1-\epsilon|I + L|W|$ where $|W|$ is the matrix with entries $|W|_{ij} = |W_{ij}|$, which implies the following.

Theorem 2.2 (Stability of Cohen–Grossberg Neural Networks) *Let* $(C, \mathbb{R}^n)$ *be the Cohen–Grossberg network given by* (9.1) *where* φ *has Lipschitz constant L. If* $|1-\epsilon| + L\rho(|W|) < 1$, *then* $(C, \mathbb{R}^n)$ *is stable.*

As mentioned in the introduction, the dynamics of most real networks are *time delayed*. That is, an interaction between two network elements will typically not happen instantaneously but will be delayed due to either the physical separation of these elements, their finite processing speeds, or be delayed due to other factors. Thus, if we represent the state of the network at time $k \geq 0$ by $\mathbf{x}^k \in X$, not only does x^{k+1} depend on x^k but also on some collection of the previous states of the network $\mathbf{x}^k, \mathbf{x}^{k-1}, \ldots, \mathbf{x}^{k-\eta-1} \in X$.

Definition 2.3 (Time-Delayed Dynamical Network) For a fixed $\eta \geq 1$ let $X^\eta = \bigoplus_{i=0}^{\eta-1} X$. Suppose $\bar{H} : X^\eta \to X$ is a continuous function given by $\mathbf{x}^{k+1} = \bar{H}(\mathbf{x}^k, \mathbf{x}^{k-1}, \ldots, \mathbf{x}^{k-\eta+1})$. Let $H : X^\eta \to X^\eta$ be the function given by

$$H(\mathbf{x}^k, \ldots, \mathbf{x}^{k-\eta+1}) = \big(\mathbf{x}^{k+1}, \mathbf{x}^k, \ldots, \mathbf{x}^{k-\eta+2}\big).$$

where $x^{k+1} = \bar{H}(\mathbf{x}^k, \ldots, \mathbf{x}^{k-\eta+1})$. The dynamical system (H, X^η) generated by iterating the function H is called a *time-delayed dynamical network.*

If (H, X^η) is a time-delayed dynamical network, then $\bar{H} : X^\eta \to X$ represents the *time-delayed interactions* of the network. Specifically, the component

$$\bar{H}_i : \bigoplus_{(j,\tau)\in I_i} X_i^\tau \to X_i \quad \text{for some } I_i \subseteq \{1, \ldots n\} \times \{0, \ldots, \eta-1\}$$

describes the dynamics of the ith network element, where $(j, \tau) \in I_i \subseteq \{1, \ldots, n\} \times \{0, \ldots, \eta - 1\}$ if the jth network element *interacts* with the ith network element with a *delay time* of τ.

An example of a time-delayed dynamical network are the time-delayed CGN networks.

Example 2.2 (Time-Delayed Cohen–Grossberg Neural Networks) For $\eta \geq 1$ let $(C, \mathbb{R}^n)$ be the time-delayed CGN network with components

$$\bar{C}_i(\mathbf{x}) = (1 - \epsilon)x_i + \sum_{j=1}^{n}\sum_{\tau=0}^{\eta-1}(W_\tau)_{ij}\varphi(x_j^\tau) + c_i, \quad 1 \leq j \leq n, \tag{9.3}$$

where each $W_\tau \in \mathbb{R}^{n\times n}$, $\epsilon, c_i \in \mathbb{R}$, and $x_j^\tau \in X_j^\tau$.

Clearly, time-delayed dynamical networks can be much more complicated than their undelayed versions. Despite this, a *time-delayed* dynamical network (H, X^η) is a dynamical network in the sense of Definition 2.1 since H is a continuous function on a product space X^η. Hence, the following corollary of Theorem 2.1 holds (see [3, 4]).

Corollary 2.1 (Stability of Time-Delayed Networks) *The time-delayed dynamical network* (H, X^η) *is stable if* $\rho(M_H) < 1$.

The main question we wish to address is not whether or not we can determine whether a network is stable but rather, under what conditions do time delays destabilize the dynamics of a network. To understand when this happens we need to compare a network's dynamics when it has delays and when it does not.

Definition 2.4 (Undelayed Network) For the time-delayed network (H, X^η) let $\mathcal{U}(H) : X \to X$ be the map

$$\mathcal{U}(H)(\mathbf{x}) = H(\mathbf{x}, \ldots, \mathbf{x}),$$

which generates the *undelayed dynamical network* $(\mathcal{U}(H), X)$.

A natural question is, if (H, X^η) is stable is $(\mathcal{U}_H, X)$ stable? Conversely, if $(\mathcal{U}_H, X)$ is stable is (H, X^η) stable? This is considered in the following example.

Example 2.3 (Lose, Gain, and Preservation of Stability) Consider the linear time-delayed dynamical network (H, X^η) and the undelayed version of this network $(\mathcal{U}(H), X)$ given by

$$H(\mathbf{x}^k, \mathbf{x}^{k-1}) = \begin{bmatrix} \alpha x_1^{k-1} + \beta x_2^k \\ \alpha x_1^k \end{bmatrix} \text{ and } \mathcal{U}(H)(\mathbf{x}) = \begin{bmatrix} \alpha x_1 + \beta x_2 \\ \alpha x_1 \end{bmatrix},$$

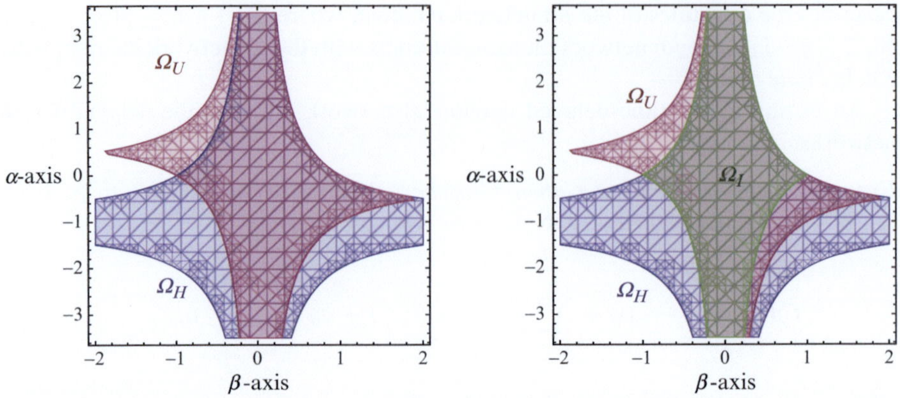

Fig. 9.1 The set $\Omega_H = \{(\alpha, \gamma) \in \mathbb{R}^2 : H \text{ is stable}\}$ and $\Omega_{\mathcal{U}} = \{(\alpha, \gamma) \in \mathbb{R}^2 : \mathcal{U}(H) \text{ is stable}\}$ are shown (*left*) where (H, X^η) is the time-delayed network in Example 2.3. The sets $\Omega_I = \{(\alpha, \gamma) \in \mathbb{R}^2 : H \text{ is intrinsically stable}\}$, Ω_H, and $\Omega_{\mathcal{U}}$ are shown (*right*)

respectively, where $\alpha, \beta \in \mathbb{R}$, $X = \mathbb{R}^2$, and $\eta = 2$. In Fig. 9.1 (left) the regions $\Omega_H = \{(\alpha, \beta) \in \mathbb{R}^2 : H \text{ is stable}\}$ and $\Omega_{\mathcal{U}} = \{(\alpha, \beta) \in \mathbb{R}^2 : \mathcal{U}(H) \text{ is stable}\}$ are shown in red and, blue respectively.

Note that neither Ω_H is contained in $\Omega_{\mathcal{U}}$ nor $\Omega_{\mathcal{U}}$ contained in Ω_H. Therefore, there are parameter values $(\alpha, \beta) \in \mathbb{R}^2$ such that when we remove the delays from (H, X^η) we destabilize (stabilize) the network. Similarly, there are parameter values $(\alpha, \beta) \in \mathbb{R}^2$ such that when we add the delays back into $(\mathcal{U}(H), X)$ we destabilize (stabilize) the network. This is a simple example of a well-known fact that changing a network's structure of delays can qualitatively change the network's stability.

Another important observation is that the parameter set $\Omega_H \supset \{(a, b) \in \mathbb{R}^2 : \rho(M_F) < 1\}$ and the parameter set $\Omega_{\mathcal{U}} \supset \{(a, b) \in \mathbb{R}^2 : \rho(M_{\mathcal{U}(H)}) < 1\}$. That is, the network (H, X^η) can be stable even if $\rho(M_H) > 1$ and $(\mathcal{U}(H), X)$ can be stable even if $\rho(M_{\mathcal{U}(H)}) > 1$.

What is perhaps even more important an unexpected is that the parameter set

$$\{(a, b) \in \mathbb{R}^2 : \rho(M_H) < 1\} = \{(a, b) \in \mathbb{R}^2 : \rho(M_{\mathcal{U}(H)}) < 1\}. \tag{9.4}$$

That is, although (H, X^η) and $(\mathcal{U}(H), X)$ are stable for different parameters, the set of parameters for which $\rho(M_H) < 1$ and $\rho(M_{\mathcal{U}(H)}) < 1$ are the same. We let this parameter set given by (9.4) be the set $\Omega_I \subset \Omega_H, \Omega_{\mathcal{U}}$ shown in Fig. 9.1 (right). Our goal is to describe the importance of this parameter region for the time-delayed network (H, X^η).

Based on Theorem 2.1, if the spectral radius $\rho(M_F) < 1$ then the network (F, X) is stable. However, the converse does not always hold. Hence, the property $\rho(F_M) < 1$ is a stronger form of stability than the standard notion of stability. To distinguish between these two forms of stability we give the following definition.

Definition 2.5 (Intrinsic Stability) The dynamical network (F, X) is called *intrinsically stable* if $\rho(M_F) < 1$.

This raises the general question as to what it means for a network to be intrinsically stable and specifically to what extent time-delays effect a network's stability if the network is intrinsically stable. To answer this question we need to consider the types of delays that can occur in a network.

Definition 2.6 (Types of Time-Delays) Suppose the time-delayed dynamical network (H, X^η) has the component functions

$$\bar{H}_i : \bigoplus_{(j,\tau)\in I_i} X_i^\tau \to X_i \text{ for some } I_i \subseteq \{1, \ldots, n\} \times \{0, \ldots, \eta - 1\}.$$

Then (H, X^η) has *single-type time-delays* if for each $i = 1, \ldots, n$ and $j = 1, \ldots, n$ there is at most one τ such that $(j, \tau) \in I_i$. Otherwise, (H, X^η) has *distributed time-delays*.

The difference between these two types of delays is illustrated in the following example.

Example 2.4 (Single and Distributed Time-Delays) Consider the two time-delayed CGN networks $(A, \mathbb{R}^8)$ and $(B, (\mathbb{R}^4)$ given by

$$A(\mathbf{x}^{k-1}, \mathbf{x}^{k-3}) = \begin{bmatrix} (1-\epsilon)x_1^{k-1} + a\tanh(x_2^{k-3}) + c_1 \\ (1-\epsilon)x_2^{k-1} + a\tanh(x_1^{k-3}) + c_2 \end{bmatrix},$$

$$B(\mathbf{x}^{k}, \mathbf{x}^{k-1}) = \begin{bmatrix} (1-\epsilon)x_1^{k} + a\tanh(x_2^{k}) - a\tanh(x_2^{k-1}) \\ (1-\epsilon)x_2^{k} + a\tanh(x_1^{k}) - a\tanh(x_1^{k-1}) \end{bmatrix},$$

respectively. The network $(A, \mathbb{R}^8)$ has single-type time delays whereas the network $(B, \mathbb{R}^4)$ has distributed time-delays. The reason $(B, \mathbb{R}^8)$ has distributed time-delays is that the component $B_1 = B_1(x_1^k, x_2^k, x_2^{k-1})$ depends on the variable x_2 over two different time-scales. Similarly, $B_2 = B_2(x_2^k, x_1^k, x_1^{k-1})$ depends on x_1 over two different time scales. Hence, the dynamics of both B_1 and B_2 incorporate distributed time-delays.

The undelayed versions of $(A, \mathbb{R}^8)$ and $(B, \mathbb{R}^4)$ are given by

$$\mathcal{U}(A)(\mathbf{x}) = \begin{bmatrix} (1-\epsilon)x_1 + a\tanh(x_2) + c_1 \\ (1-\epsilon)x_2 + a\tanh(x_1) + c_2 \end{bmatrix} \text{ and } \mathcal{U}_B(\mathbf{x}) = \begin{bmatrix} (1-\epsilon)x_1 \\ (1-\epsilon)x_2 \end{bmatrix}, \text{ respectively.}$$

Using Eq. (9.2) one can quickly calculate that

$$\rho(M_A) = \sqrt{\frac{|1-\epsilon| + \sqrt{|1-\epsilon|^2 + 4|a|}}{2}} \text{ and } \rho(M_{\mathcal{U}(A)}) = |1-\epsilon| + |a|.$$

Although the spectral radius of the delayed dynamical network has a much more complicated expression than its undelayed version, one can show that $\rho(M_A) < 1$ if and only if $\rho(M_{\mathcal{U}(A)}) < 1$. That is, $(A, \mathbb{R}^8)$ is intrinsically stable if and only if the undelayed network $(\mathcal{U}(A), \mathbb{R}^2)$ is intrinsically stable. In contrast,

$$\rho(M_B) = \frac{|1-\epsilon| + |a| + \sqrt{(|1-\epsilon| + |a|)^2 + 4|a|}}{2} \quad \text{and} \quad \rho(M_{\mathcal{U}(B)}) = |1-\epsilon|.$$

Here, $\rho(M_B) < 1$ implies that $\rho(M_{\mathcal{U}(B)}) < 1$ but the reverse implication does not hold. That is, the intrinsic stability of the delayed network $(B, \mathbb{R}^4)$ is not equivalent to the intrinsic stability of the undelayed version of the network. For instance, if $a = 1$ and $\epsilon = 1/2$, then $(B, \mathbb{R}^4)$ is not stable while $(\mathcal{U}(B), \mathbb{R}^2)$ is intrinsically stable.

The question raised by this example is, what is the difference between the networks $(A, \mathbb{R}^8)$ and $(B, \mathbb{R}^4)$ since in the first removing delays does not change whether the network is intrinsically stable but does in the second. This is answered by the following theorem (see [3, 4]).

Theorem 2.3 (Preservation of Intrinsic Stability) *If the time-delayed network (H, X^η) has single-type time delays, then (H, X^η) is intrinsically stable if and only if $(\mathcal{U}(H), X)$ is intrinsically stable.*

Adding or removing *single-type* time-delays does not destabilize a network, if the network is intrinsically stable. This is why $(A, \mathbb{R}^8)$ and $(\mathcal{U}(A), \mathbb{R}^2)$ are intrinsically stable for the same parameter values. However, since $(B, \mathbb{R}^4)$ has distributed time-delays there is no guarantee that this network will be intrinsically stable if $(\mathcal{U}(B), \mathbb{R}^2)$ is intrinsically stable. As it turns out though, we can say what happens if $(B, \mathbb{R}^4)$ is intrinsically stable (see [3, 4]).

Theorem 2.4 (Removal of General Time-Delays) *If the time-delayed network (H, X^η) is intrinsically stable, then so is its undelayed version $(\mathcal{U}_H, X)$.*

Combining the results of Theorems 2.3 and 2.4 the single-type and distributed time-delays of a network (H, X^η) can be shortened or lengthened by any finite amount and the resulting network will be stable if the original network is intrinsically stable. That is, suppose the delayed network (H, X^η) has components $x_i^{k+1} = \bar{H}_i(x_{j_1}^{k-k_1}, \ldots, x_{j_m}^{k-k_m})$. If the network $(\mathcal{H}, X^\mu)$ has components

$$x_i^{k+1} = \bar{\mathcal{H}}(\mathbf{x}) = \bar{H}_i(x_{j_1}^{k-\kappa_1}, \ldots, x_{j_m}^{k-\kappa_m}) \text{ where each } \kappa_j \text{ is a nonnegative integer,}$$

i.e. $(\mathcal{H}, X^\mu)$ is the network (H, X^η) in which the length of the time delays has been modified, then $(\mathcal{H}, X^\mu)$ will be intrinsically stable if the original network (H, X^η) is intrinsically stable. In this sense intrinsically stable networks are much more resilient to changes in a network's environment that effect time-delays than networks that are simply stable.

In addition, to determine whether a time-delayed network is stable one can check to see whether the much simple undelayed version of this network is intrinsically stable. If it is and the original delayed network has single-type time-delays, then this delayed network must also be stable. This observation is not only important in terms of simplifying the process of computing network stability but will be used in the following section to improve our estimate of a network's stability.

9.3 Network Reductions and Improved Stability Estimates

A network's stability matrix is not only useful for determining whether a network has a globally attracting equilibrium state but can also be used to characterize a network's structure of interactions.

The graph Γ_F of a network (F, X) gives a much more visual representation of the network's structure (topology). In fact, the majority of the analysis done on networks has been focused on determining properties of their underlying graph structure [10]. Here, we investigate how time delays effect a network's structure and in turn its stability.

The dynamical network (F, X) does not formally have time-delays but elements that do not directly influence one another can have an influence through other network elements. For instance, if the ith network element effects the jth which influences the kth, then there is an implicit delayed interaction from the ith to the kth element of the network. Here we investigate how it is possible to improve the stability estimates of the previous section by accounting for these implicit delays.

To do this we need to look at the graph-theoretic representation of a network's set of interactions.

Definition 3.1 (Graph of Interactions) The *graph of interactions* of the dynamical network (F, X) is the graph $\Gamma_F = (V, E, \omega)$ where

(i) the *vertex set* $V = \{v_1, \ldots, v_n\}$;
(ii) the edge e_{ij} from v_i to v_j is in the *edge set* E if $i \in I_j$; and
(iii) the edge e_{ij} is given the *weight* $\omega(e_{ij}) = (M_F)_{ij}$.

The vertex set V represents the elements of the network while the edges E represent the various interactions between these elements. The weights given to the edges by the function $\omega : E \to \mathbb{R}$ represent the strengths of these interactions.

Example 3.1 (Implicit Time-Delays) Let $(F, \mathbb{R}^4)$ be the CGN network given by

$$C(\mathbf{x}) = \begin{bmatrix} \tanh(x_2) + \tanh(x_4) + c \\ \tanh(x_1) + c \\ \tanh(x_2) + \tanh(x_4) + c \\ \tanh(x_3) + c \end{bmatrix}, \quad \text{where } c \in \mathbb{R},$$

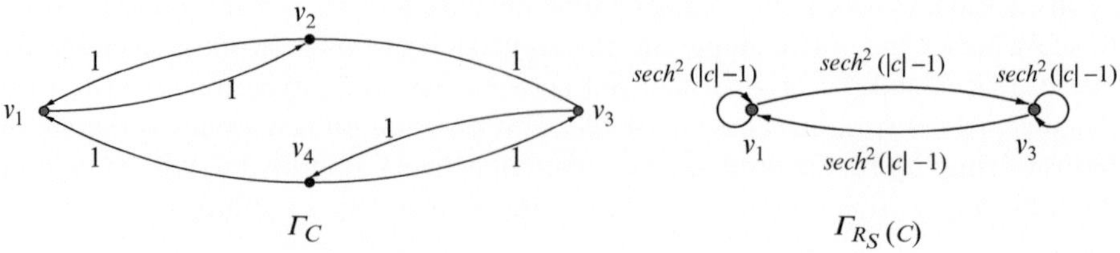

Fig. 9.2 The graphs of interactions Γ_C and $\Gamma_{\mathcal{R}_S(C)}$ of the CGN network $(C, \mathbb{R}^2)$ and its reduction $(\mathcal{R}_S(C), \mathbb{R}^2)$ over the set $S = \{v_1, v_3\}$ are shown *left and right*, respectively

which has the graph of interactions Γ_C shown in Fig. 9.2. Note that the dependencies of the network's components are given by $x_1^{k+1} = C_1(x_2^k, x_4^k)$, $x_2^{k+1} = C_2(x_1^k)$, $x_3^{k+1} = C_3(x_2^k, x_4^k)$, and $x_4^{k+1} = C_4(x_3)$.

By replacing the variables of the component functions $C_1(x_2^k, x_4^k)$ and $C_3(x_2^k, x_4^k)$ by their corresponding functions, i.e. by $x_2^k = C_2(x_1^{k-1})$ and $x_2^k = C_2(x_1^{k-1})$, the result is the time-delayed dynamical network given by

$$\mathbf{x}^{k+1} = \begin{bmatrix} x_1^{k+1} \\ x_3^{k+1} \end{bmatrix} = \begin{bmatrix} C_1(C_2(x_1^{k-1}), C_4(x_3^{k-1})) \\ C_3(C_2(x_1^{k-1}), C_4(x_3^{k-1})) \end{bmatrix} = \begin{bmatrix} \sum_{j=1,3} \tanh[\tanh(x_j) + c] + c \\ \sum_{j=1,3} \tanh[\tanh(x_j) + c] + c \end{bmatrix}.$$

By removing the delays from this network we have the *reduced version* $(\mathcal{R}_S(C), \mathbb{R}^2)$ of the dynamical network $(C, \mathbb{R}^4)$ over the index set $S = \{1, 3\}$ where

$$\mathcal{R}_S(C)(\mathbf{x}) = \begin{bmatrix} \tanh[\tanh(x_1) + c] + \tanh[\tanh(x_3) + c] + c \\ \tanh[\tanh(x_1) + c] + \tanh[\tanh(x_3) + c] + c \end{bmatrix}.$$

To create a reduced version of a dynamical network we need a set of network elements S over which the network is reduced (see [4]).

Definition 3.2 (Complete Structural Set) Let $\Gamma_F = (V, E, \omega)$. A subset $S \subseteq V$ is a *complete structural set* of Γ_F if

(i) $\Gamma_F|\bar{S}$ has no cycles; and
(ii) for all $v_i, v_j \in S$ there is at most one path from v_i to v_j in Γ_F, that contains no other vertices of S.

Note that the set $S = \{v_1, v_3\}$ is a complete structural set of the graph of interactions Γ_C corresponding to the dynamical network $(C, \mathbb{R}^4)$ considered in Example 3.1 (see Fig. 9.2). A dynamical network is reduced over a complete structural set as follows.

Definition 3.3 (Dynamical Network Reduction) Suppose S is a complete structural set of Γ_F. If $v_i \in S$ and $v_j \notin S$, replace the variable x_j of $F_i(\mathbf{x})$ by the function

$F_j(\mathbf{x})$ for each i and j. Repeat this process until only variables index by S remain. The resulting dynamical network is the *reduction* $(\mathcal{R}_S(F), X|S)$ of (F, X) induced by the set S.

The dynamical network $(\mathcal{R}_S(C), \mathbb{R}^2)$ given in Example 3.1 is the reduction of the higher-dimensional network $(C, \mathbb{R}^4)$ over the set $S = \{v_1, v_3\}$. When a network is reduced the implicit delays between elements indexed by the set S are removed. Since these are assumed to be single-type time-delays by part (ii) of Definition 3.2, if the resulting reduction is intrinsically stable then the original unreduced network must also be stable.

Theorem 3.1 (Reductions and Stability) *If at least one reduction of* (F, X) *is intrinsically stable, then* (F, X) *is stable.*

A reduction of a dynamical network may be intrinsically stable even if the network itself is not. It is therefore possible to determine whether a network is stable by reducing it. This demonstrated in the following example.

Example 3.2 (Improved Stability Estimates) We again consider the CGN network $(C, \mathbb{R}^4)$ from Example 3.1. For this network one can compute the spectral radius $\rho(M_C) = (1 + \sqrt{5})/2 > 1$. Since the network's spectral radius is greater than one we cannot directly conclude whether or not it is stable. However, its reduction $(\mathcal{R}_S(C), \mathbb{R}^2)$ has the spectral radius $\rho(M_{\mathcal{R}_S(C)}) < \text{sech}^2(|c| - 1)$. Since this is less than 1 when $|c| > 1.881$ then the original unreduced network is stable for these values of $c \in \mathbb{R}$.

The reason reductions allow for improved estimates of a network's stability is that when a network is reduced the information that is spread throughout the network is consolidated, which allows for better estimates. It is worth emphasizing that reductions are easy to compute since they are local modifications of the original network. Moreover, reductions result in a lower-dimensional networks although the components of these network have a more complicated expression.

9.4 Evolution of Network Structure

In previous sections the focus has been on the dynamics of a network with a fixed structure of interactions (topology). However, real networks are not only dynamic in terms of the behavior of their individual elements but also in terms of the underlying graph structure. For example, a biological neural network will evolve over time by adding new neurons and connections to process and store increasing amounts of information. Similarly, the Internet has an ever changing structure of connections between the servers and other physical connections that link the world's computers.

One of the central questions in network science is how a network can maintain a specific function as its structure evolves. The issue is that network growth, although important in carrying out the network's function, can have a destabilizing effect on the network's dynamics, which can lead to poor performance. In this section we describe a flexible method for modeling the growth of a network that maintains both the network's local and spectral structure. This in turn will allow us to describe a general mechanism that ensures the network remains stable as it grows.

Specifically, the goal of this section is to describe how intrinsic stability is also a natural notion for stability of a network with an evolving structure. This is done in two parts. In the first we show that we can also expand a network over any set S of its elements. We begin by describing how an expansion preserves the eigenvalues of a graph and then use this to show how an expanded network maintains its stability under expansion if it is intrinsically stable.

Similar to the interaction graphs described in Sect. 9.3, we let $G = (V, E, \omega)$ be the graph with vertices V, edges E, and edge weights given by the function ω. The graph G is *strongly connected* if there is a path from each vertex $v_i \in V$ to any other vertex $v_j \in V$. The *strongly connected components* of G are the maximal subgraphs of G that are strongly connected. Using this we can define a path of strongly connected components.

Definition 4.1 (Component Branches) For a graph $G = (V, E, \omega)$ and vertex set $S \subseteq V$ let $C_1, \ldots, C_m$ be strongly connected components of $G|\bar{S}$. If there are edges $e_0, e_1, \ldots, e_m \in E$ and two vertices $v_i, v_j \in S$ such that

(i) e_k is an edge from a vertex in C_k to a vertex in C_{k+1} for $k = 1, \ldots, m-1$;
(ii) e_0 is an edge from v_i to a vertex in C_1; and
(iii) e_m is an edge from a vertex in C_m to v_j, then we call the sequence

$$\beta = v_i, e_1, C_1, e_2, C_2, \ldots, C_m, e_m, v_j$$

a *path of components* of G with respect to S. In the case that $v_i = v_j$ then β is a *cycle of components*. We call the collection $\mathcal{B}_S(G)$ of these paths and cycles the *component branches* of G with respect to S.

The first step in expanding a graph $G = (V, E, \omega)$ is to choose a subset S over which the graph is to be expanded. Once a set S has been chosen, then its component branches $\mathcal{B}_S(G)$ are defined. It is from these branches that we build the expanded graph.

Definition 4.2 (Graph Expansion) Suppose $G = (V, E, \omega)$ and $S \subseteq V$. Let $\mathcal{X}_S(G) = (\mathcal{V}, \mathcal{E}, \mu)$ be the *evolved graph* which consists of the component branches $\mathcal{B}_S(G) = \{\beta_1, \ldots, \beta_\ell\}$ in which we *merge*, i.e. identify, each vertex $v \in S$ in any branch β_i with the same vertex v in any other branch β_j.

Example 4.1 (Expansion of a Graph) Let $G = (V, E, \omega)$ be the graph shown in Fig. 9.3a where each edge is given unit weight. For the set $S = \{v_1, v_4\} \subset V$ the graph $G|\bar{S}$ has the two components C_1 and C_2 shown in (b). The four component branches of $\mathcal{B}_S(G)$ are shown in (c), which are merged into the expanded graph $\mathcal{X}_S(G)$ in (d) by merging each of the vertices v_1 and v_4 into a single vertex, respectively.

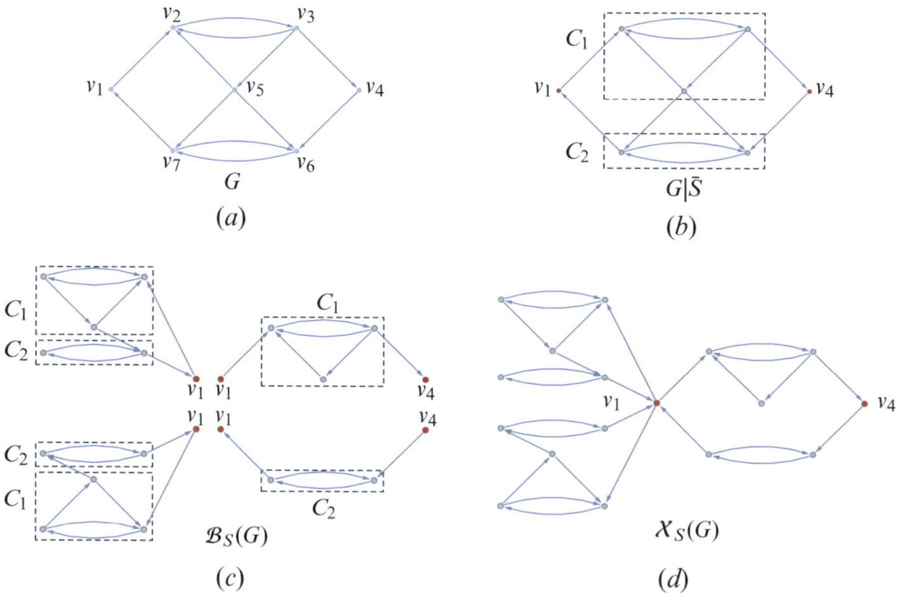

Fig. 9.3 An expansion of the graph G shown in (**a**) over the set $S = \{v_1, v_3\}$ is illustrated. The strongly connected components C_1 and C_2 of the restricted graph $G|\bar{S}$ are shown in (**b**). The four branches of $\mathcal{B}_S(G)$ are shown in (**c**). By merging the vertices v_1 and v_4, respectively, into single vertices the result is the expanded graph $\mathcal{X}_S(G)$ in (**d**)

By construction the expanded graph has the same local structure, i.e. strongly connected components, as the original graph. A natural question is to what extent the spectrum of the graph has also been preserved under expansion.

Theorem 4.1 (Spectra of Expanded Graphs) *Let $G = (V, E, \omega)$. If $S \subseteq V$, then let $C_1, \ldots, C_m$ be the strongly connected components of $G|\bar{S}$ where $\bar{S}$ is the complement of S. Then*

$$\sigma(\mathcal{X}_S(G)) = \sigma(G) \cup \sigma(C_1)^{n_1-1} \cup \sigma(C_2)^{n_2-1} \cup \cdots \cup \sigma(C_m)^{n_m-1}$$

where n_i is the number of components C_i in the evolved graph $\mathcal{X}_S(G)$ and $\sigma(C_i)^{n_i-1}$ denotes $n_i - 1$ copies of the eigenvalues of C_i.

The spectrum of an expanded graph mirrors the actual structure of the expansion in that the eigenvalues of $\mathcal{X}_S(G)$ are the eigenvalues of the original graph G plus the eigenvalues of the *new* components $C_1, \ldots, C_m$ in the expansion including multiplicities [5]. Our goal is to use this result to show that not only can we expand graphs but also certain dynamical networks can be expanded in a way that preserve their stability.

Towards this goal we observe that there is a one-to-one correspondence between the matrices $A \in \mathbb{R}^{n \times n}$ and the graphs $G = (V, E, \omega)$ where A corresponds to G if $\omega(e_{ij}) = A_{ij}$ for all $1 \leq i, j \leq n$. If A corresponds to G and $S \subseteq V$, we let A_S be the matrix that corresponds with the expansion $\mathcal{X}_S(G)$.

Definition 4.3 (Expanded Dynamical Networks) Let (F, X) be the dynamical network with components

$$F_i(\mathbf{x}) = \sum_{j=1}^{n} A_{ij} f_{ij}(x_j), \quad i = 1, \ldots, n \tag{9.5}$$

where $A \in \{0,1\}^{n \times n}$ and each $f_{ij} : X_j \to \mathbb{R}$ is a continuous function. For the subset $S \subseteq V$ of $\Gamma_F = (V, E, \omega)$, let $(\mathcal{X}_S(F), X_S)$ be the *expanded dynamical network* with components

$$\mathcal{X}_S(F)_i(\mathbf{x}) = \sum_{j=1}^{m} (A_S)_{ij} f_{S(ij)}(x_j), \quad i = 1, \ldots, m$$

where $X_S = \mathbb{R}^m$ for $A_S \in \mathbb{R}^{m \times m}$. The index $S(ij) = pq$ where the entry $(A_S)_{ij}$ corresponds to the entry A_{pq}.

Example 4.2 (Cohen–Grossberg Neural Network Expansion) Consider the CGN network $(C, \mathbb{R}^4)$ given by

$$C(\mathbf{x}) = \begin{bmatrix} (1-\epsilon)x_1 + \beta[\tanh(x_2)] + c \\ (1-\epsilon)x_2 + \beta[\tanh(x_1) + \tanh(x_3)] + c \\ (1-\epsilon)x_3 + \beta[\tanh(x_4)] + c \\ (1-\epsilon)x_4 + \beta[\tanh(x_1) + \tanh(x_3)] + c \end{bmatrix}$$

where $\epsilon, \beta, c \in \mathbb{R}$. For $\Gamma_C = (V, E, \omega)$ we let $S \subset V$ be all elements of the network that influence the dynamics of at least two other network elements. These are the vertices highlight red in the graph Γ_C in Fig. 9.4a.

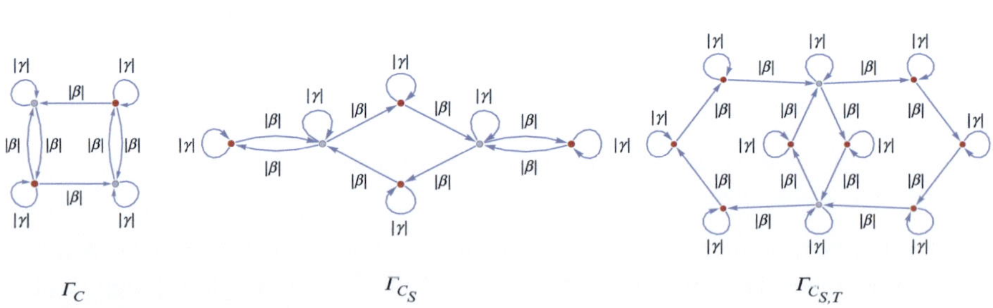

Fig. 9.4 The graphs of interactions Γ_C, Γ_{C_S}, and $\Gamma_{C_{S,T}}$ corresponding to the sequence of expansions $(C, \mathbb{R}^4)$, $(C_S, \mathbb{R}^6) = (\mathcal{X}_S(C), \mathbb{R}^6)$, and $(C_{S,T}, \mathbb{R}^{10}) = (\mathcal{X}_T(C_S), \mathbb{R}^{10})$ are shown. Here the parameter $\gamma = 1 - \epsilon$. The vertices over which the networks are expanded are highlighted *red*

The expanded dynamical network $(C_S, \mathbb{R}^6)$ where we let $C_S = \mathcal{X}_S(C)$ is given by

$$C_S(\mathbf{x}) = \begin{bmatrix} (1-\epsilon)x_1 + \beta[\tanh(x_4) + \tanh(x_5)] + c \\ (1-\epsilon)x_2 + \beta[\tanh(x_3) + \tanh(x_6)] + c \\ (1-\epsilon)x_3 + \beta[\tanh(x_1)] + c \\ (1-\epsilon)x_4 + \beta[\tanh(x_2)] + c \\ (1-\epsilon)x_5 + \beta[\tanh(x_1)] + c \\ (1-\epsilon)x_6 + \beta[\tanh(x_2)] + c \end{bmatrix}.$$

The graph of interactions Γ_{C_S} is shown in Fig. 9.4b. One can continue this process of expansion by selecting those elements T of $(C_S, \mathbb{R}^6)$ that influence the dynamics of at least two other network elements. The result of expanding the dynamical network $(C_S, \mathbb{R}^6)$ over T is the dynamical network $(C_{S,T}, \mathbb{R}^{10}) = (\mathcal{X}_T(C_S), \mathbb{R}^{10})$ whose graph of interactions is shown in Fig. 9.4c. That is, a network can be repeatedly expanded over any subset of network elements.

A network's growth can have a destabilizing effect on a network. For instance, cancer is the uncontrolled growth of cells, which can lead to the failure of certain types of biological networks. The following result gives a partial answer to the general question of how a dynamical network can remain stable as its structure evolves.

Theorem 4.2 (Stability of Expanded Dynamical Networks) *Let (F, X) be a dynamical network and S a subset of the vertices of Γ_F. The evolved dynamical network $(\mathcal{X}_S(F), X_S)$ is intrinsically stable if and only if (F, X) is intrinsically stable.*

If a network evolves in a way that can be modeled by an expansion or sequence of expansions, then it will remain stable if it is intrinsically stable. On the other hand, it is possible for a network to destabilize itself as it grows if the network is stable but not intrinsically stable.

Example 4.3 (Loss of Stability) Consider the stable linear network $(L, \mathbb{R}^3)$ and its expansion $(\mathcal{X}_S(L), \mathbb{R}^4)$ given by

$$L(\mathbf{x}) = \begin{bmatrix} 0 & -1 & 3/4 \\ 0 & 0 & 1/2 \\ -1/2 & 0 & 3/2 \end{bmatrix} \begin{bmatrix} x_1 \\ x_2 \\ x_3 \end{bmatrix} \quad \text{and} \quad \mathcal{X}_S(L)(\mathbf{x}) = \begin{bmatrix} 0 & -1 & 0 & 3/4 \\ 0 & 0 & 1/2 & 0 \\ -1/2 & 0 & -3/2 & 0 \\ -1/2 & 0 & 0 & -3/2 \end{bmatrix} \begin{bmatrix} x_1 \\ x_2 \\ x_3 \\ x_4 \end{bmatrix},$$

respectively, where $S = \{v_1, v_2\}$. The dynamical network $(L, \mathbb{R}^3)$ is stable but not intrinsically stable. When it is expanded into the network $(\mathcal{X}_S(L), \mathbb{R}^4)$ it loses its stability becoming unstable.

9.5 Concluding Remarks

We have described a number of network transformations that allow us to simplify the analysis of a network's dynamics. For a time-delayed dynamical network we have described how one can remove the network's single-type time-delays. The resulting undelayed network *inherits* the stability of the more complicated delayed system if this network is intrinsically stable, which is not the case, in general, if the network is only stable. Hence, in place of the much more difficult analysis of the delayed network we can analyze the stability of the smaller simplified undelayed network.

This analysis can be further simplified if we reduce the network by removing some set of the network's implicit delays. Once these implicit delays are removed, one can determine whether this smaller *reduced network* is intrinsically stable. If it is, then the original network is stable. If not, then one can repeat this process by further reducing the network. If after some number of reductions the network becomes intrinsically stable, then the original network is stable. In this way, network reductions can be used to prove a network is stable when a direct approach to proving that the network is stable fails.

Similarly, if a network is intrinsically stable, then any expansion of the network inherits this property. Since *network expansions* are a way of modeling the growth of a network with an evolving structure of interactions, this demonstrates how a network can grow while preserving its stability. Thus, a network that is intrinsically stable is more robust with respect to changes in its interactions due to time delays and network growth. Since time delays and growth are typically unavoidable in real networks the implication is that it is preferable to design an intrinsically stable network, e.g. an intrinsically stable computer, electrical, or transportation network, rather than one that is simply stable.

Overall, the theory of transformations of multidimensional dynamical systems and dynamical networks allows us to improve and often simplify the analysis of these objects. Moreover, it provides a new and powerful tool for data analysis and visualization of multidimensional systems and in particular of dynamical networks [4].

Acknowledgements The work of Leonid Bunimovich is partially supported by the NSF grant DMS-1600568.

References

1. Alpcan, T., Basar, T.: A globally stable adaptive congestion control scheme for internet-style networks with delay. IEEE/ACM Trans. Netw. **13**, 6 (2005)
2. Bunimovich, L.A., Webb, B.Z.: Isospectral graph transformations, spectral equivalence, and global stability of dynamical networks. Nonlinearity **25**, 211–254 (2012)
3. Bunimovich, L.A., Webb, B.Z.: Restrictions and stability of time-delayed dynamical networks. Nonlinearity **26**, 2131–2156 (2013)

4. Bunimovich, L.A., Webb, B.Z.: Isospectral Transformations: A New Approach to Analyzing Multidimensional Systems and Networks. Springer Monographs in Mathematics. Springer, Berlin (2014)
5. Bunimovich, L.A., Webb, B.Z.: Mechanisms for network growth that preserve spectral and local structure (2016). arXiv:1608.06247 [nlin.AO]
6. Cao, J.: Global asymptotic stability of delayed bi-directional associative memory neural networks. Appl. Math. Comput. **142**(2–3), 333–339 (2003)
7. Chena, S., Zhaoa, W., Xub, Y.: New criteria for globally exponential stability of delayed Cohen–Grossberg neural network. Math. Comput. Simul. **79**, 1527–1543 (2009)
8. Cheng, C.-Y., Lin, K.-H., Shih, C.-W.: Multistability in recurrent neural networks. SIAM J. Appl. Math. **66**(4), 1301–1320 (2006)
9. Cohen, M., Grossberg S.: Absolute stability and global pattern formation and parallel memory storage by competitive neural networks. IEEE Trans. Syst. Man Cybern. **SMC-13**, 815–821 (1983)
10. Newman, M.E.J.: Networks an Introduction. Oxford University Press, Oxford (2010)
11. Tao, L., Ting, W., Shumin, F.: Stability analysis on discrete-time Cohen–Grossberg neural networks with bounded distributed delay. In: Proceedings of the 30th Chinese Control Conference, Yantai, 22–24 July (2011)
12. Wang, L., Dai, G.-Z.: Global stability of virus spreading in complex heterogeneous networks. SIAM J. Appl. Math. **68**(5), 1495–1502 (2008)

Chapter 10
Discovering, Constructing, and Analyzing Synchronous Clusters of Oscillators in a Complex Network Using Symmetries

Louis M. Pecora, Francesco Sorrentino, Aaron M. Hagerstrom, Thomas E. Murphy, and Rajarshi Roy

10.1 Introduction

In 1986 an article appeared in the journal *Izvestiya-Vysshikh-Uchebnykh-Zavedenii,-Radiofizika* called "Stochastic synchronization of oscillations in dissipative systems" [1]. Along with a few other papers around that time and a little later [2–5] it sets the stage for studying what has now become a well-defined

L.M. Pecora (✉)
U. S. Naval Research Laboratory, Washington, DC 20375, USA
e-mail: louis.pecora@nrl.navy.mil

F. Sorrentino
Department of Mechanical Engineering, University of New Mexico, Albuquerque, NM 87131, USA

A.M. Hagerstrom
Department of Physics, University of Maryland, College Park, MD 20742, USA

Institute for Research in Electronics and Applied Physics, University of Maryland, College Park, MD 20742, USA

T.E. Murphy
Department of Electrical and Computer Engineering, University of Maryland, College Park, MD 20742, USA

Institute for Research in Electronics and Applied Physics, University of Maryland, College Park, MD 20742, USA

R. Roy
Department of Physics, University of Maryland, College Park, MD 20742, USA

Institute for Research in Electronics and Applied Physics, University of Maryland, College Park, MD 20742, USA

Institute for Physical Science and Technology, University of Maryland, College Park, MD 20742, USA

I.S. Aranson et al. (eds.), *Advances in Dynamics, Patterns, Cognition*, Nonlinear Systems and Complexity 20, DOI 10.1007/978-3-319-53673-6_10

dynamical situation, exact (or near exact) synchronization of chaotic oscillators in a network. The topic remains of great interest to the nonlinear dynamics community and it is a good example of the many contributions that Misha Rabinovich has made to the study of dynamics. The synchronization of two coupled chaotic systems has been extended to the synchronization of all (identical) oscillators in various networks. The latter is often referred to as global synchronization (GS). The conditions for global synchronization are often related to the network structure through the master stability function [6]. Perhaps more commonplace is partial or cluster synchronization (CS), in which patterns or sets of synchronized subgroups of nodes (oscillators) occur. In cluster synchronization nodes in the same cluster synchronize with each other, but not with nodes in other clusters. This more complex patterning phenomenon has been established in several models of networks [7–15].

With only a few exceptions most of the studies of cluster synchronization have focused on particular clusters or cluster types, especially those which are engineered to guarantee clusters form. However, it is possible to find general approaches that expose many types of CS in networks of oscillators, even for those so large or complex that it would be beyond human ability to analyze for CS in a lifetime. We present a comprehensive treatment of cluster synchronization in systems that are coupled using various versions of the network (graph) adjacency matrix (A), which represents the network by setting $A_{ij} = 0$ in the position where nodes i and j are not connected and $A_{ij} = 1$ when they are. Our method uses the tools of computational group theory to reveal the symmetries of networks and predict the patterns of synchronization that can arise as well as analyze the stability of the clusters [16]. We also show that these clusters can undergo bifurcations to less synchronized clusters without causing the whole network to desynchronize. We called this isolated desynchronization, but it is the equivalent of a transition to a chimera state [17]. Of equal importance is that we established the above behavior in an electro-optical experiment [16, 18], which shows that such cluster behavior can appear in real systems (a rough version of structural stability) and be analyzed by the group theory approach.

We also show that this approach can be extended to networks that are coupled using the network (graph) Laplacian L, where $L_{ij} = A_{ij} - \delta_{ij} \sum_{j=1}^{N} A_{ij}$ and δ_{ij} is the Kronecker delta. This is important because not only do Laplacian coupled systems show the same symmetries as their adjacency matrix counterparts, but they also may show additional CS patterns that are not the result of symmetries, but are possible because the clusters in these cases still have nodes whose equations of motion yield invariant flows for the dynamics of nodes in the same cluster. We will show that these cases can still be aided by computational group theory and the stability analysis mentioned above can be extended to the Laplacian coupled case.

We note that some other papers and books have been written on general analysis of CS using applications of group theory and symmetries to network dynamics, including types of cluster synchronization [19–23] as well as using balanced input approaches [20, 24]. These are important and interesting studies of the relation of symmetries and input sets to the phenomenon of CS, however, they did not attempt to move the calculations into the realm of larger and/or more complex networks using computation and very little attention was paid to stability calculations, which we show here.

10.2 Theory

10.2.1 Symmetries of Networks

In Fig. 10.1 we show a network of five oscillators, which we consider to be identical along with eight edges (couplings between oscillators), which we also consider to be identical. Two symmetries or permutations are displayed in the figure and there are eight symmetries in all (four rotations including the identity and four mirror planes). There are permutation symmetries that map any of the outer nodes into each other, but no permutation maps the center into any of the others. Since the edges (the couplings) are identical this implies the equations of motion for the network are the same for the outer nodes, but different for the central node. This means that if the outer nodes are started in exactly the same state, i.e. same initial conditions, they will all follow the same trajectory forever. This situation is called *flow invariant dynamics*. The resulting dynamics will have the outer nodes in the same synchronization cluster and the inner node in a separate cluster by itself. The dynamical trajectories of the outer node and the inner nodes will be different in general. An exception is when the coupling is through a Laplacian coupling, which we will talk about later.

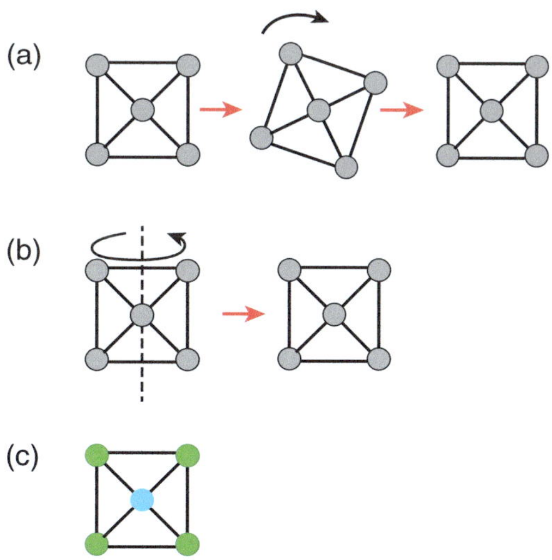

Fig. 10.1 An example of a simple network with symmetries along with dynamical consequences. (**a**) Shows a 5-node network that has a rotation symmetry (fourfold). (**b**) Is an example of a mirror or out of plane symmetry. (**c**) Shows the expected cluster configuration of the oscillators at the nodes that results from the symmetries

Generally, we want to do for more complex networks what we did above for the 5-node network. In terms of the system's state space we have found a subspace (flat submanifold, actually) in which the motion will be confined and one component of the submanifold will contain the motion of the four outer nodes and another component will contain the motion of the center node (strictly speaking the outer node manifold is $4m$ and the center node manifold is m dimensional, where m is the dimension of the node's dynamical variables). The subspace containing the synchronization motion is often referred to as the synchronization manifold and it contains the trajectories of the synchronization clusters. We put these statements into more precise mathematical form below.

10.2.2 The Dynamical Equations

The following set of general dynamical equations describe a network of N coupled identical oscillators

$$\dot{\mathbf{x}}_i(t) = \mathbf{F}(\mathbf{x}_i(t)) + \sigma \sum_j A_{ij}\mathbf{H}(\mathbf{x}_j), \; i = 1, \ldots, N, \tag{10.1}$$

where $\mathbf{x}_i$ is the n-dimensional state vector of the ith oscillator, $\mathbf{F}$ describes the dynamics of each oscillator, A is a coupling matrix that describes the connectivity of the network, σ is the overall coupling strength, and $\mathbf{H}$ is the output function of each oscillator. Equation (10.1) or its equivalent forms provide the dynamics for many networks of oscillators including all those in [6–8, 10–12, 15, 21–23, 25]. This includes some cases of time delays in the coupling functions. In [25] it is pointed out that it is only necessary for the form of the equations of motion or, more importantly, the variational equations to have the form of Eq. (10.1) near the synchronization manifolds. The form of Eq. (10.1) also applies to discrete time systems or more general coupling schemes [26]. And the same form emerges in the more general case for the variational equations where the vector field and the coupling combine into one function, for example $\mathbf{F}(\mathbf{x}_i(t), \{\mathbf{x}_j(t)\})$, where $\{\mathbf{x}_j(t)\}$ is the input of all nodes connected to node i, so long as the nodes are treated as having the same basic dynamics [26]. Further extensions of Eq. (10.1) have been studied [9, 13, 14, 27]. Existing systems which can be modeled by equations of the same form as Eq. (10.1) is large [28, 29]. These include genetic networks, circadian networks, ecology, neuronal networks, cortical networks, consensus problems, opinion formation, power grids, and concentration of metabolites in a cell, for example.

For simplicity, we take all nodes to have identical dynamics and be bidirectionally coupled to other nodes in the network by couplings of the same weight, i.e. A_{ij} is taken to be the (symmetric) adjacency matrix of 1's and 0's with the factor of σ controlling the weight of the couplings. Generalizations are possible for the coupling. We take the vector field $\mathbf{F}$ to be identical for all nodes.

The cases of symmetries in the Laplacian coupling schemes are encompassed by our analysis as shown here. This is because the row sums are not affected by the symmetry operations. Laplacian coupling schemes are usually used to allow for global synchronization. We consider the Laplacian case below when other clusters show up beyond the symmetry-related ones. We think the adjacency matrix coupling may be more representative of networks that form naturally (e.g., neurons) where row sums will not necessarily appear in feedback.

The symmetries of the network form a (mathematical) group $\mathcal{G}$. Each symmetry g of the group can be described by a permutation matrix R_g that re-orders the nodes in a way that leaves the dynamical equations unchanged as we noted in Sect. 10.2.1. This implies that applying a symmetry permutation R_g to the equations of motion Eq. (10.1) demands that we must have that each R_g commute with A. This property is crucial and we will see that it leads to possible simplifications of variational equations for the stability of the synchronous clusters. The set of symmetries (or automorphisms) [19, 30] of a network can be quite large, even for small networks, but they can be calculated from A using widely available discrete algebra routines [31, 32]. Figure 10.2a shows two graphs generated by randomly removing six edges from an otherwise fully connected 12-node network. Although the graphs appear similar and exhibit no obvious symmetries, the first instance has no symmetries (other than the identity permutation), while the other has 384 symmetries. So for even a moderate number of nodes (12) finding the symmetries can become impossible by inspection.

Thus, the computational group theory software (e.g., Sage [31]) can quickly find the symmetries of sizable networks and display the clusters of synchronization (in group theory, the orbits of the group—those nodes that will be permuted only among themselves by the R_g permutations). These are displayed in colors in Fig. 10.2a. We note that there are two clusters in Fig. 10.2a which are always permuted together (red and green), that is, when one is permuted the other must be permuted, too, to keep the adjacency matrix invariant. This will have repercussions mentioned below. The software can also show the decomposition of the symmetry group into subgroups allowing a path of potential bifurcations as parameters are varied and the stability of a cluster is lost. This leads directly to the question of finding the stability of the clusters and it turns out group theory and the software provide help with that calculation, too.

10.2.3 Stability and the Variational Equations

The stability can be found as follows. We first form the variational equation shown in Eq. (10.2) below:

$$\delta\dot{\mathbf{x}}(t) = \left[\sum_{m=1}^{M} E^{(m)} \otimes D\mathbf{F}(\mathbf{s}_m(t)) + \sigma A \sum_{m=1}^{M} E^{(m)} \otimes D\mathbf{H}(\mathbf{s}_m(t))\right] \delta\mathbf{x}(t), \qquad (10.2)$$

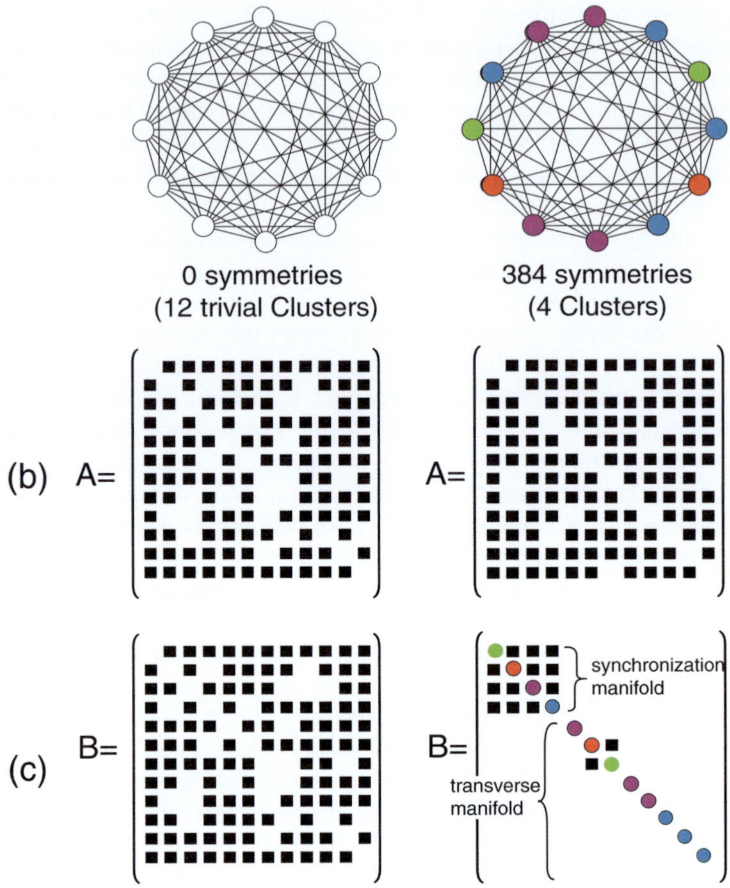

Fig. 10.2 Two randomly generated networks with varying amounts of symmetry and associated coupling matrices. (**a**) Nodes of the same color are in the same synchronization cluster. The *colors* show the maximal symmetry the network dynamics can have given the graph structure. (**b**) A graphic showing the structure of the adjacency matrices of each network (*black squares* are 1, *white squares* are 0). (**c**) Block diagonalization B of the coupling matrices A for each network. *Colors* denote the cluster, as in (**a**). *Black squares* are numerically nonzero, *white spaces* are zeros

where the Nn-dimensional vector $\delta\mathbf{x}(t) = [\delta\mathbf{x}_1(t)^T, \delta\mathbf{x}_2(t)^T, \ldots, \delta\mathbf{x}_N(t)^T]^T$ and $E^{(m)}$ is an N-dimensional diagonal matrix such that

$$E_{ii}^{(m)} = \begin{cases} 1, & \text{if} \quad i \in C_m, \\ 0, & \text{otherwise}, \end{cases} \tag{10.3}$$

$i = 1, \ldots, N$ and C_m is the set of the mth cluster's indices. One might think that we now just solve for the Lyapunov exponents or Floquet multipliers of the system from Eq. (10.2) and that would reveal stability, but there is a fundamental problem with that approach. We want to understand the stability of perturbations away from the synchronous state, but there are perturbations in which the perturbation on each node in a cluster is the same, meaning that the perturbation does not move the system away from the synchronous state. These perturbations keep the system in the synchronization manifold. We only want perturbations that move us away from the synchronization manifold. These are called transverse perturbations. This is where the theory of group representations will help.

Note that Fig. 10.2a is in node coordinates, that is the permutations R_g are the usual matrices with 1's and 0's that move the node vectors around. The coupling matrix is also in the node coordinates. Figure 10.2b shows the structure of the perturbation coupling matrix in Eq. (10.2). The R_g's form a representation of the group and from the theory of group representations there is one transformation T that will cause all the R_g's to be simultaneously block diagonalized so that each block only operates on a subspace that carries an irreducible representation (IRR) of the group [33]. There are corollaries of Schur's lemma [19, 30, 33] which state that since all R_g's commute with A, if we use T to transform to a set of IRR coordinates A will be block diagonalized to align with the blocks of the transformed R_g's. This block diagonalization of A is exactly what is needed to simplify the variational equations and separate the synchronization manifold perturbations from the transverse perturbations. The calculation of T can be done by using projection matrices [16, 34].

Applying T to Eq. (10.2) we get the variational matrix equation shown in Eq. (10.4), where $\boldsymbol{\eta}(t) = T \otimes I_n \delta\mathbf{x}(\mathbf{t})$, $J^{(m)}$ is the transformed $E^{(m)}$, and B is the block diagonalization of the coupling matrix A,

$$\dot{\boldsymbol{\eta}}(t) = \left[\sum_{m=1}^{M} J^{(m)} \otimes D\mathbf{F}(\mathbf{s}_m(t)) + \sigma B \otimes I_n \sum_{m=1}^{M} J^{(m)} \otimes D\mathbf{H}(\mathbf{s}_m(t)) \right] \boldsymbol{\eta}(t), \tag{10.4}$$

In Eq. (10.4) we have linearized about synchronized cluster states $\{\mathbf{s}_1, \ldots, \mathbf{s}_M\}$, $\boldsymbol{\eta}(t)$ is the vector of variations of all nodes transformed to the IRR coordinates and $D\mathbf{F}$ and $D\mathbf{H}$ are the Jacobians of the nodes' vector field and coupling function, respectively. We note that this analysis holds for any node dynamics, steady state, periodic, chaotic, etc.

Figure 10.2c shows the structure of the coupling matrices B after A is transformed by the block diagonalizing matrix T. When there are no symmetries T is merely the identity matrix as shown in (c). However, the T matrix is nontrivial when there are symmetries and this leads to a true block diagonalization and simplification of the coupling matrix as shown for the 384-symmetry network in (c). The 4×4 block in the upper-left-hand part of B is block related to the synchronization manifold. It can be used to calculate the Lyapunov exponents of the synchronized motion. The remaining blocks, several of which are trivial (1-dimensional) are related to the perturbations of the transverse manifold and govern the stability of the synchronization clusters. Each block is related to a cluster or a few clusters and determines the stability of that (those) clusters. The 2×2 block exists because the red and green marked clusters are intertwined [16], that is any permutation of one must permit a similar permutation of the other to keep the adjacency matrix invariant. This means that when one of the clusters becomes unstable so does the other one. Let's see how all this applies to an experimental situation and a particular network.

10.3 An Electro-Optic Experiment

We tested these theoretical results on an experimental system. This is an important first step to determining whether phenomena from a pristine theoretical system would possibly be seen in a more realistic situation. Figure 10.3 shows the setup of the electro-optic experiment. A simple explanation of the system's functioning is that light from a light-emitting diode passes through a polarizing beam splitter. One part becomes circularly polarized and is reflected back from a spatial light modulator (SLM). The SLM reflection surface phase shifts the light by an amount that can be controlled voltages on different cells of a rectangular array behind the SLM. This light then recombines with the other half of the beam and interferes with it constructively or destructively. The camera sees bright and dark square patches from the SLM. The voltages on the patches of the SLM and, hence, the phases can be changed at any time. This change can be made dependent on the input from the camera and controlled by a computer. In this way, the feedback and voltage changes make a dynamical system that changes with each time step. This is modeled by an iterated map and creates an experimental system in which, through the computer, each patch can represent a node (oscillator) and we can control which nodes influence each other. The map is written as

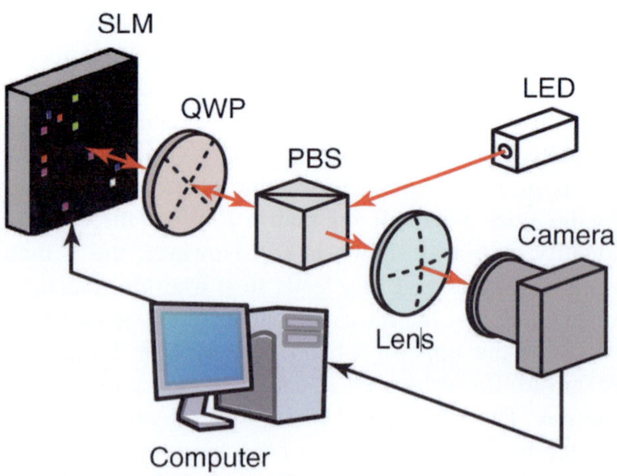

Fig. 10.3 The electro-optical experiment. *QWP* quarter-wave polarizer, *PBS* polarized beam splitter, *LED* Light-emitting diode

$$x_i^{t+1} = \left[\beta \mathcal{I}(x_i^t) + \sigma \sum_j A_{ij} \mathcal{I}(x_j^t) + \delta \right] \bmod 2\pi \tag{10.5}$$

where x_i^t is the phase of the ith node or patch on the SLM at the time t, β is the self feedback strength, and the offset δ is introduced to suppress the trivial solution $x_i = 0$. Equation (10.5) is a discrete time equivalent of Eq. (10.1). Depending on the values of β, σ, and δ, Eq. (10.5) can show constant, periodic, or chaotic dynamics. There are no experimentally imposed constraints on the adjacency matrix A_{ij}, which makes this system an ideal platform to explore synchronization in complex networks. For more details see [16].

We constructed a network as shown in Fig. 10.4a. This had 32 symmetries and 5 clusters with one being a trivial (single node) cluster. Two of the clusters are intertwined (the red and blue nodes). We ran the experiment and calculated the average synchronization error. This was determined by examining the time-average total variance of each cluster about the instantaneous mean of the cluster. We also block diagonalized the coupling matrix as detailed above. This leads to Fig. 10.4b. The synchronization manifold is the upper-left-hand corner block (5 × 5). We used this to calculate the maximum Lyapunov exponent for each cluster to compare with the synchronization error from the experiment as we varied the feedback parameter β in Eq. (10.5). Note that there is a 2 × 2 block in the transverse variational part for the red and blue intertwined nodes. Those nodes' clusters will both become

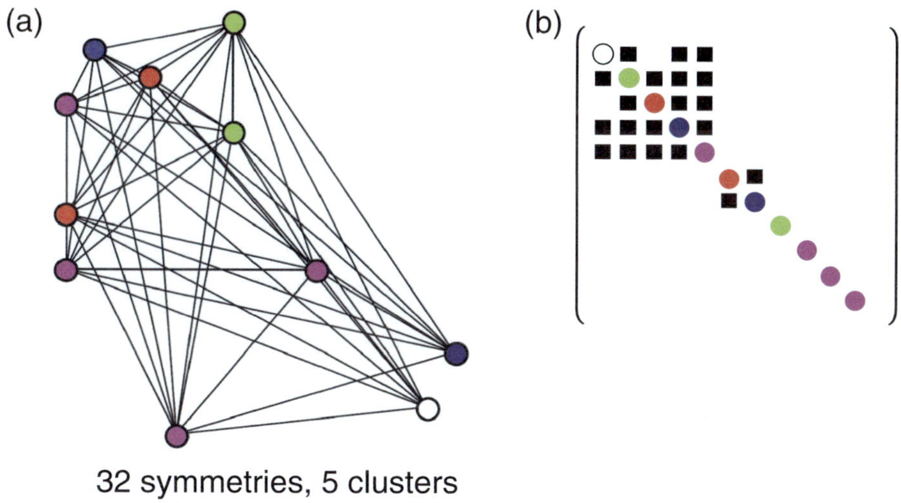

Fig. 10.4 Experimental network. (**a**) The network constructed for the spatial light modulator experiment. Nodes in the same cluster are colored the same. (**b**) The block diagonalization of the coupling matrix for (**a**). Note the synchronization manifold block in the *upper-left-hand corner* and the intertwined clusters block for the *red and blue nodes* just diagonally below the synchronization block

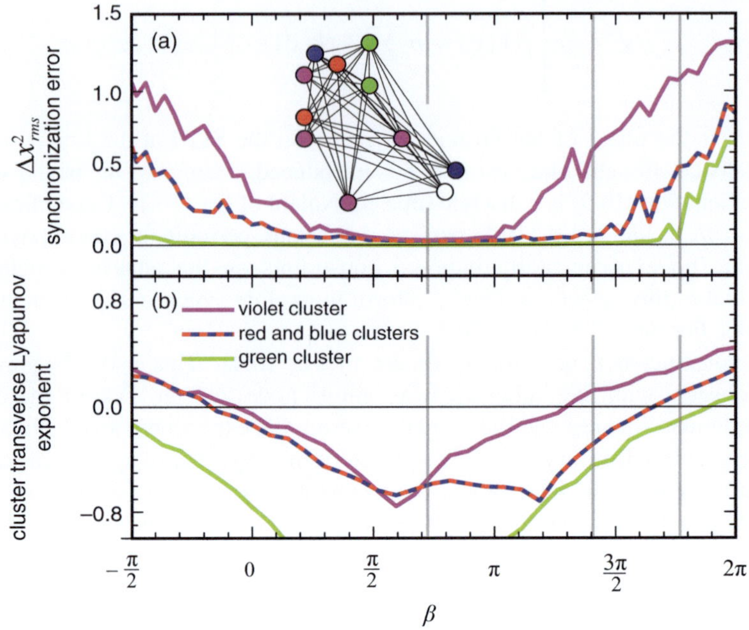

Fig. 10.5 Experimental results and Lyapunov exponents for the clusters in the experiment. (**a**) is the synchronization error (mismatch) for each cluster. (**b**) is the maximum Lyapunov exponent for each cluster or cluster group (*red and blue* in this case)

unstable when the maximum Lyapunov exponent for that block goes above 0.0. For more information on these measurements and calculation see [16].

Figure 10.5a shows the average synchronization error for each cluster or cluster group (the red and blue clusters). In the experiment all clusters are well synchronized between $\beta = 0.3\pi$ and 1.1π. The Lyapunov exponents calculations agree with this for the most part. The differences in range come in at the extremes where a cluster is about to desynchronize and, most likely, this leads to more attractor bubbling in the experiment where the clusters are not perfectly matched, which shows up in the synchronization error at the extremes.

From the experiment we see that as β is decreased from 0.3 the violet cluster loses stability first. It splits into two synchronized clusters with two nodes in each. This is a symmetry breaking bifurcation. Further decrease in β leads to the red-blue clusters becoming unstable. The green cluster remains synchronized almost all the way to $\beta = 2\pi$. As beta is increased from π the violet cluster loses stability and breaks into two synchronized clusters in a symmetry-breaking bifurcation as before. Further increase of β causes the red-blue cluster to destabilize and then the green. The approximate values of β at which these bifurcations take place and the same order is seen in the transitions of the clusters' Lyapunov exponents predicted from theory.

10.4 Laplacian Coupling and Synchronization Beyond Symmetries

Above we considered only synchronization clusters which are formed because of symmetries of the adjacency matrix. In the case of Laplacian coupling if we remain with undirected (two-way, equal) coupling, we still can get synchronization clusters, but they are not the result of symmetries. Instead they result from the combination or merging of symmetry-based clusters. Below we outline this phenomenon. For a more detailed explanation, we refer to [35].

We consider the case of Laplacian coupling where the diagonal of the coupling matrix contains minus the row sums of the original adjacency matrix (we actually use the negative of the Laplacian as usually defined in graph theory). It is well known that it is possible to have global synchronization in this case, where all nodes are synchronized to all other nodes. This comes about because the diagonal of the coupling matrix cancels the sum of the off-diagonal leaving the remaining vector field without the coupling terms for each node. This leads naturally to the question, "are there synchronous clusters that can be formed in the Laplacian system beyond the global synchronization case?" The answer is yes. Below we outline how this is possible.

Just as in the global synchronization case what matters is that the equations of motion are the same for all nodes in a cluster. This can happen if the inputs to each node in the cluster are the same. The diagonal term in the Laplacian coupling is a self-coupling that cancels any inputs that are in sync with the node under consideration. If the other inputs from other nodes in other clusters to all nodes in the considered cluster are the same, then the cluster can achieve synchronization. Or to put it in dynamics terms, when set to the same initial conditions the nodes in the cluster are flow invariant because they all have the same equations of motion in the synchronized state. We will show how these clusters can be found or constructed from the symmetry-based clusters using a 5-node network with symmetries. The Laplacian matrix for this 5-node system is

$$L = \begin{pmatrix} -3 & 1 & 0 & 1 & 1 \\ 1 & -3 & 1 & 0 & 1 \\ 0 & 1 & -3 & 1 & 1 \\ 1 & 0 & 1 & -3 & 1 \\ 1 & 1 & 1 & 1 & -4 \end{pmatrix} \qquad (10.6)$$

Figure 10.6 shows symmetry-based clusters and their merged Laplacian partners. Row 1 shows the most symmetric case where only the center node is dynamically different. As mentioned above we can merge the center node and the outer nodes to form a cluster that will have flow invariant dynamics in the fully (globally) synchronous cases. This can be seen from the structure of Eq. (10.1). All the nodes will have equations of motion in which the coupling term will not be present in

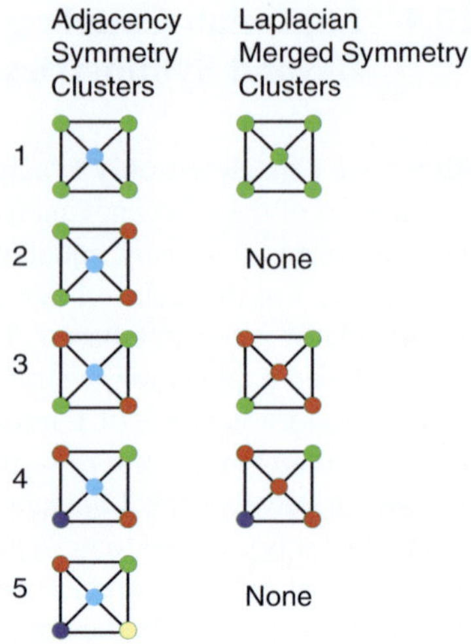

Fig. 10.6 Symmetry clusters for adjacency matrix and allowed merged clusters for Laplacian matrix cases. The *left column* contains the clusters predicted from the symmetries of the 5-node network. The *right column* shows which merged clusters from the *left column* are allowed. In rows 3 and 4 we have left out other merged clusters that are related to the ones shown by a symmetry analysis of the original network, which makes their analysis equivalent

the synchronous state because of the negative self-coupling from the Laplacian diagonal. So all will have the isolated vector field $\mathbf{F}(\mathbf{x})$. This makes the synchronous state flow invariant.

A more interesting case is from row 3 in Fig. 10.5. Here there are two outer clusters and the central node is a (trivial) cluster. We can merge the central node with either synchronized outer pair. Then each node in the diagonal triple of merged nodes has two inputs from the green node. The outer nodes have one input from a red node and the center node has two. But the center has a Laplacian self-coupling term that is one larger. The red node contributions all cancel and we are left with the following equation of motion for each red cluster node,

$$\dot{\mathbf{x}}_i(t) = \mathbf{F}(\mathbf{x}_i(t)) + \sigma[2\mathbf{H}(\mathbf{s}_g) - 2\mathbf{H}(\mathbf{s}_r)], \tag{10.7}$$

where $\mathbf{s}_g$ and $\mathbf{s}_r$ are the synchronized dynamical variables of the green and red clusters, respectively. In row 4 of Fig. 10.5 a similar process can merge the center node with the outer synchronized red nodes. As in the row 3 case this yields equations of motion similar to Eq. (10.7) except with a green and dark blue node replacing the two greens in Eq. (10.5).

We can ask the question, in row 2 of Fig. 10.5 can we merge the central node with one of the other clusters? We can see that will not work since the central node has two green inputs and the outer nodes have one (recall the red inputs all cancel properly). This case cannot be flow invariant and the mergers of clusters in row 3 either lead to non-flow invariant cases or symmetric cases already discovered.

Although the above analysis works for simple cases, is there a way to analyze possible mergers so that more complicated cases can be analyzed easily and even by the group theory software? The answer is, yes. We show how this can be done.

We use the example of merging in row 3 of Fig. 10.5. We note the general result in Laplacian coupled systems that inputs which are synchronous with a node cancel with the nodes self-coupling. This means that the synchronized dynamics are the same as though nodes within the same cluster are not coupled. We can model this with a Laplacian in which those nodes are not coupled and the diagonal components are adjusted accordingly,

$$L = \begin{pmatrix} -2 & 1 & 0 & 1 & 0 \\ 1 & -3 & 1 & 0 & 1 \\ 0 & 1 & -2 & 1 & 0 \\ 1 & 0 & 1 & -3 & 1 \\ 1 & 0 & 1 & 0 & -2 \end{pmatrix}, \tag{10.8}$$

where we start indexing the nodes in the lower-right-hand corner and proceed counter-clockwise with the center node being number 5. We call the matrix in Eq. (10.8) the *dynamically equivalent* coupling matrix since the dynamics for Eqs. (10.8) and (10.6) are the same. However, if we draw the equivalent network for the dynamically equivalent Laplacian (10.8) we get the network shown in Fig. 10.7. It is easy to see that the central node can now be permuted to either outside node and so the symmetries of Eq. (10.8) give the correct merged cluster. This generalizes to the algorithm of testing symmetry cluster merging: (1) Construct the dynamically equivalent Laplacian. (2) Analyze this using the Sage [31] software. (3) If the merged cluster is one of the symmetry clusters of the dynamically equivalent Laplacian, then the merged cluster is allowed. If not, it is not allowed. This test also correctly yields a negative outcome for the attempted merging of the center node in row 2 as described above.

We only briefly note here that we can also develop the variational equations that describe the merged cluster network. We start with the block diagonalized variational matrix from the original Laplacian symmetries and reduce them further by a vector space analysis that models the decrease in the number of synchronized clusters and the subsequent increase in the transverse perturbations which model the breakup of this merged cluster. Details for this analysis are in [35].

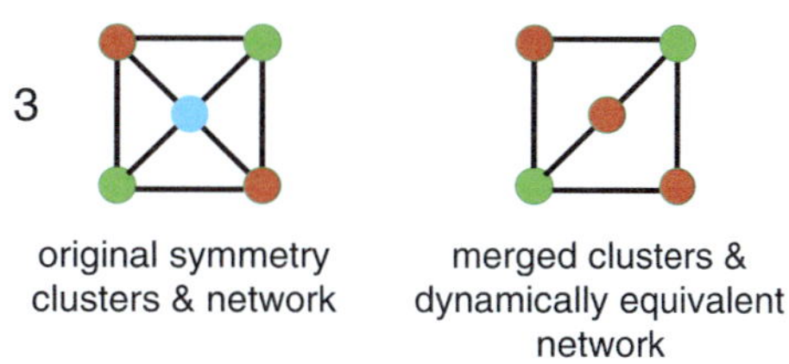

Fig. 10.7 Symmetry-based clusters and their merged Laplacian partners

10.5 Conclusions and Remarks

As we mentioned in our original paper on this work [16] the existence of symmetry-based synchronous clusters subsumes several types of phenomena seen in special cases. One is what is called remote synchronization, where nodes not directly connected have identical synchronous trajectories [22]. Another is the existence of chimera states in symmetric networks in which the coherent population is identically synchronized, as in the recent experiments on globally coupled networks in [36]. In these chimera states, the full symmetry of the network is broken in a phenomenon we originally called isolated desynchronization [16]. Here, the full symmetry is the globally synchronized state; through isolated desynchronization, a subgroup cluster desynchronizes from the globally synchronized state, resulting in a chimera state. These chimera states are permitted according to the symmetry analysis; whether they actually occur, of course, depends on the particular system and bifurcations present. We note that we have not explored the symmetry-breaking bifurcations in these systems. This will be a rich area of research much as it was in the simpler networks studied in the comprehensive book of Golubitsky, Stewart, and Schaeffer [19].

We only cover a small part of our work on Laplacian coupling and cluster construction. The interested reader should examine our paper on that topic [35]. Basically, we show how we can form new, merged clusters from the symmetry clusters that will also support synchronization. We develop an analysis (as above) that will indicate whether possible cluster mergings will yield allowed new merged clusters that synchronize. We also develop in detail there how to derive the stability variational equation for the new system with merged clusters from the original block-diagonalized equation from the pure symmetry case.

Not all possible clusters will be found using only symmetry, especially for directed networks where inputs to each node is really what matters. In this case the better test is to group nodes into clusters that include only those nodes that have the same inputs, thereby guaranteeing that they will have the same dynamics and the synchronized state will be flow invariant. However, note that finding common inputs is not trivial since we start not knowing which nodes are in which clusters. In a sense the solution to the problem requires the answer beforehand. However, there is a nice algorithm (scaling linearly with the size of the network) for finding the smallest number of clusters given by Belykh and Hasler [37]. But finding a bifurcation route to subclusters is not trivial and remains an unsolved problem. In addition, a simplification of the variational matrix as above remains to be developed and is another nontrivial situation. For more information, the reader should consult [24, 37–40].

Finally, the experiments detailed above suggest that there is robustness in this analysis, i.e. small variations in coupling strengths, parameters, etc. do not necessarily destroy the clusters. That is, the symmetries and clusters have some structural stability and we can consider the network to have approximate symmetries. Recently, we have studied this in order to quantify it and describe it more rigorously. The result is that structural stability will, in many cases, exist in general, although the ranges of parameters for which the cluster and symmetry results hold will vary with the dynamics of the nodes and the makeup of the network. This work will appear in [41].

References

1. Afraimovich, V.S., Verichev, N.N., Rabinovich, M.I.: Stochastic synchronization of oscillations in dissipative systems. Izvestiya-Vysshikh-Uchebnykh-Zavedenii,-Radiofizika **29**, 1050–1060 (1986). Original Russian reference
2. Pikovsky, A.: On the interaction of strange attractors. Z. Phys. B **55**, 149–154 (1984)
3. Fujisaka, H., Yamada, T.: Stability theory of synchronized motion in coupled-oscillator systems. Prog. Theor. Phys. **69**, 32–47 (1983)
4. Volkovskii, A.R., Rul'kov, N.F.: Experimental study of bifurcations at the threshold for stochastic locking. Sov. Tech. Phys. Lett. **15**, 249–251 (1989)
5. Pecora, L.M., Carroll, T.L.: Synchronization in chaotic systems. Phys. Rev. Lett. **64**, 821–824 (1990)
6. Pecora, L.M., Carroll, T.L.: Master stability functions for synchronized coupled systems. Phys. Rev. Lett. **80**, 2109–2112 (1998)
7. Zhou, C., Kurths, J.: Hierarchical synchronization in complex networks with heterogeneous degrees. Chaos **16**, 015104 (2006)
8. Do, A.-L., Höfener, J., Gross, T.: Engineering mesoscale structures with distinct dynamical implications. New J. Phys. **14**, 115022 (2012)
9. Dahms, T., Lehnert, J., Schöll, E.: Cluster and group synchronization in delay-coupled networks. Phys. Rev. E **86**, 016202 (2012)
10. Fu, C., Deng, Z., Huang, L., Wang, X.: Topological control of synchronous patterns in systems of networked chaotic oscillators. Phys. Rev. E **87**, 032909 (2013)
11. Kanter, I., Zigzag, M., Englert, A., Geissler, F., Kinzel, W.: Synchronization of unidirectional time delay chaotic networks and the greatest common divisor. Europhys. Lett. **93**, 6003 (2011)
12. Rosin, D.P., Rontani, D., Gauthier, D.J., Schöll, E.: Control of synchronization patterns in neural-like boolean networks. Phys. Rev. Lett. **110**, 104102 (2013)
13. Sorrentino, F., Ott, E.: Network synchronization of groups. Phys. Rev. E **76**, 056114 (2007)
14. Williams, C.R.S., Murphy, T.E., Roy, R., Sorrentino, F., Dahms, T., Schöll, E.: Experimental observations of group synchrony in a system of chaotic optoelectronic oscillators. Phys. Rev. Lett. **110**, 064104 (2013)
15. Belykh, V.N., Osipov, G.V., Petrov, V.S., Suykens, J.A.K., Vandewalle, J.: Cluster synchronization in oscillatory networks. Chaos **18**, 037106 (2008)
16. Pecora, L.M., Sorrentino, F., Hagerstrom, A.M., Murphy, T.E., Roy, R.: Cluster synchronization and isolated desynchronization in complex networks with symmetries. Nat. Commun. **5**, 1–8 (2014)
17. Abrams, D.M., Strogatz, S.H.: Chimera states for coupled oscillators. Phys. Rev. Lett. **93**, 174102 (2004)
18. Hagerstrom, A.M., Murphy, T.E., Roy, R., Hovel, P., Omelchenko, I., Schöll, E.: Experimental observation of chimeras in coupled-map lattices. Nat. Phys. **8**, 658–661 (2012)
19. Golubitsky, M., Stewart, I., Schaeffer, D.G.: Singularities and Groups in Bifurcation Theory, vol. II. Springer, New York (1985)
20. Golubitsky, M., Stewart, I.: The Symmetry Perspective: From Equilibrium to Chaos in Phase Space and Physical Space. Berkhäuser, Basel (2002)
21. D'Huys, O., Vicente, R., Erneux, T., Danckaert, J., Fischer, I.: Synchronization properties of network motifs: influence of coupling delay and symmetry. Chaos **18**, 037116 (2008)

22. Nicosia, V., Valencia, M., Chavez, M., Díaz-Guilera, A., Latora, V.: Remote synchronization reveals network symmetries and functional modules. Phys. Rev. Lett. **110**, 174102 (2013)
23. Russo, G., Slotine, J.-J.E.: Symmetries, stability, and control in nonlinear systems and networks. Phys. Rev. E **84**, 041929 (2011)
24. Judd, K.: Networked dynamical systems with linear coupling: synchronisation patterns, coherence and other behaviours. Chaos **23**, 043112 (2013)
25. Motter, A.E., Myers, S.A., Anghel, M., Nishikawa, T.: Spontaneous synchrony in power-grid networks. Nat. Phys. **9**, 191–197 (2013)
26. Fink, K.S., Johnson, G., Carroll, T., Mar, D., Pecora, L.: Three-oscillator systems as universal probes of coupled oscillator stability. Phys. Rev. E **61**, 5080–5090 (2000)
27. Irving, D., Sorrentino, F.: Synchronization of dynamical hypernetworks: dimensionality reduction through simultaneous block-diagonalization of matrices. Phys. Rev. E **86**, 056102 (2012)
28. Arenas, A., Díaz-Guilera, J.K.A., Moreno, Y., Zhou, C.: Synchronization in complex networks. Phys. Rep. **469**, 93–153 (2008)
29. Newman, M.: Networks, An Introduction, chap. 18. Oxford University Press, Oxford (2011)
30. Tinkham, M.: Group Theory and Quantum Mechanics. McGraw-Hill, New York (1964)
31. Stein, W.: SAGE: Software for Algebra and Geometry Experimentation (2013). http://www.sagemath.org/sage/, http://sage.scipy.org/
32. The GAP Group, GAP: Groups, Algorithms, and Programming, Version 4.4 (2005). http://www.gap-system.org
33. Sagan, B.E.: The Symmetric Group. Wadsworth Brooks, Pacific Grove, CA (1991)
34. Walen, A., Brennany, S., Sauer, T., Schiff, S.: Observability and controllability of nonlinear networks: the role of symmetry. Phys. Rev. X **5**, 011005-1–17 (2015)
35. Sorrentino, F., Pecora, L.M., Hagerstrom, A.M., Murphy, T.E., Roy, R.: Complete characterization of stability of cluster synchronization in complex dynamical networks. Sci. Adv. **5**, 011005-1–17 (2015)
36. Hart, J.D., Bansal, K., Murphy, T.E., Roy, R.: Experimental observation of chimera and cluster states in a minimal globally coupled network. Chaos **26**, 094801 (2016)
37. Belykh, I., Hasler, M.: Mesoscale and clusters of synchrony in networks of bursting neurons. Chaos **21**, 016106 (2011)
38. Kamei, H., Cock, P.: Computation of balanced equivalence relations and their lattice for a coupled cell network. SIAM J. Appl. Dyn. Syst. **12**, 352–382 (2013)
39. Golubitsky, M., Stewart, I., Török, A.: Coupled cell networks with multiple arrows. SIAM J. Appl. Dyn. Syst. **4**, 78–100 (2005)
40. Schaub, M.T., O'Clery, N. Billeh, Y.N., Delvenne, J.-C. Lambiotte, R., Barahona, M.: Graph partitions and cluster synchronization in networks of oscillators. Chaos **26**, 094821 (2016)
41. Sorrentino, F., Pecora, L.: Approximate cluster synchronization in networks with symmetries and parameter mismatches. CHAOS **26**, 094823 (2016). http://doi.org/10.1063/1.4961967

Chapter 11
When Repulsive Inhibition Promotes Synchrony of Bursting Neurons: Help from the Enemy

Reimbay Reimbayev, Kun Zhao, and Igor Belykh

Dedicated to Mikhail I. Rabinovich on the occasion of his 75th birthday

11.1 Introduction

Neurons can generate complex patterns of bursting activity. Over the last 30 years, much work has been dedicated to the classification of bursting rhythms and the bifurcation transitions between them (see [1–13] and the references therein).

When coupled in a network, bursting neurons can attain different forms of synchrony: burst synchronization when only the envelopes of the spikes synchronize, complete synchrony, anti-phase bursting, and other forms of phase-locking [14–17]. Synchrony has been broadly observed in pathological brain states, especially during epilepsy and Parkinson's tremors [18, 19]. Epilepsy is characterized by two behaviors, short bursts of synchronized neuronal activity and long events called seizures [19]. The emergence of a specific cooperative rhythm depends on the intrinsic properties of coupled neurons, a type and strength of synaptic coupling, and network circuitry (see, for example, [20–44]).

Fast excitatory synapses are known to facilitate bursting synchrony (see [15, 17, 35]). However, fast non-delayed inhibition is known to promote pairwise anti-phase synchronization in purely inhibitory networks [20] whereas fast excitation induces synchrony as long as the coupling strength exceeds a threshold value [15–17, 44]. Slow or time-delayed inhibitory and excitatory synapses reverse their roles such that slow or delayed inhibitory connections favor neural synchrony [21, 23, 28, 33, 34]. At the same time, synchronization in a pair of reciprocally coupled neurons with fast non-delayed inhibitory neurons is typically unstable.

R. Reimbayev • K. Zhao
Department of Mathematics and Statistics, Georgia State University, 30 Pryor Street, Atlanta, GA 30303, USA

I. Belykh (✉)
Department of Mathematics and Statistics and Neuroscience Institute, Georgia State University, 30 Pryor Street, Atlanta, GA 30303, USA
e-mail: ibelykh@gsu.edu

I.S. Aranson et al. (eds.), *Advances in Dynamics, Patterns, Cognition*, Nonlinear Systems and Complexity 20, DOI 10.1007/978-3-319-53673-6_11

More specifically, it has been shown that fast non-delayed inhibition is always repulsive in the two-coupled network of spiking (non-bursting) cells [34], unless each cell has at least two slow intrinsic variables [33]. Recently, it was shown that fast non-delayed reciprocal inhibition can promote synchrony in some bursting cells such as the leech heart interneuron model and Purkinje neuron model, provided that the inhibitory connections are weak and the initial conditions are chosen close enough, within the spiking phase of bursting [38]. However, this fragile synchronous rhythm has a small basin of attraction and is largely dominated by much stronger co-existing anti-phase bursting.

The network architecture also plays an important role in synchronization of an inhibitory network. For example, it was shown that even weak common inhibition of a bursting network with strong repulsive inhibitory connections by an external pacemaker neuron can induce synchronization within the network. This common inhibition can win out over the much (e.g., 100 times) stronger repulsive connections, provided that the pacemaker's duty cycle, the fraction of the period during which the neuron bursts, is sufficiently long [36]. Inhibitory connections also play various roles in the emergence of synchronous and asynchronous rhythms in neuronal motifs [32, 37, 40, 45, 46]. For example, the presence of a single reciprocally connected pair provides dynamical relaying in neuronal motifs that yields zero-lag synchrony despite long conduction delays [45, 46].

In this chapter, we review our recent results [47, 48] that report a counterintuitive find that fast non-delayed *repulsive* inhibitory connections can robustly promote synchronization, when added to an excitatory network of square-wave bursting neurons. This synergistic effect is caused by the ability of inhibition to effectively switch the type of network behavior from square-wave [1] to plateau bursting [6]. Square-wave bursting [1] was named after its shape during a burst which resembles a square wave. Plateau bursting is characterized by spikes of decreasing size that turn into a plateau towards the end of the active phase of bursting [6]. Square-wave bursters are difficult to synchronize [15] and their spike synchronization requires strong excitatory coupling, whereas plateau bursters with smaller spikes are more prone to synchrony. The added inhibition causes plateau bursting so that weaker excitatory coupling is sufficient to induce synchrony in the excitatory–inhibitory network. This effect is generic and observed in different models of bursting neurons. In this study, we choose the Hindmarsh–Rose neuron model and Sherman β-cell model as individual units of networks. Both models are known to generate different types of complex bursting behavior [3, 30, 49, 50]. The emergence of bursting rhythms in isolated and coupled Hindmarsh–Rose neurons has received a particularly great deal of attention in the literature [3, 13, 17, 44]. This includes major contributions made by M.I. Rabinovich (see a review [11]) who introduced the senior author of this chapter (I.B.), a graduate student at that time, to the exciting field of neural modeling and its implications in central pattern generators and brain networks.

While many studies use reduced neuronal models such as phase or relaxation oscillators where the spikes are ignored, our results promote the use of the detailed biophysical models, taking into account neuronal spikes and bursts. The discovered synergistic effect is due to nonlinear interactions of spikes; as a result, it is not observed in networks of the reduced models. Yet, there is experimental evidence that the onset and self-termination of seizures is accompanied by the transition between different types of network bursting activities [12, 51] where the spikes play an important role. Remarkably, the transition to abnormal synchrony corresponds to switching to plateau-like bursting [12].

We use the stability analysis to reveal the general mechanism of the induced synchronization and demonstrate that there is an optimal balance between the excitatory and inhibitory couplings that trigger synchronized bursting. These results are applicable to synchronization in a pair of connected neurons as well as to large networks with mixed excitatory–inhibitory connections. We discover universal scaling laws for the onset and loss of stable synchronization where the synchronization conditions are fully controlled by the number of excitatory and inhibitory inputs each neuron receives, regardless of the network size and topology. The independence of the synchronization conditions in purely excitatory networks of bursting neurons from the details of network architecture, except for the in-degree of each neuron, was reported in [17]. In this work, we show that the inhibition-induced synchrony is also controlled by the number of inhibition inputs to each neuron; however, the scaling law for the synchrony loss is different and involves a ratio of excitatory and inhibitory inputs. These general laws are drastically different from those in linearly coupled networks with positive (attractive) and negative (repulsive) coupling where the synchronization conditions are controlled by the structure of negative connections via the eigenvalues of the corresponding Laplacian matrix [52–54].

The layout of this chapter is as follows. First, in Sect. 11.2.1, we introduce a two-cell network model and the Hindmarsh–Rose model as its individual unit. We show that the uncoupled cell model exhibits square-wave bursting and discuss the generation mechanism. Then, we introduce the self-coupled system that governs the type of synchronous bursting. We show that the self-coupled system switches from square-wave to plateau bursting with an increase in the excitatory and/or inhibitory couplings. We then demonstrate that this transition between different types of bursting induces stable synchronization in the two-cell coupled networks. In Sect. 11.2.2, we present the variational equations for the stability of the synchronous solution and explain the main synchronization mechanism. In Sect. 11.2.3, we provide additional support for explaining synchrony loss, caused by overly strong inhibition. In Sect. 11.2.4, we present the universal scaling laws for the stability of synchronization in large networks. In Sect. 11.3, we study a two-cell network of Sherman model to demonstrate that the synergistic effect is generic and observed in different bursting models capable of switching from square-wave to plateau bursting. In Sect. 11.4, a brief discussion of the obtained results is given.

11.2 Networks of Hindmarsh–Rose Models

We begin with the Hindmarsh–Rose neuron model [55] which represents a class of phenomenological models of spiking and bursting neurons. Without direct relation to specific biophysical mechanisms, these models aim at reproducing the characteristic features of the bursting behavior.

11.2.1 Two-Cell Network

The simplest network of two bursting Hindmarsh-Rose neuron models with excitatory and inhibitory connections reads:

$$\dot{x}_i = ax_i^2 - x_i^3 - y_i - z_i + g_{\rm exc}(V_{\rm exc} - x_i)\Gamma(x_j) + g_{\rm inh}(V_{\rm inh} - x_i)\Gamma(x_j),$$
$$\dot{y}_i = (a+\alpha)x_i^2 - y_i, \quad \dot{z}_i = \mu(bx_i + c - z_i), \quad i,j = 1,2. \tag{11.1}$$

Here, x_i represents the membrane potential of neuron i, and variables y_i and z_i take into account the transport of ions across the membrane through fast and slow ion channels, respectively. The cells are identical and the symmetrical synaptic connections are fast and instantaneous. The parameters $g_{\rm exc}$ and $g_{\rm inh}$ are the excitatory and inhibitory coupling strengths, respectively. The reversal potentials $V_{\rm exc} = 2 > x_i(t)$ and $V_{\rm inh} = -2 < x_i(t)$ for all values of $x_i(t)$ and any t, i.e. the synapses are excitatory and inhibitory, respectively. The fast synaptic coupling is modeled by the sigmoidal function $\Gamma(x_j) = 1/[1 + \exp\{-\lambda(x_j - \Theta_{\rm s})\}]$ [29]. The synaptic threshold $\Theta_{\rm s} = -0.25$ is set to ensure that every spike in the single cell burst can reach the threshold [17]. As a result, a spike arriving from a presynaptic cell j activates the synapse current (through $\Gamma(x_j)$ switching from 0 to 1) entering the postsynaptic cell i. Hereafter, the parameters are chosen and fixed as follows: $a = 2.8$, $\alpha = 1.6$, $\lambda = 10$, $c = 5$, $b = 9$, $\mu = 0.001$ [17, 44].

From a neuroscientist's perspective, such a network can be viewed as the interaction between two excitatory neurons with direct excitatory and tertiary synapses [56] where the latter excites the presynaptic terminal of an inhibitory interneuron, allowing inhibition of the other excitatory cell (see Fig. 11.1) [57]. From a physicist's perspective, this is a network of two pulse-coupled oscillators with attractive (excitatory) and repulsive (inhibitory) connections.

We use this two-cell network to demonstrate the synergistic effect and clearly describe its stability mechanism. We will then show that the same results carry over to larger networks whose architecture always supports Dale's law [58] such that synaptic (outgoing) connections from a neuron to other cells are either all excitatory or inhibitory.

Each cell in the network (11.1) receives one inhibitory and one excitatory input from the other cell, therefore the network system (11.1) has an invariant manifold $D = \{x_1 = x_2 = x(t),\ y_1 = y_2 = y(t),\ z_1 = z_2 = z(t)\}$, that defines complete synchronization between the cells.

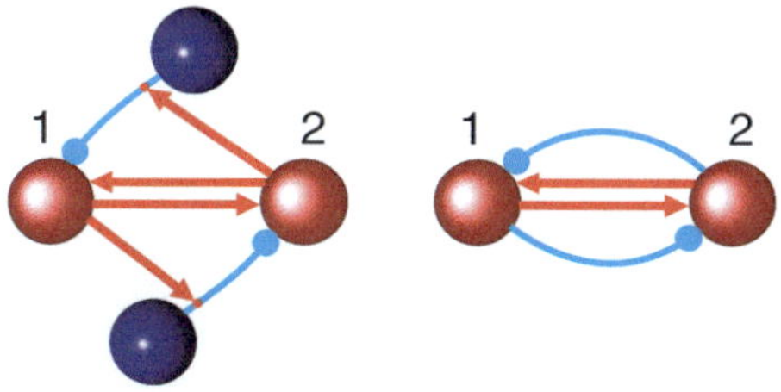

Fig. 11.1 (*Left*). Possible interactions between two excitatory neurons 1 and 2 with direct excitatory and tertiary synapses. The tertiary synapses mediate inhibition by exciting the presynaptic terminals of inhibitory interneurons at their somas. This network can be viewed as a pair of neurons effectively coupled through both excitatory and inhibitory connections (*right*). Excitatory (inhibitory) connections are depicted by arrows (*circles*). The dynamics of the two-cell network is studied in Fig. 11.3

Note that the synchronous dynamics differs from that of the uncoupled cell as the former is governed by the self-coupled system with extra coupling terms:

$$\dot{x} = ax^2 - x^3 - y - z + k_{\text{exc}}\, g_{\text{exc}}(V_{\text{exc}} - x)\Gamma(x) + k_{\text{inh}}\, g_{\text{inh}}(V_{\text{inh}} - x)\Gamma(x),$$
$$\dot{y} = (a + \alpha)x^2 - y, \qquad \dot{z} = \mu(bx + c - z). \tag{11.2}$$

As a result, the synchronous dynamics and the type of bursting depend on the coupling strengths g_{exc} and g_{inh}. This property is a key ingredient of the synergistic effect reported in this chapter. There are critical coupling strengths g_{exc} and g_{inh} at which square-wave bursting in the self-coupled system (11.2) turns into plateau-type bursting, depicted in Fig. 11.2. This happens via the disappearance of a homoclinic bifurcation (HB) in the 2-D fast subsystem ($\mu = 0$) of system (11.2) that governs the type of synchronized bursting. Figure 11.2 illustrates the bifurcation mechanism of this transition from square-wave to plateau bursting. According to the Izhikevich classification [6], square-wave bursting corresponds to fold/homoclinic bursting where the burst termination is determined by a homoclinic loop to a saddle in the fast subsystem. Increasing synaptic coupling in the self-coupled system (11.2), whether excitatory or inhibitory, eventually leads to the disappearance of this homoclinic bifurcation and induces plateau bursting (fold/fold bursting in the Izhikevich classification). This can be achieved by strong excitation (see Fig. 11.3b) or by weaker inhibition (see Fig. 11.3c). The fast (x, y)-subsystem of the self-coupled system (11.2) has the nullcline $z = h(x) \equiv -\alpha x^2 - x^3 + g_{\text{exc}}(V_{\text{exc}} - x)\Gamma(x) + g_{\text{inh}}(V_{\text{inh}} - x)\Gamma(x)$. The excitatory (inhibitory) coupling moves the nullcline $z = h(x)$ to the right (left) (see Fig. 11.2). Remarkably, a small shift of the right branch of $z = h(x)$ towards the synaptic threshold $x = \Theta_s$ (to the left) caused by weaker inhibition effectively decreases the divergence inside the limit cycle of the fast system, forming the spiking manifold. This causes the limit cycle to shrink in size and makes the homoclinic orbit disappear. At the same time, a much larger amount of excitation is necessary to shift the right branch of $z = h(x)$ to a far right

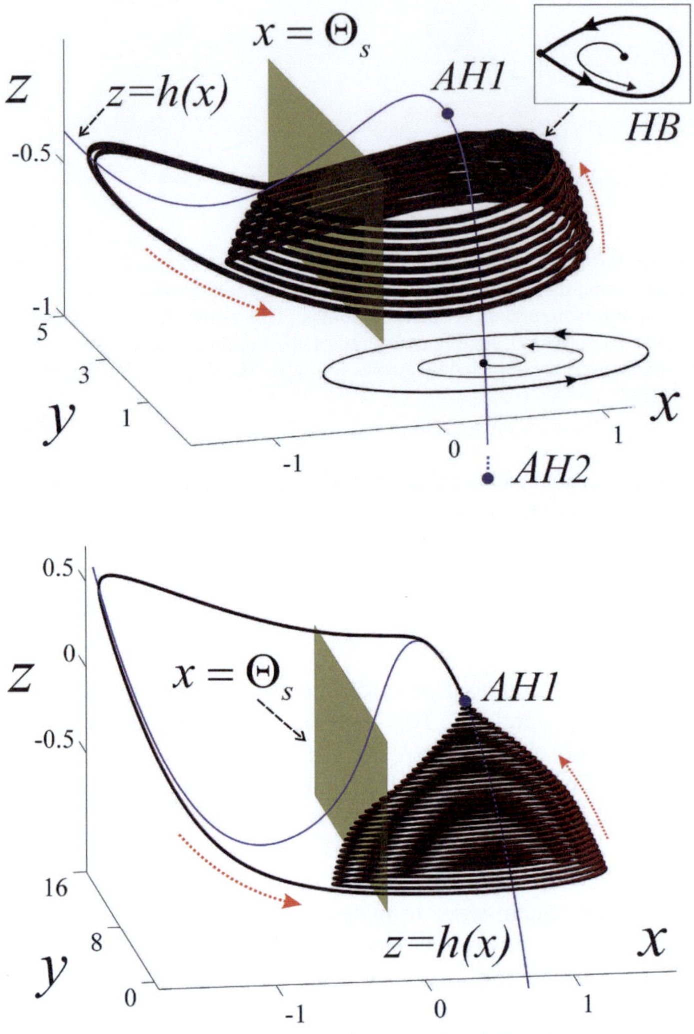

Fig. 11.2 Transition from square-wave to plateau bursting in the self-coupled system (11.2), controlling the type of synchronous bursting. (*Top*). Square-wave burster in the uncoupled network (11.1). The right branch of the fast nullcline $z = h(x)$ contains two points $AH1$ and $AH2$ corresponding to supercritical Andronov–Hopf bifurcations. A limit cycle of the fast system ($\mu = 0$) is born from the Andronov–Hopf bifurcation $AH2$ and grows in size as z increases. This family of limit cycles constitutes the spiking manifold which terminates at the homoclinic bifurcation HB of the saddle point of the fast system, located on the middle branch of $z = h(x)$. The red (*dotted*) curve schematically indicates the route for bursting trajectories. The plane $x = \Theta_s$ displays the synaptic threshold. (*Bottom*) Plateau bursting induced by the combination of excitatory and inhibitory coupling ($g_{exc} = 0.6$ and $g_{inh} = 0.25$), corresponding to point (c) in Fig. 11.3. The added inhibition leads to the disappearance of the homoclinic bifurcation such that the spiking manifold extends further up and disappears as the limit cycle shrinks to zero amplitude and disappears via the reverse Andronov–Hopf bifurcation $AH1$

region where the divergence is small enough for a similar switch from square-wave to plateau bursting via the disappearance of the homoclinic orbit (see the HB curve in Fig. 11.3(top)); the curve is calculated using the bifurcation analysis software CONTENT [59].

Figure 11.3 presents the synergistic effect in the two-cell network of HR neurons (11.1) and reveals that there is a broad interval of inhibitory strengths over which the repulsive inhibition complements attractive excitation in promoting neural synchrony. Notice that the onset of spike (complete) synchronization through boundary $E1$ is accompanied by or close to the transition from square-wave to plateau bursting, indicated by the curve HB. The two curves practically coincide up to the values of $g_{\text{exc}} \approx 0.8$ such that a significant reduction of the synchronization threshold for g_{exc} as much as ten times, observed at the lower values of g_{exc} is governed by this transition between the two types of bursting. This transition occurs in both the purely excitatory (Fig. 11.3b) and mixed excitatory–inhibitory connections (Fig. 11.3c). The addition of inhibition to the purely excitatory network, whose synchrony requires a much stronger coupling, makes the cells switch to plateau bursting with smaller spikes which can be synchronized by the weaker excitatory coupling. The blue (dark) synchronization region, bounded by curves $E1$ and $E2$, corresponds to synchronized bursting and indicates a synergistic balance between the excitation and inhibition. Overly strong inhibition destroys synchrony (through boundary $E2$) and leads to anti-phase bursting, as expected (Fig. 11.3d).

It is worth noticing that switching to synchronized plateau bursting also shifts the plateau part of the burst to the right from the synaptic threshold (see Fig. 11.2). Due to the choice of the synaptic sigmoidal function $\Gamma(x_j)$ in (11.1), the coupling between the cells remains continuous during this part of the burst while being pulsatile in the first half of the burst where the spikes cross the synaptic threshold Θ_s. This might not be the case in cortical networks where the coupling is always pulsatile. Figure 11.3e indicates the region between the stability boundary $E1$, corresponding to the onset of induced synchrony, and the HB curve, indicating the transition to synchronized plateau bursting. This region corresponds to synchronized square-wave bursting where all the spikes cross the synaptic threshold Θ_s, making the coupling pulsatile for all times. We have also performed numerical simulations of the network (11.1) with the sigmoidal function $\Gamma(x_j)$, replaced by the Heaviside function $H(x_j)$, representing realistic fast pulse-coupling. The obtained stability diagrams are similar to the ones of Fig. 11.3 with a slight expansion of the left stability zone bounded by $E1$ along the x and y axes, up to the synchronization coupling threshold $g_{\text{exc}} = 1.35$ in the purely excitatory network (cf. the synchronization threshold $g_{\text{exc}} = 1.28$ in the network with the sigmoidal function $\Gamma(x_j)$). This increase in the coupling comes from the fact that the Heaviside-type pulse-coupling has a weaker impact, compared to the sigmoidal-type coupling. As a result, larger values of g_{exc} and g_{inh} are required to achieve the same effect.

Why is plateau bursting easier to synchronize? Why does inhibition not desynchronize plateau oscillations as it seems to have an apparent destabilizing effect?

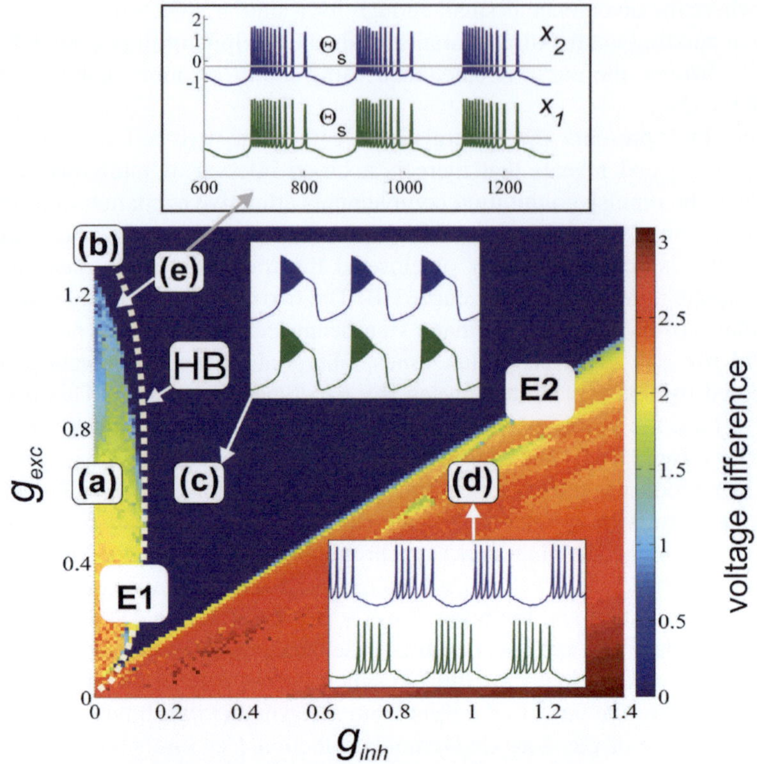
x2
x1
Θs
Θs
600
800
1000
1200
(b)
(e)
HB
(a)
(c)
(d)
E1
E2
g_exc
g_inh
voltage difference
0
0.2
0.4
0.6
0.8
1
1.2
1.4
0.5
1.5
2
2.5
3

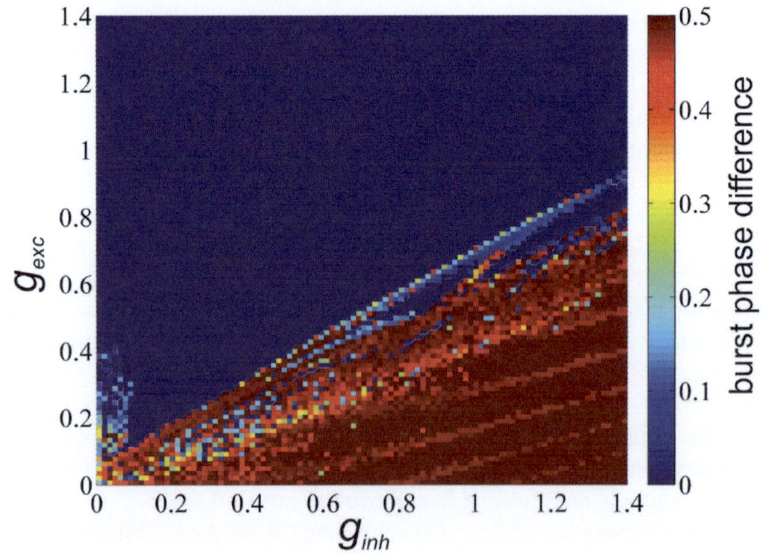
g_exc
g_inh
burst phase difference
0
0.2
0.4
0.6
0.8
1
1.2
1.4
0.1
0.3
0.5

The following subsection offers answers to these questions and reveals the stability mechanism associated with the transition from square-wave to plateau bursting in the system (11.1).

11.2.2 Stability Mechanism

To explain the synchronization mechanism, we use the stability equations for the infinitesimal transverse perturbations $\xi_{12} = x_1 - x_2$, $\eta_{12} = y_1 - y_2$, $\zeta_{12} = z_1 - z_2$ [17]:

$$\begin{aligned} \dot{\xi}_{12} &= (2ax - 3x^2)\xi_{12} - \eta_{12} - \zeta_{12} - \Omega(x)\,\xi_{12}, \\ \dot{\eta}_{12} &= 2(a+\alpha)x\xi_{12} - \eta_{12}, \\ \dot{\zeta}_{12} &= \mu(b\xi_{12} - \zeta_{12}), \end{aligned} \tag{11.3}$$

where $\Omega(x) = S_1 + S_2$ with

$$\begin{aligned} S_1 &= (g_{\text{exc}} + g_{\text{inh}})\,\Gamma(x), \\ S_2 &= (g_{\text{exc}}(V_{\text{exc}} - x) + g_{\text{inh}}(V_{\text{inh}} - x))\,\Gamma_x(x). \end{aligned} \tag{11.4}$$

In (11.3), $x(t)$ is the synchronous solution defined via the self-coupled system (11.2). The stability of the zero equilibrium $\{\xi_{12} = 0,\ \eta_{12} = 0,\ \zeta_{12} = 0\}$ of the linearized system (11.3) corresponds to the stability of the synchronous solution in the original network. The function $\Omega(x)$ represents the contribution of the excitatory and inhibitory coupling; it favors the stability of synchronization when it becomes positive and has a destabilizing impact when it is negative [17]. More specifically, the coupling term $-\Omega(x)\,\xi_{12}$ aims at stabilizing the zero equilibrium of system (11.3) when it is positive and tends to destabilize the zero equilibrium when it is negative. The two terms S_1 and S_2, composing $\Omega(x)$, heavily depend on whether the voltage $x(t)$ exceeds the synaptic threshold Θ_s. The first term S_1 contains the sigmoidal synaptic function $\Gamma(x)$ and becomes significant for $x(t) \geq \Theta_s$. Once turned on, the term $S_1 > 0$ makes $\Omega(x) > 0$ for $x(t) \geq \Theta_s$ (see Fig. 11.4) and favors the stability for both excitatory and inhibitory coupling as $g_{\text{exc}} + g_{\text{inh}} > 0$.

Fig. 11.3 Synchronization in the two-cell network (11.1) as a function of excitation (g_{exc}) and inhibition (g_{inh}). (*Top panel*). The color bar indicates the voltage difference $|x_1 - x_2|$, averaged over the last three bursting periods. The *blue* (*dark*) zone (c) corresponds to the zero voltage difference (complete synchronization), appearing from random initial conditions. Observe the effect when a small increase of inhibition from 0 dramatically lowers the synchronization threshold from 1.28 to 0.11. Note that the inhibition desynchronizes the cells in the absence of excitation ($g_{\text{exc}} = 0$), independent from the coupling strength and initial conditions. Bifurcation curve HB (*white dotted line*) corresponds to the transition to synchronized plateau bursting. (*Bottom panel*). Burst synchronization. The color bar indicates the phase difference between the bursts, $\Delta\phi = \phi_1 - \phi_2$, averaged over the last three bursting periods. The normalized phase $0 \leq \phi_i \leq 1$ of the ith bursting cell ($i = 1, 2$) is initiated and reset every cycle at the beginning of the burst. The normalized phase difference $\Delta\phi$ ranges from 0 (burst synchrony, *blue* (*dark*) color) to 0.5 (anti-phase bursting, *red* (*lighter*) color). Notice a similar effect of burst synchronization, induced by repulsive inhibition

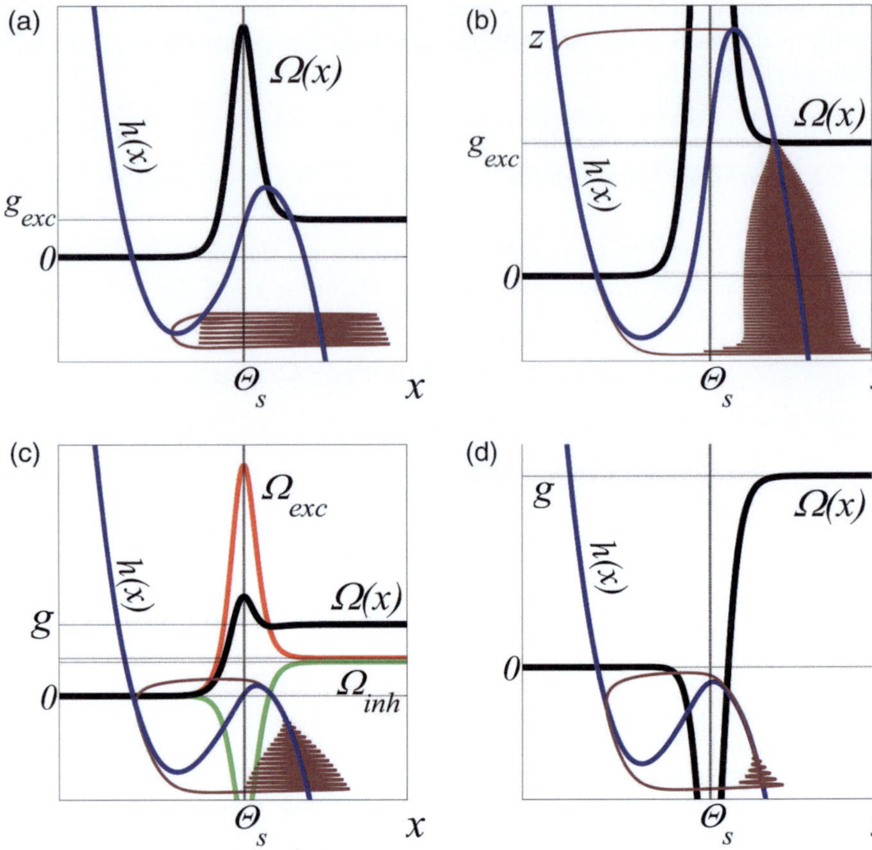

Fig. 11.4 Stability function $\Omega(x)$ for synchronized bursting. Panels (**a**), (**b**), (**c**), and (**d**) correspond to the points (**a**), (**b**), (**c**), and (**d**) in Fig. 11.3. (**a**). $g_{\text{exc}} = 0.6$, $g_{\text{inh}} = 0$: Unstable square-wave synchronous bursting (*light brown*) and the fast nullcline $h(x)$ of the self-coupled system, together with $\Omega(x)$ superimposed on its own scale. The impact of $\Omega(x)$ is not sufficient to stabilize the subthreshold part of the spikes where the coupling is insignificant (to the left from the threshold Θ_s). (**b**). $g_{\text{exc}} = 1.28$, $g_{\text{inh}} = 0$: The increased excitation makes the impact of $\Omega(x)$ stronger; more importantly it changes the type of synchronous bursting. Notice that the spikes have shifted to the right and moved to the region where the strong coupling is present. (**c**) $g_{\text{exc}} = 0.6$, $g_{\text{inh}} = 0.25$: The *red* curve represents the contribution of the excitatory coupling $\Omega_{\text{exc}} = g_{\text{exc}}\Gamma(x) + g_{\text{exc}}(V_{\text{exc}} - x)\Gamma_x(x)$, the *light green* curve corresponds to that of the inhibitory coupling $\Omega_{\text{inh}} = g_{\text{inh}}\Gamma(x) + g_{\text{inh}}(V_{\text{inh}} - x)\Gamma_x(x)$, and the *thick black line* indicates the combined curve $\Omega(x) = \Omega_{\text{exc}} + \Omega_{\text{inh}}$. Adding the inhibition decreases the impact of $\Omega(x)$ (cf. with (**a**) where $\Omega(x)$ equals Ω_{exc} in (**c**)). At the same time, it induces plateau bursting, with the spikes in the region above the threshold, where the coupling is sufficiently strong to synchronize them. (**d**). $g_{\text{exc}} = 0.6$, $g_{\text{inh}} = 0.9$: Strong inhibition destabilizes synchronous plateau bursting. $\Omega(x)$ has a negative drop in the region, covering the upper knee of the nullcline. As a result, the cells diverge when slowly crawling up this part of the nullcline. Note that synchronous plateau bursting of the self-coupled system is unstable and does not represent the dynamics observed in the network; the cells get locked into anti-phase square-wave bursting (cf. Fig. 11.3d)

The second term, S_2, can change sign; the term due to the excitatory coupling $g_{exc}(V_{exc} - x)$ is positive and therefore attractive, whereas the inhibitory one $g_{inh}(V_{inh} - x)$ is negative and repulsive. It contains the derivative $\Gamma_x(x)$ which has a peak around Θ_s and rapidly decaying tails (in the case of the Heaviside function $H(x_i)$, $\Gamma_x(x)$ turns into the delta function). Therefore, the term S_2 switches and remains on for the values of x, close to the threshold Θ_s when the spikes cross the threshold. It becomes decisive for the overall sign of $\Omega(x)$ in a region around the threshold Θ_s, giving a distinct bell shape to $\Omega(x)$ (see Fig. 11.4).

When $x(t)$ drops below the threshold Θ_s, the cells are practically uncoupled. Our Lyapunov function based analysis of synchronization in excitatory networks [17, 44] suggests that the spikes are the most unstable part of the synchronous solution such that their stabilization via the synaptic coupling yields complete synchronization. The above-threshold part of the synchronous solution lies in the stability zone as the coupling function $\Omega(x) > 0$, for any combination of g_{exc} and g_{inh}. Therefore, this part of the solution can be stabilized by making the coupling stronger. At the same time, the subthreshold part of the synchronous spikes is difficult to stabilize as the contribution of the term S_2 rapidly decays to zero below from the threshold. Moreover, only excitatory coupling can stabilize the synchronous trajectory in the subthreshold region as it yields the positive peak of the bell-shaped curve $\Omega(x)$ (see Fig. 11.4a). The addition of inhibition lowers this peak and can make it negative (see Fig. 11.4d), making the region around the threshold less stable. Figure 11.4a and b show that increasing $\Omega(x)$ (via increasing g_{exc}) induces synchrony in the purely excitatory network. However, it requires fairly strong excitation to stabilize the synchronous solution, especially its subthreshold part. Figure 11.4c demonstrates that adding the inhibition has a two-fold effect. It lowers the stabilizing impact of $\Omega(x)$ around and below the synaptic threshold; however, it helps switching the type of synchronous bursting via (11.2), making the spikes shorter and moving them towards the stability region, controlled by the synchronizing term S_1. Increasing inhibition typically switches synchronous square-wave bursting to plateau bursting which places the spikes of synchronous bursting into the stability (above-threshold) region that can be in turn effectively stabilized by the excitatory coupling via S_1. Therefore, the combination of $g_{exc} + g_{inh}$ synergistically induces synchronized bursting within a wide region of parameters g_{exc} and g_{inh}. Its right stability boundary $E2$ (cf. Fig. 11.3) corresponds to synchrony loss and is defined by the mutual arrangements between the graphs of $\Omega(x)$ and the nullcline $h(x)$ (Fig. 11.4d). This happens when the upper knee of $h(x)$ falls inside the instability zone where $\Omega(x)$ is negative (cf. Fig. 11.4d). The following subsection contains an additional argument for predicting the slope of boundary $E2$. This estimate $g_{exc} = 0.78\, g_{inh}$ (see Sect. 11.2.3) coincides remarkably well with the numerically calculated boundary $E2$ in Fig. 11.3.

It is important to restate that the dynamics and type of synchronous bursting $x(t)$ are controlled by the self-coupled system (11.2) and depend on both g_{exc} and g_{inh}. This property allows the inhibition to induce plateau bursting in the self-coupled system (11.2). The synchronous bursting observed in the self-coupled system (11.2) does not necessarily represent the emergent network dynamics. This synchronous solution can be unstable, especially when g_{inh} is overly strong as in Fig. 11.4d. Therefore, the network generates a different stable rhythm; this is typically anti-phase square-wave bursting as in Fig. 11.3d (cf. the two insets for the consistency).

While the onset of inhibition-induced synchronization is typically governed by the transition from square-wave to plateau bursting, the addition of inhibition can also induce synchronized square-wave bursting in a smaller region of parameters (Fig. 11.3e). However, the synchronization mechanism is essentially the same; the inhibition decreases the subthreshold part of the spikes, without changing the type of bursting, and thus facilitates synchronization. Although, fairly strong excitation is required, making the synergistic effect less pronounced.

11.2.3 Slope of Synchrony Loss Boundary E2

This subsection provides additional support for explaining synchrony loss, caused by overly strong inhibition via the stability boundary $E2$ (see Fig. 11.3). In addition to the stability argument based on the variational equations (see Sect. 11.2.2), we use a more straightforward approach to predict the slope of the boundary $E2$ in the two-cell network.

Note that the combined action of two excitatory and inhibitory synapses in the network (11.1) essentially amounts to that of one synaptic connection with strength g_{syn} and synaptic reversal potential E_{syn}. The corresponding system reads:

$$
\begin{aligned}
\dot{x}_i &= ax_i^2 - x_i^3 - y_i - z_i + g_{\text{syn}}(E_{\text{syn}} - x_i)\Gamma(x_j),\\
\dot{y}_i &= (a+\alpha)x_i^2 - y_i,\\
\dot{z}_i &= \mu(bx_i + c - z_i),\;\; i,j = 1,2. \qquad (11.5)
\end{aligned}
$$

The synaptic reversal potential E_{syn} changes in the range $[-2,\ 2]$, allowing us to vary the type of the connection from purely inhibitory when $E_{\text{syn}} = -2 < x_i$ for all $x_i(t)$, to purely excitatory when $E_{\text{syn}} = 2 > x_i(t)$. In this setting, changing the coupling strengths g_{exc} and g_{inh} in the network (11.1) with fixed $V_{\text{exc}} = 2$ and $V_{\text{inh}} = -2$ is equivalent to changing the values of g_{syn} and E_{syn} in the network (11.5). Figure 11.5 shows robust synchronization in an interval of g_{syn} and E_{syn}. Here, the left stability boundary, indicating the drop of the synchronization threshold from 1.28 with decreasing E_{syn} from 2, corresponds to the boundary $E1$ in Fig. 11.3. The vertical stability boundary for synchrony loss at $E_{\text{syn}} = -0.25$ corresponds to the boundary $E2$ in Fig. 11.3. The origin of this almost vertically rising boundary, starting roughly at $E_{\text{syn}} = -0.25$ is of no mystery if one realizes that this is also the synaptic threshold $\Theta_s = -0.25$. It is not a coincidence that these two values appear equal. Note that the synaptic connection becomes purely inhibitory when $x_i(t)$ exceeds the reversal potential E_{syn}. Therefore, the part of the synchronous solution lying above E_{syn} (mainly, the above-threshold part of the spikes) cannot be robustly stabilized. At the same time, when $x_i(t)$ is below E_{syn}, the synapse is

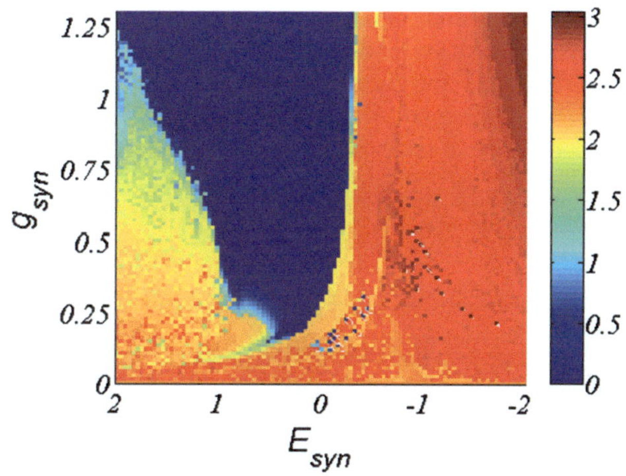

Fig. 11.5 Role of E_{syn} in synchronization of the two-cell network (11.5.) The stability diagram and color coding are similar to those of Fig. 11.3. Decreasing the reversal potential E_{syn} from 2 first dramatically lowers the synchronization threshold. Dropping E_{syn} below -0.25 makes the connection essentially inhibitory such that synchronization cannot be achieved for any value of g_{syn}: note the vertically rising stability boundary around $E_{\text{syn}} = -0.25$

excitatory. As Fig. 11.5 suggests, when E_{syn} is chosen as low as Θ_{s}, the excitatory action of the synapse is non-existent as the synapse is practically off below the synaptic threshold Θ_{s}.

This is the key observation for predicting the slope of the stability boundary $E2$ in the original network (11.1). We return to the network (11.1) and notice that for the overall impact of the excitatory and inhibitory connections to be robustly synchronizing, the overall input to the ith cell, $g_{\text{exc}}(V_{\text{exc}} - x_i)\Gamma(x_j) - g_{\text{inh}}(V_{\text{inh}} - x_i)\Gamma(x_j)$ must remain positive. Rewriting this condition yields $\frac{g_{\text{exc}}V_{\text{exc}}+g_{\text{inh}}V_{\text{inh}}}{g_{\text{exc}}+g_{\text{inh}}} - x_j > 0$, as $\Gamma(x_j) \geq 0$. Notice that the first term plays a role of the reversal potential E_{syn} in the network (11.5). Therefore, according to Fig. 11.5, $\frac{g_{\text{exc}}V_{\text{exc}}+g_{\text{inh}}V_{\text{inh}}}{g_{\text{exc}}+g_{\text{inh}}}$ cannot exceed $E_{\text{syn}} \approx \Theta_{\text{s}} = -0.25$ for synchronization to remain stable. This yields the following condition on the stability boundary $g_{\text{exc}} = \frac{\Theta_{\text{s}}-V_{\text{inh}}}{V_{\text{exc}}-\Theta_{\text{s}}} g_{\text{inh}}$, written in terms of the parameters of the original network (11.1). Plugging in the values of the parameters $V_{\text{inh}} = -2$, $V_{\text{exc}} = 2$, and $\Theta_{\text{s}} = -0.25$, one gets $g_{\text{exc}} = 0.78 g_{\text{inh}}$. This condition predicts the slope of the boundary line $E2$ remarkably well. This argument also carries over to larger networks and supports the scaling law for synchrony loss: $g_{\text{exc}} = \alpha \frac{k_{\text{inh}}}{k_{\text{exc}}} g_{\text{inh}}$, reported in the following subsection.

11.2.4 Larger Networks: The Scaling Laws

The discovered inhibition-induced synchronization phenomenon is also present in larger networks of square-wave bursters. We demonstrate that the structure of the added inhibitory connections is not important and only the number of inhibitory inputs controls the onset of synchronization, independent from all other details of their network topology. In the context of complex dynamical networks, this unexpected result indicates the drastically different roles of network topology in synchronization of linearly [53, 54] and synaptically coupled networks with attractive and repulsive connections.

Towards this goal, we consider a network of n bursting Hindmarsh–Rose neuron models with excitatory and inhibitory connections:

$$\dot{x}_i = ax_i^2 - x_i^3 - y_i - z_i + g_{\text{exc}}(V_{\text{exc}} - x_i)\sum_{j=1}^{n} c_{ij}\Gamma(x_j) + g_{\text{inh}}(V_{\text{inh}} - x_i)\sum_{j=1}^{n} d_{ij}\Gamma(x_j),$$

$$\dot{y}_i = (a+\alpha)x_i^2 - y_i, \quad \dot{z}_i = \mu(bx_i + c - z_i), \; i,j = 1,\ldots,n. \tag{11.6}$$

The connectivity matrices $C = (c_{ij})$ and $D = (d_{ij})$ define the structure of excitatory and inhibitory connections, respectively; both mutual and unidirectional coupling are allowed. It is required that all row-sums of C and D are equal to k_{exc} and k_{inh}, the property that implies a network where each cell has k_{exc} inputs from excitatory neurons and k_{inh} from inhibitory ones. This constraint is chosen to ensure the existence of complete synchrony and to allow the use of the stability conditions to reveal the synchronization mechanism. Other notations and parameters are identical to those of the two-cell system (11.1).

Figure 11.6 demonstrates a synergistic effect in the n-cell network (11.6). It shows that the size of the left desynchronization zone, bounded by the g_{exc} axis and boundary $E1$ (cf. Fig. 11.3), scales down vertically and horizontally by k_{exc} and k_{inh} times, respectively. As a result, the stability boundaries $E1$ for the onset of synchrony are nearly identical for networks of different sizes and topologies, provided that k_{exc} and k_{inh} are uniform for each cell. In support of this claim, we have analyzed a series of different regular and random networks (11.1) with uniform numbers of excitatory (k_{exc}) and inhibitory (k_{inh}) synapses per neuron. For all simulated networks, numerical results are consistent with the scaling law above. Figure 11.6 demonstrates two representative pairs of networks yielding the largest and smallest regions of inhibition-induced synchronization for all possible network topologies (11.1) with the given number of excitatory and inhibitory inputs. Figure 11.7 summarizes the numerical simulations of different networks with different topologies and shows how the synchronization effect of added inhibition scales with the size of the network.

Fig. 11.6 Stability diagrams for network synchronization, similar to that of Fig. 11.3. The color bar indicates the mean voltage difference $\sum_{i=1}^{n-1}\sum_{j>i}^{n} \frac{2}{n(n-1)}(x_i - x_j)$, calculated and averaged over the last three bursting periods. Notice the nearly identical diagrams for pairs of ten-cell irregular and five-cell regular networks with $k_{\text{exc}} = 4$ and $k_{\text{inh}} = 4$ (*two top rows*) and $k_{\text{exc}} = 2$ and $k_{\text{inh}} = 4$ (*two bottom rows*). Excitatory (inhibitory) connections are depicted by arrows (*circles*). Excitatory (inhibitory) neurons in the ten-cell irregular networks [with only outgoing excitatory (inhibitory) connections] are denoted by *light* (*dark*) circles. The height and width of the *left* instability zone, adjacent to the g_{exc}-axis and corresponding to desynchronized square-wave bursting are inversely proportional to k_{exc} and k_{inh}, respectively (also compare with Fig. 11.3)

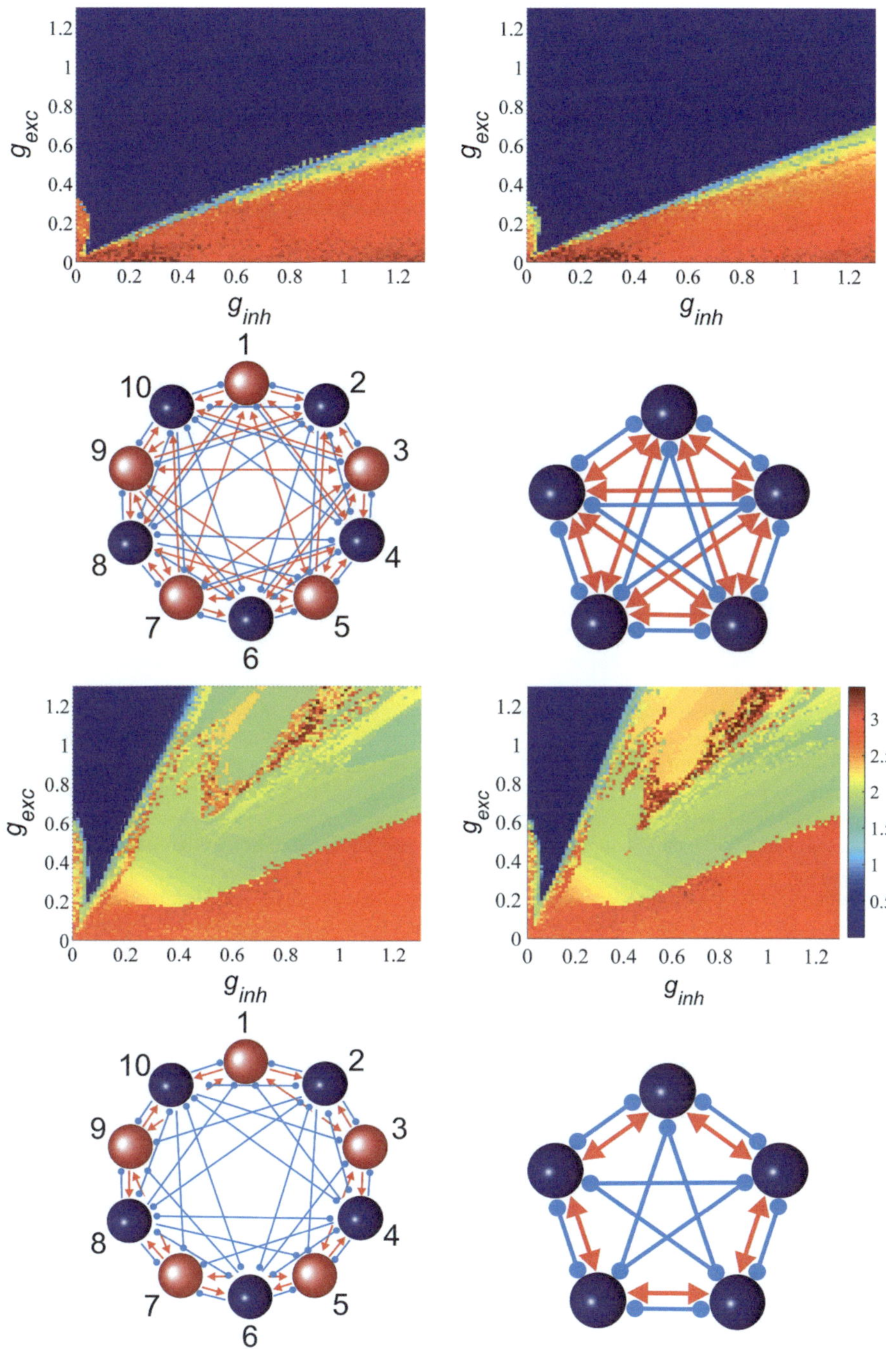

g_{exc}
g_{inh}
1
2
3
4
5
6
7
8
9
10

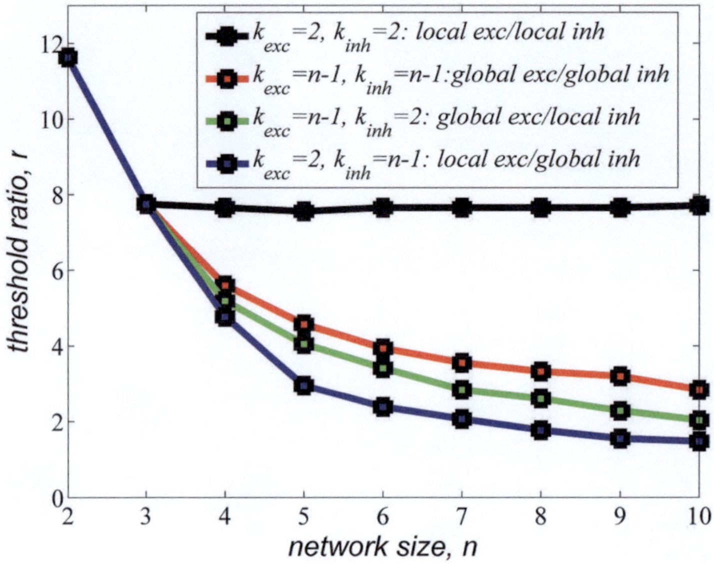

Fig. 11.7 Ratio of the synchronization threshold in an excitatory network without inhibition and the minimum synchronization threshold achieved by adding inhibition, as a function of the network size n, for different values of k_{exc} and k_{inh}. The ratio of the synchronization threshold reduction, induced by added inhibition is as large as 12 for the two-cell network (compare with Fig. 11.3.) The four curves represent four types of network topology: rings of cells with local excitatory and inhibitory connections ($k_{\text{exc}} = 2$ and $k_{\text{inh}} = 2$); all-to-all networks with both global excitatory and inhibitory connections ($k_{\text{exc}} = n - 1$ and $k_{\text{inh}} = n - 1$); networks with global excitatory and local inhibitory connections ($k_{\text{exc}} = n - 1$ and $k_{\text{inh}} = 2$); and rings of cells with local excitatory and all-to-all inhibitory connections ($k_{\text{exc}} = 2$ and $k_{\text{inh}} = n - 1$). Notice that the addition of global inhibition to a locally coupled excitatory network (local excitation/global inhibition) yields the smallest reduction in the synchronization threshold for $n > 3$ (*lowest line*), and therefore has the worst synchronization properties. At the same time, the addition of local inhibition to the same locally coupled excitatory network yields the highest reduction ratio for $n > 3$ (*top line*) and indicates a non-trivial synergistic effect of the combined inhibitory and excitatory topologies. Also observe that global inhibition promotes synchronization more significantly than local inhibition when added to a globally coupled excitatory network, as the global excitation/global inhibition configuration has a higher synchronization threshold reduction ratio (*second line from the top*), compared to that of the global excitation/local inhibition configuration (*third line from the top*)

To show that the scaling laws carry over to larger networks with random coupling matrices, we have simulated a 100-cell random network where each cell receives four excitatory $k_{\text{exc}} = 4$ and four inhibitory $k_{\text{inh}} = 4$ connections (Fig. 11.8). The network consists of 80 excitatory and 20 inhibitory cells such that the excitatory (inhibitory) cells only have excitatory (inhibitory) outgoing connections, thereby abiding by Dale's law. Both excitatory and inhibitory coupling strengths are mismatched by adding $\Delta g_{ij} \cdot q$ to g_{exc} and g_{inh} for each existing connection (i, j). The mismatch parameter Δg_{ij} is expressed as a percentage of g_{exc} and g_{inh} and kept equal to 5%; the values of the parameter q are chosen randomly from the

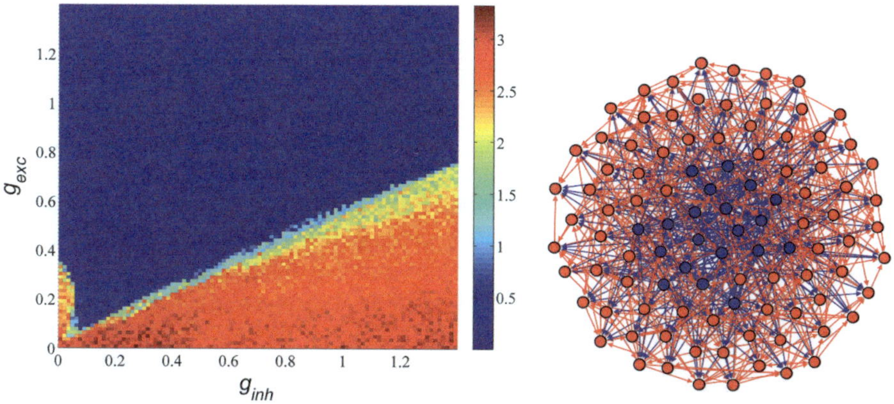

Fig. 11.8 (*Top*) Induced synchronization in a 100-cell randomly generated network with uniform $k_{\text{exc}=4}$ and $k_{\text{inh}=4}$. (*Bottom*) The network has 80 excitatory (*red/light*) and 20 inhibitory (*blue/dark*) cells. The excitatory connections are marked by *red* arrowed lines; the inhibitory coupling is indicated by *blue* arrows. Both excitatory and inhibitory coupling strengths are heterogeneous, with randomly distributed mismatch up to 10%. The *color bar* indicates the mean voltage difference as in Fig. 11.6. The stability diagram is similar to those of the two *left* diagrams in Fig. 11.6, corresponding to the 5- and 10-cell networks with $k_{\text{exc}} = 4$ and $k_{\text{inh}} = 4$. Complete spike synchronization is impossible in this mismatched network; however, an approximate synchronization with small voltage differences (offsets between the spikes) is robustly present. Various shades of *blue* and the non-homogeneous structure of the synchronization stability zone correspond to slight voltage offsets due to the parameter mismatch

interval $(-1,\ 1)$ for each excitatory and inhibitory connection (i,j), yielding a 10% maximum mismatch. The stability diagram supports the scaling law and has a structure similar to the two left diagrams in Fig. 11.6, all corresponding to different network topologies with the uniform number of connections $k_{\text{exc}} = 4$ and $k_{\text{inh}} = 4$.

To target realistic biological networks with non-uniform numbers of excitatory and inhibitory inputs per neuron, we have simulated a 100-cell network, similar to that of Fig. 11.8, but with an *average* number of inputs $k_{\text{exc}} = 4$ and $k_{\text{inh}} = 4$. This heterogeneous network has been generated from the network topology of Fig. 11.8 with $k_{\text{exc}} = 4$, $k_{\text{inh}} = 4$, and non-mismatched g_{exc} and g_{inh} by randomly choosing a pair of cells and changing their in-degrees of the excitatory and inhibitory inputs by subtracting one incoming connection of each type from one cell and adding these connections to the other cell. As a result, one half of the cells have $k_{\text{exc}} = 3$ and $k_{\text{inh}} = 3$, while the other half have $k_{\text{exc}} = 5$ and $k_{\text{inh}} = 5$, yielding the average $k_{\text{exc}} = 4$ and $k_{\text{inh}} = 4$. This effective mismatch between the overall strength of the incoming connections to each neuron is larger than the 10% maximum mismatch used in the previous example (Fig. 11.8); however, the stability diagram for approximate synchronization is quite similar to Fig. 11.8, except for the appearance of a more irregular structure of the synchronization stability zone due to the increased coupling mismatch. Increasing the heterogeneity mismatch between the connections even further shall eventually make approximate spike synchronization impossible; however, we expect induced burst synchronization to persist.

To explain the scaling law, we shall return to the transversal variational Eqs. (11.3) written for $n-1$ difference variables $\xi_{ij} = x_j - x_i$, $\eta_{ij} = y_j - y_i$, $\zeta_{ij} = z_j - z_i$, $i, j = 1, \ldots, n$. The equations for the purely excitatory networks were given in [17] where an analog of the Master Stability Function [52] for synaptically coupled networks (11.1) was used to analyze the stability of the most unstable transverse mode. Unfortunately, the Master Stability Function cannot be applied to mixed excitatory–inhibitory networks in general as it requires simultaneous diagonalization of both the excitatory (C) and inhibitory (D) connectivity matrices. This is impossible in general unless the two matrices commute [60]. In the latter case, the stability equation for the most unstable transverse synchronous mode is the Eq. (11.3) with a new stability function $\Omega^{\text{new}}(x) = (k_{\text{exc}} g_{\text{exc}} + k_{\text{inh}} g_{\text{inh}})\,\Gamma(x) - g_{\text{exc}}(V_{\text{exc}} - x)\Gamma_x(x)(k_{\text{exc}} + \gamma_2^{\text{exc}}) - g_{\text{inh}}(V_{\text{inh}} - x)\Gamma_x(x)(k_{\text{inh}} + \gamma_2^{\text{inh}})$, where γ_2^{exc} and γ_2^{inh} are the second largest eigenvalues of the (commuting) Laplacian connectivity matrices for the excitatory and inhibitory networks, $\mathrm{C}^{\mathrm{L}} = \mathrm{C} - k_{\text{exc}}\mathbf{I}$ and $\mathrm{D}^{\mathrm{L}} = \mathrm{D} - k_{\text{inh}}\mathbf{I}$, respectively. The first term in $\Omega^{\text{new}}(x)$ accounts for the number and strength of excitatory and inhibitory inputs. The last two terms, containing the partial derivative Γ_x and the networks structure via γ_2^{exc} and γ_2^{inh}, only matter for the stability/instability of synchronization in the region of $x(t)$, close to the synaptic threshold Θ_{s}, similar to the two-cell network case. The shift of the nullcline $h(x)$ and switching from square-wave to synchronous plateau bursting are governed by $k_{\text{exc}} g_{\text{exc}}$ and $k_{\text{inh}} g_{\text{inh}}$ via the self-coupled system (11.2). As a result, the spikes of the synchronous bursting solution leave the bell-shaped zone (similar to Fig. 11.4c) such that the contribution of the last two terms in $\Omega^{\text{new}}(x)$ becomes insignificant for synchronization. This yields the scaling law when the minimum strength of added inhibition g^*_{inh}, sufficient to induce plateau bursting synchrony is inversely proportional to k_{inh}, regardless of the network size and structure (compare, for example, $g^*_{\text{inh}} \approx 0.14$ in the two-cell network of Fig. 11.3 and $g^*_{\text{inh}} \approx 0.035 = 0.14/4$ in the networks of Fig. 11.6 with $k_{\text{inh}} = 4$, all calculated at the level $g_{\text{exc}} = 0.2$). Notice that the 5-cell networks of Fig. 11.6 correspond to the commuting excitatory and inhibitory connectivity matrices: global excitation/global inhibition and local excitation/global inhibition. In the case where the connectivity matrices do not commute (the 10-cell networks of Fig. 11.6 and the 100-node network of Fig. 11.8), the eigenvalues of the connectivity matrices cannot be used and the stability function $\Omega^{\text{new}}(x)$ cannot be derived. A modification of the Connection Graph method [61] that uses graph theoretical reasoning instead of the spectrum of the connectivity matrices can be used to write down a set of similar stability functions. However, the stability argument is essentially the same, the induced synchronization is governed by the transition to plateau bursting that is in turn controlled by the self-coupled system. Consequently, the same scaling law for the inverse dependence of the induced synchronization threshold on g_{exc} and k_{inh} also holds for realistic non-commuting coupling configurations. Our results also indicate that the loss of stable synchrony via the right (inclined) boundary (similar to boundary $E2$ in Fig. 11.3) is governed by a simple condition $g_{\text{exc}} = \alpha \frac{k_{\text{inh}}}{k_{\text{exc}}} g_{\text{inh}}$, where α is a scaling factor, uniform for different topologies with the same ratio $k_{\text{inh}}/k_{\text{exc}}$. As in the two-cell network yielding the slope $g_{\text{exc}} = 0.78\, g_{\text{inh}}$, this condition is determined by the shift of the nullcline $h(x)$ such that the upper knee of $h(x)$ moves close to the synaptic threshold Θ_s and falls into the instability zone (as in Fig. 11.4c).

11.3 Coupled Sherman Models

In this section, we demonstrate that a similar synergistic effect is observed in a network of physiologically based Hodgkin–Huxley-type models such as the pancreatic β-cell Sherman model [30], exhibiting square-wave and plateau-type bursting. Similarly to system (11.1), we consider the simplest network of two coupled Sherman models with both excitatory and inhibitory connections:

$$\begin{aligned}
\tau \frac{dV_i}{dt} &= F(V_i, n_i, S_i) + g_{\text{exc}}(E_{\text{exc}} - V_i)\Gamma(V_j) + g_{\text{inh}}(E_{\text{inh}} - V_i)\Gamma(V_j),\\
\tau \frac{dn_i}{dt} &= G(V_i, n_i) \equiv n_\infty(V_i) - n_i,\\
\tau_s \frac{dS_i}{dt} &= H(V_i, S_i) \equiv S_\infty(V_i) - S_i, \quad i, j = 1, 2.
\end{aligned} \tag{11.7}$$

Here, V_i represents the membrane potential of the ith cell. Function $F(V_i, n_i, S_i) = -[I_{Ca}(V_i) + I_K(V_i, n_i) + I_S(V_i, S_i)]$ defines three intrinsic currents: fast calcium, I_{Ca}, potassium, I_K, and slow potassium, I_S, currents:

$$I_{Ca} = \bar{g}_{Ca}\, m_\infty(V_i)\,(V_i - E_{Ca}), \quad I_K = \bar{g}_K\, n_i\,(V_i - E_K), \quad I_S = \bar{g}_S\, S_1\,(V_i - E_K).$$

The gating variables for n_i and S_i are the opening probabilities of the fast and slow potassium currents, respectively, and

$$\begin{aligned}
m_\infty(V_i) &= [1 + \exp((-20 - V_i)/12)]^{-1}\\
n_\infty(V_i) &= [1 + \exp((-16 - V_i)/5.6)]^{-1}\\
S_\infty(V_i) &= [1 + \exp((-35.245 - V_i)/10)]^{-1}.
\end{aligned}$$

Other intrinsic parameters are $\tau = 20$, $\tau_S = 10{,}000$, $\bar{g}_{Ca} = 3.6$, $E_{Ca} = 25\,\text{mV}$, $\bar{g}_K = 10$, $E_K = -75\,\text{mV}$, $\bar{g}_S = 4$. The reversal potentials $E_{\text{exc}} = 10\,\text{mV}$ and $E_{\text{inh}} = -75\,\text{mV}$ make the synapses excitatory and inhibitory, respectively, as $E_{\text{exc}} > V_i$ ($E_{\text{inh}} < V_i$) for all values of $V_i(t)$. The synaptic coupling function is modeled by the sigmoidal function $\Gamma(V_j) = 1/[1+\exp\{-10(V_j - \Theta_s)\}]$. The synaptic threshold is chosen as follows: $\Theta_{\text{syn}} = -40\,\text{mV}$. Other notations are similar to those of the network of Hindmarsh–Rose neurons (11.1).

The presence of the large parameter $\tau_S = 10{,}000$ on the left-hand side of the S-equation makes the system (11.7) slow-fast such that the (V_i, n_i)-equations represent the 2-D fast "spiking" subsystem for the ith cell; the S_i-equation corresponds to the slow 1-D "bursting" system. Therefore, we use the standard decomposition into

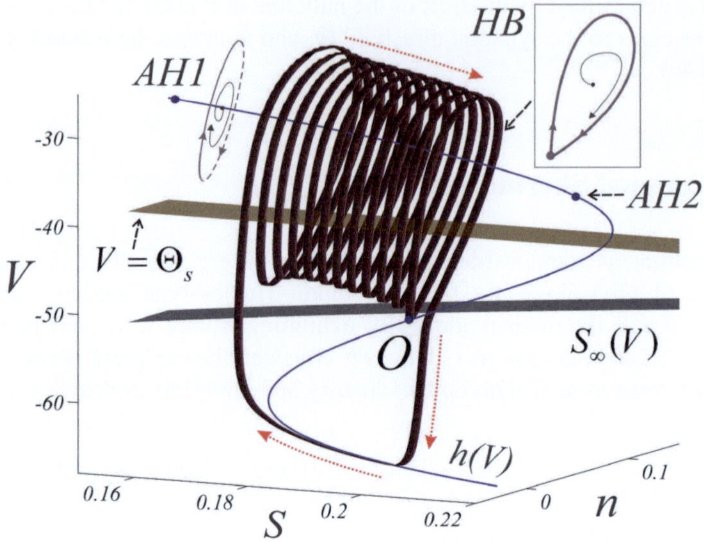

Fig. 11.9 Square-wave burster of the uncoupled Sherman models (11.7). The fast system displays a supercritical Andronov–Hopf bifurcation at $S = S_{AH1}$ and a homoclinic bifurcation (loop) at $S = S_{\mathrm{HB}}$. The spiking manifold is composed of limit cycles in the fast system and terminates at the homoclinic bifurcation HB. The intersection of the fast ($h(V)$) and slow ($S_\infty(V)$) nullclines indicates a unique saddle point O of the full system. The *red dotted curve* shows the route for bursting in the full system. The plane $V = \Theta_{\mathrm{s}}$ displays the synaptic threshold above which the presynaptic cell can influence the postsynaptic one

fast and slow subsystems; the types of bursting that can exist in the uncoupled cell systems (11.1) with $g_{\mathrm{exc}} = 0$ and $g_{\mathrm{inh}} = 0$ are defined by the S-parameter sequences of phase portraits of the 2-D fast system. This analysis has been performed for a similar pancreatic cell [49] and revealed different types of bursting such as square-wave, plateau, and pseudo-plateau bursting [50]. Figure 11.9 illustrates the standard sequence of phase portraits in the uncoupled systems (11.7) with $g_{\mathrm{exc}} = 0$ and $g_{\mathrm{inh}} = 0$, giving rise to square-wave bursting.

The equilibrium point on the upper branch of the nullcline $h(V)$ in the 2-D fast subsystem undergoes a supercritical Andronov–Hopf bifurcation for $S = S_{AH1}$, softly giving rise to a stable limit cycle that encircles the unstable point and forms the spiking manifold for $S_{AH1} < S < S_{\mathrm{HB}}$. Its upper edge is defined by a homoclinic bifurcation at $S = S_{\mathrm{HB}}$. Here, the stable limit merges into a stable homoclinic loop and disappears. For the given location of the slow nullcline $S_\infty(V)$, the trajectories jump down to the lower branch of the fast nullcline, creating square-wave (fold/homoclinic) bursting. It is worth noticing the similarities of the phase portraits' sequences and bifurcations leading to square-wave bursting in the networks (11.1) and (11.7).

Similarly to the network of Hindmarsh–Rose neurons (11.1), the network (11.7) has an invariant manifold $D = \{V_1 = V_2 = V(t),\ n_1 = n_2 = n(t),\ S_1 =$

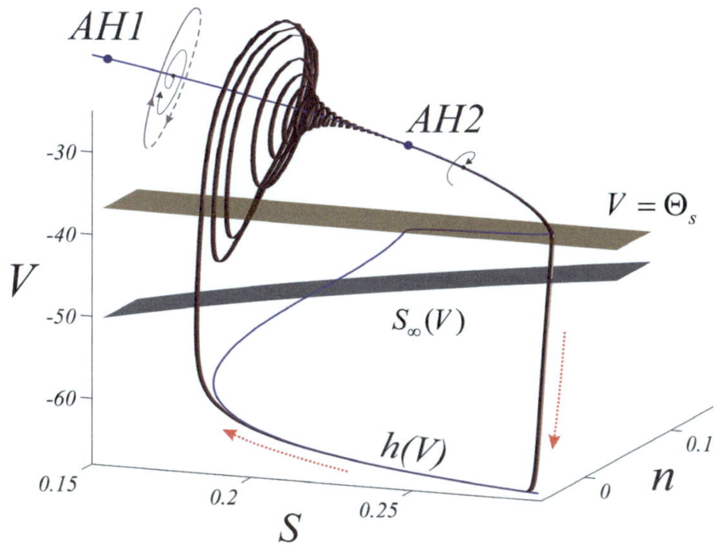

Fig. 11.10 Plateau-type burster of the self-coupled model (11.8), governing the synchronous network dynamics. Note the disappearance of a homoclinic bifurcation in the fast system due to the synaptic coupling. The stable limit cycle of the fast system disappears through a reverse Andronov-Hopf bifurcation at $S = S_{AH2}$, ending the spiking manifold. The *red dotted curve* shows the route for plateau-type bursting. The non-smooth part of the fast nullcline at $V = \Theta_s$ is due to the synaptic coupling, turning on when the trajectory jumps up to the spiking manifold and crosses the threshold Θ_s. The coupling strengths $g_{\rm exc} = 0.14$ and $g_{\rm inh} = 0.06$ correspond to the point *b* in the 2-D diagram of Fig. 11.11

$S_2 = S(t)\}$, that defines complete synchronization between the cells. Synchronous dynamics on the manifold *D* is defined by the self-coupled system:

$$\begin{aligned}
\tau \frac{dV}{dt} &= F(V,n,S) + g_{\rm exc}(E_{\rm exc} - V)\Gamma(V) + g_{\rm inh}(E_{\rm inh} - V)\Gamma(V) \\
\tau \frac{dn}{dt} &= G(V,n) \\
\tau_s \frac{dS}{dt} &= H(V,S),
\end{aligned} \tag{11.8}$$

where changing coupling strengths $g_{\rm exc}$ and $g_{\rm inh}$ can induce transitions from square-wave bursting into plateau-type bursting, depicted in Figs. 11.9 and 11.10. This also happens through the disappearance of the homoclinic bifurcation in the self-coupled system (11.2) due to increased coupling strengths. While excitation alone is able to transform square-wave into plateau-type bursting at some high values of $g_{\rm exc}$, inhibition does so more effectively and its addition lowers the combined coupling strength $g_{\rm exc} + g_{\rm inh}$.

Figure 11.11a proves that the synergistic effect observed in the network of Hindmarsh–Rose models is generic and indicates that the addition of inhibition to an

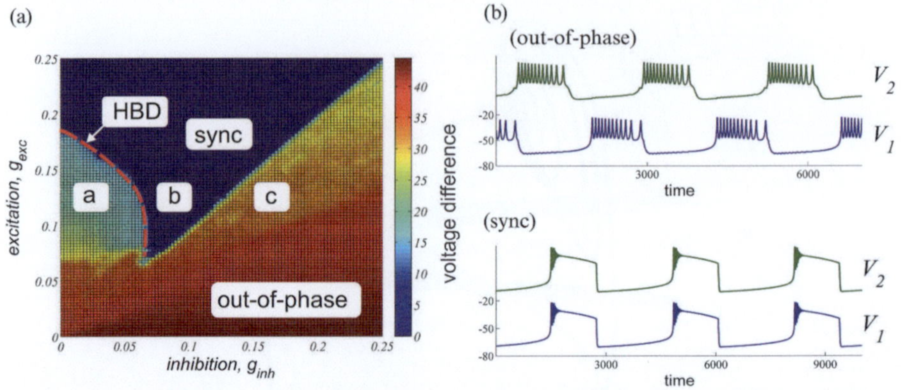

Fig. 11.11 (**a**) The stability diagram for synchronization in the two-cell network (11.7). The color bar indicates the maximum voltage difference $V_1 - V_2$, calculated over the last two bursts in the established rhythm. *Blue* (*dark*) zone corresponds the zero voltage difference and shows the synchronization region. Note the unexpected effect when an increase of the inhibitory coupling from 0 to 0.07 significantly lowers the synchronization threshold from about 0.18 to 0.07. Notice that the inhibition desynchronizes the cells in the absence of excitation ($g_{exc} = 0$). The *red dashed curve* indicates the disappearance of the homoclinic bifurcation (HBD) in the 2-D fast subsystem; it corresponds to the transition from square-wave to plateau bursting and practically coincides with the stability boundary between desynchronized and synchronized bursting. (**b**) *Top*: Typical out-of-phase voltage traces, corresponding to the *red* ("out-of-phase") zone. *Bottom*: synchronization of plateau bursting in the *blue* ("sync") parameter region

excitatory network induces synchronization in a fairly wide range of the inhibitory strength g_{inh}. Note that increasing g_{inh} first lowers the synchronization threshold and weaker excitation synchronizes the cells (e.g., from $g_{exc} = 0.18$ in the absence of inhibition to $g_{exc} = 0.07$ for $g_{inh} = 0.07$). At the same time, the inhibition cannot induce robust synchronization by itself (see the x-axis in Fig. 11.11a, which corresponds to the desynchronizing role of inhibition in the absence of excitation).

11.4 Conclusions

Different types of bursting have significantly different synchronization properties. While square-wave bursters are known for their high resistance to spike synchronization, elliptic and plateau-like bursters are much easier to synchronize and require a weaker coupling strength. Typically, fast non-delayed excitation promotes synchronization of bursters while fast non-delayed inhibition desynchronizes them. Although, counterexamples of synchronizing fast non-delayed inhibition in the weak coupling case have been reported [38, 39].

In this chapter, we have discussed the synergistic effect of combined attractive excitation and repulsive inhibition in promoting bursting synchrony. Remarkably, the addition of the inhibitory coupling lowers the synchronization threshold much more significantly than strengthening the present excitatory connections. The effect is generic and observed in several other Hodgkin-Huxley-type models of square-wave bursting cells. The effect is also independent from the choice of the synaptic interaction model, ranging from the instantaneous pulsatile coupling to a fast dynamical synapse [23]. Our preliminary results show that inhibition also promotes burst synchrony in realistic networks with a highly heterogeneous structure of connections, where spike or approximate synchrony is impossible. Our study has potential implications for understanding the emergence of abnormal synchrony in epileptic brain networks. An epileptic patient is normally (i.e., except for during a seizure) in a desynchronized state which might correspond to the instability region to the left of the $E1$-border in Fig. 11.3. Our results suggest that promoting presumably desynchronizing inhibition in an attempt to prevent the patient's seizures can have a counterproductive effect and induce abnormal synchronous firing in the excitatory–inhibitory brain network. Brain networks have been also shown to evolve their functional topology during epileptic seizures [19]. In light of this, our results on the role of network connectivity, identifying network topologies with the highest and lowest resilience of abnormal synchronized bursting can give insights into how seizures self-terminate and into how to control epileptic networks. Outside of Neuroscience, negative pairwise repulsive interactions were previously shown to have a positive effect on synchronization in *linearly* coupled networks, where negative interactions by themselves tend to destabilize synchronous states, but can compensate for other instabilities [54]. However, this intriguing phenomenon, where the structure of negative connections heavily affects the synchronization, is conceptually different from the one reported in this study. Apart from synchronization, a counterintuitive role of inhibition was reported in [62], demonstrating that the addition of inhibitory nodes to an excitatory network of 1-D discrete-time oscillators causes self-sustaining dynamics.

Acknowledgements This work was supported by the National Science Foundation (USA) under Grant No. DMS1616345 (to I.B. and R.R.), the US Army Research Office Network Science Division under Grant No. W911NF-15-1-0267 (to I.B.) and GSU Brains and Behavior program. R.R. acknowledges support as a GSU Brains and Behavior Fellow.

References

1. Rinzel, J.: Lecture Notes in Biomathematics, vol. 71, pp. 251–291. Springer, Berlin (1987)
2. Terman, D.: SIAM J. Appl. Math. **51**, 1418 (1991)
3. Wang, X.-J.: Physica (Amsterdam) **62D**, 263 (1993)
4. Bertram, R., Butte, M.J., Kiemel, T., Sherman, A.: Bull. Math. Biol. **57**, 413 (1995)
5. Belykh, V.N., Belykh, I., Colding-Joergensen, M., Mosekilde, E.: Europ. Phys. J. E **3**, 205 (2000)
6. Izhikevich, E.M.: Int. J. Bifurcation Chaos **10**, 1171 (2000)
7. Golubitsky, M., Josic, K., Kaper, T.J.: An unfolding theory approach to bursting in fast-slow systems In: Broer, H.W., Krauskopf, B., Vegter, G. (eds.) Global Analysis of Dynamical Systems: Festschrift dedicated to Floris Takens on the occasion of his 60th birthday, pp. 277–308. Institute of Physics Publications (2001)

8. Guckenheimer, J., Oliva, R.A.: SIAM J. Appl. Dyn. Syst. **1**, 105 (2002)
9. Shilnikov, A., Cymbalyuk, G.: Phys. Rev. Lett. **94**, 048101 (2005)
10. Belykh, V.N., Belykh, I., Mosekilde, E.: Int. J. Bifurcation Chaos **15**, 3567 (2005)
11. Rabinovich, M.I., Varona, P., Selverston, A.I., Abarbanel, H.D.I.: Rev. Modern Phys. **78**, 1213 (2006)
12. Frohlich, F., Bazhenov, M.: Phys. Rev. E **74**, 031922 (2006)
13. Shilnikov, A.L., Kolomiets, M.L.: Int. J. Bifurcation Chaos **18**(8), 2141 (2008)
14. Abarbanel, H.D.I., Selverston, A.I., Huerta, R., Bazhenov, M.V., Sushchik, M.M., Rubchinskii, L.L., Rabinovich, M.I.: Usp. Fiziol. Nauk **166**, 363 (1996)
15. Izhikevich, E.M.: SIAM Rev. **43**, 315 (2001)
16. Dhamala, M., Jirsa, V.K., Ding, M.: Phys. Rev. Lett. **92**, 028101 (2004)
17. Belykh, I., de Lange, E., Hasler, M.: Phys. Rev. Lett. **94**, 188101 (2005)
18. Netoff, T.I., Schiff, S.J.: J. Neurosci. **22**(16), 7297 (2002); Schindler, K., Elger, C.E., Lehnertz, K.: Clin. Neurophysiol. **118**, 1955 (2007)
19. Lehnertz, K., Ansmanna, G., Bialonski, S., Dickten, H., Geier, C., Porz, S.: Physica (Amsterdam) **267D**, 7 (2014)
20. Wang, X.-J., Rinzel, J.: Neural Comput. **4**, 84 (1992)
21. van Vreeswijk, C., Abbott, L.F., Bard Ermentrout, G.: J. Comput. Neurosci. **1**, 313 (1994)
22. Kopell, N., Ermentrout, G.B.: In: Fiedler, B. (ed.) Handbook of Dynamical Systems, vol. 2, pp. 3–54. Elsevier, Amsterdam (2002)
23. Golomb, D., Rinzel, J.: Phys. Rev. E **48**, 4810 (1993)
24. Terman, D., Kopell, N., Bose, A.: Physica (Amsterdam) **117D**, 241 (1998)
25. Ernst, U., Pawelzik, K., Geisel, T.: Phys. Rev. Lett. **74**, 1570 (1995)
26. Sadeghi, S., Valizadeh, A.: J. Comput. Neurosci. **36**, 55 (2014)
27. Matias, F.S., Carelli, P.V., Mirasso, C.R., Copelli, M.: Phys. Rev. E **84**, 021922 (2011)
28. Elson, R.C., Selverston, A.I., Abarbanel, H.D.I., Rabinovich, M.I.: J. Neurophysiology **88**, 1166 (2002)
29. Somers, D., Kopell, N.: Biol. Cybern. **68**, 393 (1993)
30. Sherman, A.: Bull. Math. Biol. **56**, 811 (1994)
31. Bazhenov, M., Huerta, R., Rabinovich, M.I., Sejnowski, T.: Physica (Amsterdam) **116D**, 392 (1998)
32. Canavier, C.C., Baxter, D.A., Clark, J.W., Byrne, J.H.: Biol. Cybern. **80**, 87 (1999)
33. Rubin, J., Terman, D.: Neural Comput. **12**, 597 (2000)
34. Rubin, J., Terman, D.: SIAM J. Appl. Dyn. Sys. **1**, 146 (2002)
35. Kopell, N., Ermentrout, G.B.: Proc. Natl. Acad. Sci. USA **101**, 15482 (2004)
36. Belykh, I., Shilnikov, A.: Phys. Rev. Lett. **101**, 078102 (2008)
37. Shilnikov, A., Gordon, R., Belykh, I.: Chaos **18**, 037120 (2008)
38. Jalil, S., Belykh, I., Shilnikov, A.: Phys. Rev. E **81**, 045201(R) (2010)
39. Jalil, S., Belykh, I., Shilnikov, A.: Phys. Rev. E **85**, 036214 (2012)
40. Wojcik, J., Clewley, R., Shilnikov, A.: Phys. Rev. E **83**, 056209-6 (2011)
41. Rabinovich, M.I., Torres, J.J., Huerta, R., Abarbanel, H.D.I.: Phys. Rev. E **60**, R1130 (1999)
42. Nowotny, T., Huerta, R., Rabinovich, M.I.: Chaos **18**, 037119 (2008)
43. Nowotny, T., Rabinovich, M.I.: Phys. Rev. Lett. **98**, 128106 (2007)
44. Belykh, I., Hasler, M.: Chaos **21**, 016106 (2011)
45. Vicente, R., Gollo, L.L., Mirasso, C.R., Fischer, I., Pipa, G.: Proc. Natl. Acad. Sci. USA **105**, 17157 (2008)
46. Gollo, L.L., Mirasso, C., Sporns, O., Breakspear, M.: PLoS Comput. Biol. **10**(4), e1003548 (2014)
47. Belykh, I., Reimbayev, R., Zhao, K.: Phys. Rev. E **91**, 062919 (2015)
48. Reimbayev, R., Belykh, I.: Int. J. Bifurcation Chaos **24**, 1440013 (2014)
49. Tsaneva-Atanasova, K., Osinga, H., Riel, T., Sherman, A.: J. Theor. Biol. **264**, 1133 (2010)
50. Stern, J., Osinga, H., LeBeau, A., Sherman, A.: Bull. Math. Biol. **70**, 68 (2008)
51. Krishnan, G.P., Bazhenov, M.: J. Neurosci. **24**, 8870 (2011)

52. Pecora, L.M., Carroll, T.L.: Phys. Rev. Lett. **80**, 2109 (1998); Pecora, L.M.: Phys. Rev. E **58**, 347 (1998)
53. Leyva, I., Sendina-Nadal, I., Almendral, J.A., Sanjuań, M.A.F.: Phys. Rev. E **74**, 056112 (2006)
54. Nishikawa, T., Motter, A.E.: Proc. Natl. Acad. Sci. USA **107**, 10342 (2010)
55. Hindmarsh, J.L., Rose, M.: Proc. R. Soc. Lond. Ser. B **221**, 87 (1984)
56. Ren, M., Yoshimura, Y., Takada, N., Horibe, S., Komatsu, Y.: Science **316**, 758 (2007)
57. Connors, B.W., Cruikshank, S.J.: Nat. Neurosci. **10**, 808 (2007)
58. Dale, H.H.: Proc. R. Soc. Med. **28**, 319 (1934)
59. CONTENT is availabe on its main developer's webpage (Yu.A. Kuznetsov): http://www.staff.science.uu.nl/~kouzn101/CONTENT
60. Sorrentino, F.: New J. Phys. **14**, 033035 (2012); Irving, D., Sorrentino, F.: Phys. Rev. E **86**, 056102 (2012)
61. Belykh, V.N., Belykh, I.V., Hasler, M.: Physica (Amsterdam) **195D**, 159 (2004)
62. Larremore, D.B., Shew, W.L., Ott, E., Sorrentino, F., Restrepo, J.G.: Phys. Rev. Lett. **112**, 138103 (2014)

Part III
Brain

Chapter 12
The Variational Principles of Cognition

Karl Friston

12.1 Introduction

This chapter offers a teleological explanation of dynamics; namely, what do they do in the context of adaptive behavior, perception, and cognition. The premise here is that the brain is trying to optimize something (specifically variational free-energy) and uses a generalized gradient descent to perform this optimization. In other words, one can understand neuronal dynamics as optimizing a quantity through the method of steepest ascent—described with a (complicated) set of ordinary differential equations. It is these equations that give rise to the itinerant (wandering) dynamics that have been the focus of several groundbreaking contributions from Mikhail Rabinovich [2, 7, 48–50, 58].

In what follows, we will see how the optimization of free-energy leads naturally to optimal action and perception. Crucially, the nature of this optimization rests on the brain's internal or generative model of the world that it navigates. This model includes prior beliefs about the causal structure and dynamics in this world, which constrain both perception and action. This adds a second level of dynamics that reflect our prior expectations about the trajectories of states and their attractors in our environment. It is at this point we call on itinerant dynamics and winnerless competition [2] to furnish prior beliefs about sensorimotor trajectories. In particular, we will look at action-observation in the context of handwriting and how it rests on stable heteroclinic channels [49].

This chapter comprises two parts. In the first, we provide a heuristic overview of the free-energy principle, motivating it from basic principles. We will consider the underlying imperative that applies to all biological systems; namely, to conserve

K. Friston (✉)
The Wellcome Trust Centre for Neuroimaging, University College London, Queen Square, London WC1N 3BG, UK
e-mail: k.friston@ucl.ac.uk

I.S. Aranson et al. (eds.), *Advances in Dynamics, Patterns, Cognition*,
Nonlinear Systems and Complexity 20, DOI 10.1007/978-3-319-53673-6_12

themselves by minimizing surprise; and how this calls upon the minimization of free-energy. We then unpack the free-energy principle in terms of its implications for action and perception. This leads to active inference and perceptual inference, of the sort considered by the Bayesian brain hypothesis. We illustrate the key aspects of this treatment with a few selected examples and conclude by thinking about the timescales over which the dynamics of free-energy minimization may be manifest. The second part of this chapter presents a particular example in greater detail. This example considers handwriting in terms of itinerant expectations about sequences of movements. Not only does it provide a plausible account of sensorimotor execution but touches upon the cognitive neuroscience of action-observation and how we represent ourselves and others.

12.2 The Free-Energy Principle

In recent years, there has been growing interest in free-energy formulations of brain function [15, 22], not just from the neuroscience community, where has caused some puzzlement [56] but from fields as far apart as psychotherapy [10] and social politics [30]. The free-energy principle has been described as a unified brain theory [36] and may have broader implications for how we interact with our environment [11, 12, 34, 35]. This section describes the origin of the free-energy formulation, its underlying premises, and the implications for how we represent and interact with the world. Table 12.1 (see below) provides a glossary of the quantities that we will be dealing with.

The free-energy principle is a simple postulate that has complicated ramifications. It says that all agents or biological systems (like us) must minimize free-energy. This postulate is as simple and fundamental as Hamilton's law of Least Action and the celebrated H-theorems in statistical physics [42]. The principle was originally formulated as a computational account of perception [22] that borrows heavily from statistical physics and machine learning [20, 33, 43]. However, its explanatory scope includes action and behavior [25] and may be linked, at a fundamental level, to our very existence [24]: In brief, the free-energy principle takes well-known statistical ideas and applies them to problems in population (ensemble) dynamics and self-organization [3, 31, 37, 46]. In applying these ideas, many aspects of our brains, how we perceive and the way we act become understandable as necessary and self-evident (possibly self evidencing) attributes of biological systems [24, 35]. To see this consider the following problem:

How, in a changing and unpredictable world, do biological agents resist a natural tendency to disorder and thermodynamic equilibrium? All the physics that we know, such as the fluctuation theorem (which generalizes the second law of thermodynamics; [18]), suggests that random fluctuations in our environment will ultimately change our physical states to the point we cease to exist (i.e., we should gently decompose, dissipate, or evaporate). And yet, biological systems seem to violate these laws, maintaining precise physiological states for long periods of time [6].

Table 12.1 Generic variables and quantities in the free-energy formation of active inference, under the Laplace assumption (i.e., generalized predictive coding)

Variable	Description
$m \in \mathcal{M}$	**Generative model or agent:** In the free-energy formulation, each agent or system is taken to be a model of the environment in which it is immersed. This model is used to predict sensory signals
$a \subset \vartheta$	**Action:** These variables are states of the world that correspond to the movement or configuration of an agent (i.e., its effectors)
$\tilde{s}(t) = s \oplus s' \oplus s'' \oplus \ldots \in S$	**Sensory signals:** These generalized sensory signals or samples comprise the sensory states, their velocity, acceleration, and temporal derivatives to high order. In other words, they correspond to the trajectory of an agent's sensations
$\mathcal{L}(\tilde{s}\|m) = -\ln p(\tilde{s}\|m)$	**Surprise:** This is a scalar function of sensory samples and reports the improbability of sampling some signals, under a generative model of how those signals were caused. It is sometimes called (sensory) surprisal or self-information. In statistics it is known as the negative log-evidence for the model
$H(S\|m) \propto \int dt \mathcal{L}(\tilde{s}(t)\|m)$	**Entropy:** Sensory entropy is, under ergodic assumptions, proportional to the long-term time average of surprise
$\mathcal{G}(\tilde{s}, \vartheta) = -\ln p(\tilde{s}, \vartheta\|m)$	**Gibbs energy:** This is the negative log of the density specified by the generative model; namely, surprise about the joint occurrence of sensory samples and their causes
$\mathcal{F}(\tilde{s}, \tilde{\mu}) = \mathcal{G}(\tilde{s}, \tilde{\mu}) + \frac{1}{2}\ln\|\mathcal{G}_{\tilde{\mu}\tilde{\mu}}\| \geq \mathcal{L}(\tilde{s}\|m)$	**Free-energy:** This is a scalar function of sensory samples and a recognition density, which upper bounds surprise. It is called free-energy because it is the expected Gibbs energy minus the entropy of the variational density. Under a Gaussian (Laplace) assumption about the form of the variational density, free-energy reduces to the simple function of Gibbs energy shown
$\mathcal{S}(\tilde{s}, \tilde{\mu}) = \int dt \mathcal{F}(\tilde{s}, \tilde{\mu}) \geq H(S\|m)$	**Free-action:** This is a scalar functional of sensory samples and a variational density, which upper bounds the entropy of sensory signals. It is the time or path integral of free-energy
$q(\vartheta) = \mathcal{N}(\tilde{\mu}, C)$ $\tilde{\mu} = \mu \oplus \mu' \oplus \mu'' \oplus \ldots$ $C = \mathcal{G}_{\tilde{\mu}\tilde{\mu}}^{-1}$	**Variational density**: This is also known as a variational ensemble or recognition density and becomes (approximates) the conditional density over hidden causes of sensory samples, when free-energy is minimized. Under the Laplace assumption, it is specified by its conditional expectation and covariance

(continued)

Table 12.1 (continued)

Variable	Description
$\boldsymbol{\vartheta} = \{\mathbf{u}, \boldsymbol{\varphi}, a\}$ $\vartheta = \{u, \varphi\}$ $u = \{x, v\}$ $\varphi = \{\theta, \gamma\}$	**True (bold) and hidden (italics) causes:** These quantities cause sensory signals. The true quantities exist in the environment and the hidden homologues are those assumed by the generative model of that environment. Both are partitioned into time-dependent variables and time-invariant parameters
$\theta \subset \varphi \subset \vartheta$	**Hidden parameters:** These are the parameters of the mappings (e.g., equations of motion) that constitute the deterministic part of a generative model
$\gamma \subset \varphi \subset \vartheta$	**Log-precisions:** These parameters control the precision (inverse variance) of fluctuations that constitute the random part of a generative model
$x(t) = x^{(1)} \oplus x^{(2)} \oplus x^{(3)} \ldots \subset u \subset \vartheta$	**Hidden states:** These hidden variables encode the hierarchical states in a generative model of dynamics in the world
$v(t) = v^{(1)} \oplus v^{(2)} \oplus v^{(3)} \ldots \subset u \subset \vartheta$	**Hidden causes:** These hidden variables link different levels of a hierarchical generative model
$g(x^{(i)}, v^{(i)}, \theta)$ $f(x^{(i)}, v^{(i)}, \theta)$	**Deterministic mappings:** These are equations at the i-th level of a hierarchical generative model that map from states at one level to another and map hidden states to their motion within each level. They specify the deterministic part of a generative model
$\omega^{(i,v)}$ $\omega^{(i,x)}$	**Random fluctuations:** These are random fluctuations on hidden causes and the motion of hidden states. Gaussian assumptions about these fluctuations furnish the probabilistic part of a generative model
$\tilde{\Pi}^{(i,v)} = R^{(i,v)} \otimes \Pi(\gamma^{(i,v)})$ $\tilde{\Pi}^{(i,x)} = R^{(i,x)} \otimes \Pi(\gamma^{(i,x)})$	**Precision matrices:** These are the inverse covariances among (generalized) random fluctuations on the hidden cases and motion of hidden states
$R^{(i,v)}$ $R^{(i,x)}$	**Roughness matrices:** These are the inverse of a matrix encoding serial correlations among (generalized) random fluctuations on the hidden cases and motion of hidden states
$\tilde{\varepsilon}^{(i,v)} = \tilde{v}^{(i-1)} - \tilde{g}^{(i)}$ $\tilde{\varepsilon}^{(i,x)} = \mathcal{D}\tilde{x}^{(i)} - \tilde{f}^{(i)}$	**Prediction errors:** These are the prediction errors on the hidden causes and motion of hidden states evaluated at their current conditional expectation
$\xi^{(i,v)} = \tilde{\Pi}^{(i,v)}\tilde{\varepsilon}^{(i,v)}$ $\xi^{(i,x)} = \tilde{\Pi}^{(i,x)}\tilde{\varepsilon}^{(i,x)}$	**Precision-weighted prediction errors:** These are the prediction errors weighted by their respective precisions

In other words, they occupy a small number of states with a high probability and avoid a large number of other states. In short, they appear to resist thermodynamic imperatives. Mathematically, we can summarize this remarkable capacity by saying biological agents maintain a low entropy distribution on the states that they could occupy. Entropy is just the average surprise or negative log probability of an agent being in a particular state (see Table 12.1). In short, the question we need to address is how biological systems minimize their average surprise (or more exactly surprisal or self information). Surprise here just means something unexpected, like reaching into one's pocket and discovering your wallet is not there. One might think that exotic phenomena from theories of pattern-formation and self-organization may provide a sufficient explanation for the emergence of orderly (unsurprising) state-transitions. However, they do not. These patterns certainly have beautiful and intrinsic structures that unfold over short periods of time; but self-organization *per se* cannot explain the ability of biological agents to avoid surprise indefinitely. However, there is a solution that is almost tautological in its simplicity:

The solution lies in noting that surprise in ensemble dynamics is exactly the same as the (negative log) evidence for a model in statistics: $\mathcal{L} = -\ln p(\tilde{s}|m)$ (see Table 12.1). The conceptual link between surprise and log-evidence rests on assuming that every agent or person is a model of their environment or, more specifically, the sensory data to which they are exposed. This means that to minimize average surprise (entropy), each agent should maximize the evidence for its model of sensory exchanges with the world. Model optimization of this sort is a solved problem in statistics and machine learning (e.g., [41, 43]). In fact, most forms of statistical inference rest on comparing the evidence for one model relative to another, given some data. So what does this mean for our brains?

It suggests that we are obliged to optimize our model of the world through evolution, neurodevelopment, and learning. In other words, we are statistical engines that make inferences about the world, given the (sensory) data available to us. The idea that we are *inference machines* is very old and was most clearly articulated by the renowned physicist Helmholtz [32]. Indeed, perception has been explicitly equated with hypothesis testing [29] and the brain has been referred to as a Helmholtz machine [15]. More recent incarnations of this idea appear as the Bayesian brain hypothesis [38, 40] as instantiated in schemes like predictive coding [44, 51]. All these explanations borrow from Helmholtz's idea that the brain makes inferences about its sensations. A large body of work in theoretical neuroscience provides a plausible and compelling account of perception and the architecture of the wet-ware (brain) required to make these inferences [1, 4]. The ensuing perspective on biological systems says something quite profound: It says that all biological organisms can be regarded as a model of the environmental niches (econiches) they inhabit [13]. In this sense, each species represents the product of evolutionary model optimization and each phenotype (including our brain) is a physical model or transcription of causal structure in its econiche. However, we have overlooked one small problem: Optimizing models is not easy and, in most situations, evaluating surprise or model evidence is an intractable problem. This is where the free-energy comes in.

Free-energy was introduced (in the context of quantum physics) by Richard Feynman [20] to solve the sort of difficult integration problems inherent in computing model evidence. It has been exploited in statistics and machine learning (e.g., [45]) as a very efficient way of measuring and maximizing model-evidence (i.e., minimizing surprise). The idea is quite simple, instead of trying to minimize something that cannot be measured, one simply creates a bound that can be measured, which is always bigger than the unknown quantity. One then minimizes the unknown quantity by minimizing the bound. So, what is this bound?

In physics and statistics it is free-energy (recent statistical treatments of evolution consider a related quantity called free-fitness; [54]). Its construction is simple (see Fig. 12.1): The free-energy bound is constructed by adding a non-negative (Kullback–Leibler divergence) quantity to surprise. The clever thing is that adding this term renders the free-energy easily computable. This Kullback-Leibler divergence measures the difference between two probability distributions; the first is called a variational density and is an arbitrary probability distribution used to create the bound. The second is the posterior or conditional density on the causes of our sensations (for example, the presence of an object in our field of view). The posterior density is the probability of causes after seeing their consequences. Minimizing the bound reduces the difference between the variational and the posterior density. When they are identical, free-energy becomes surprise or negative log-evidence. This means to evaluate surprise, we have to make (Bayesian) inferences about what caused our sensations. This is the Bayesian brain hypothesis, where minimizing free-energy entails Bayes-optimal perception. In short, free-energy converts an intractable mathematical problem into a simple optimization problem. This statistical device furnishes another important perspective on how we, as organisms, work. It suggests that we minimize surprise by optimizing an upper bound on surprise. In other words, everything we do can be cast in terms of optimization. This is self-evidently true in many contexts, certainly in fields like reinforcement learning and economics [9, 14, 52, 55] but also fields like evolutionary biology, where adaptive fitness is optimized.

12.2.1 The Bayesian Brain

The Bayesian brain hypothesis makes complete sense in this context. If our imperative is to reduce surprise, then we need to some reference or expectations against which to measure surprise.These expectations depend upon some model of the world and its current state. The probabilistic state of the world we infer is the variational density above (Fig. 12.1) and, when things are working properly, corresponds to the true but unknown posterior density. In the brain, this variational density (or more precisely, its sufficient statistics like its mean or average) may be encoded by neuronal activity or connection strengths among different parts of the brain. This leads to an understanding of perceptual inference and learning as changing synaptic activity and connectivity, respectively, to minimize free-energy. There are many schemes that have been proposed to implement this optimization. Among the more popular is predictive coding. Under some simplifying assumptions

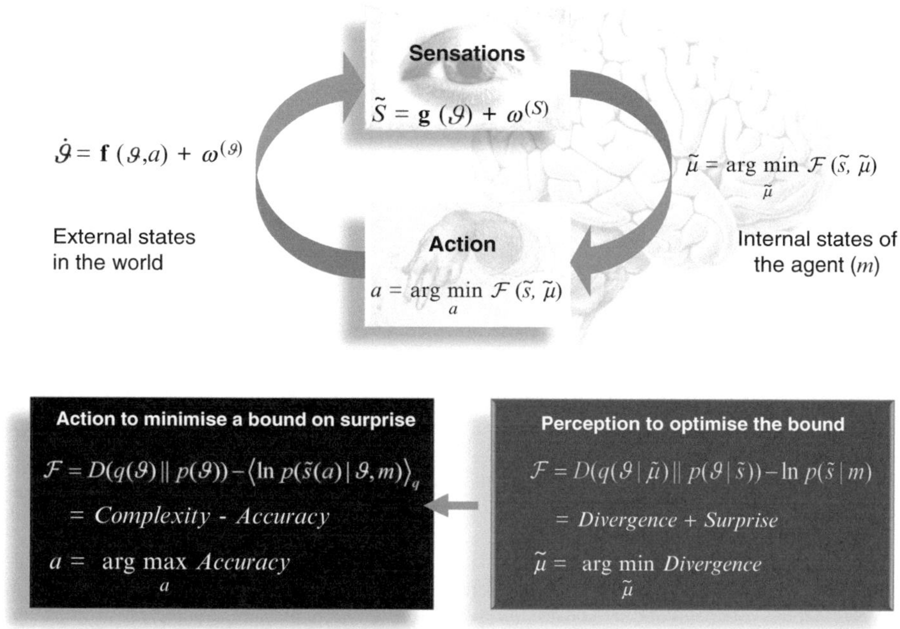

Fig. 12.1 This schematic shows the dependencies among the quantities that define the free-energy of an agent or brain, denoted by m. These include, its generalized internal states and sensory signals (generalized states include their generalized motion; i.e., velocity, acceleration, etc.). The environment is described by equations, which specify the motion of its states, when depend on action. Both internal brain states and action minimize free-energy, which is a function of sensory input and the internal states. Internal states encode a variational density on the causes of sensory input. These comprise states of the world and the amplitude of random fluctuations. The lower panels provide the key equations behind the free-energy formulation. The right equality shows that optimizing brain states, with respect to the internal states, makes the variational density an approximate conditional density on the causes of sensory input. Furthermore, it shows that free-energy is an upper bound on surprise. This is because the first term of the equality is a divergence between the variational density and the true conditional or posterior density. Because this divergence can never be less than zero, minimizing free-energy renders it a proxy for surprise. At the same time, the variational density becomes the posterior density. The left equality shows that action can only reduce free-energy by selectively sampling sensory data that are predicted under the variational density

about the shape of the probability densities involved, the free-energy reduces to the sum of squared prediction error (see Fig. 12.2). In short, minimizing free-energy corresponds to reducing prediction errors. The hierarchical scheme depicted in Fig. 12.3 represents a fairly plausible architecture that the brain might use to suppress prediction errors and thereby reduce free-energy. Crucially, this scheme is based upon a gradient ascent of free-energy (squared prediction error) and, as such, can be cast as a set of ordinary differential equations. It is these equations of motion that we suppose provide a model for neuronal dynamics that will be used in the second part of this chapter.

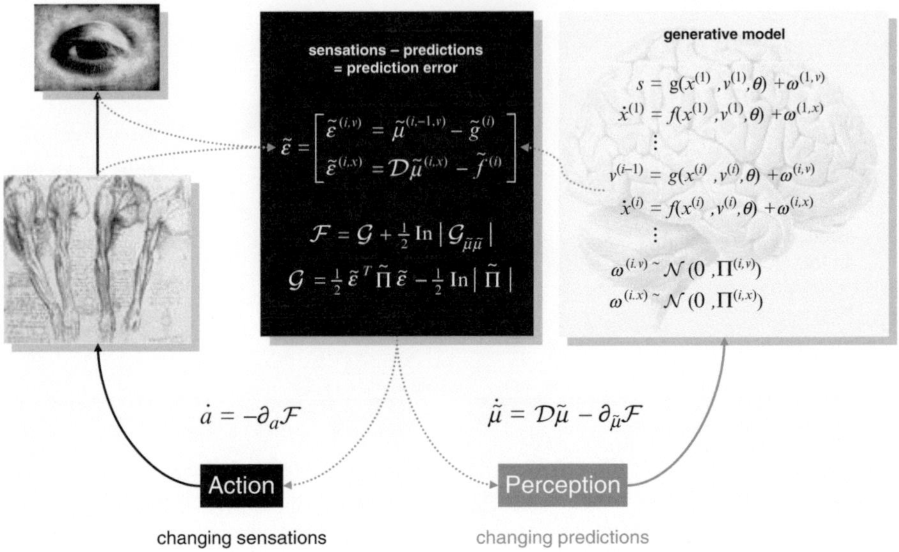

Fig. 12.2 This schematic illustrates the bilateral role of free-energy (i.e., prediction error) in driving action and perception: **Action:** Acting on the environment by minimizing free-energy enforces a sampling of sensory data that is consistent with the current representation (i.e., changing sensations to minimize prediction error). This is because free-energy is a mixture of complexity and accuracy (the first expression for free-energy in Fig. 12.1). Crucially, action can only affect accuracy. This means the brain will reconfigure its sensory epithelia to sample inputs that are predicted by its representations; in other words, to minimize prediction errors. The equation above action simply states that action performs a gradient decent on (i.e., minimizes) free-energy. **Perception:** Optimizing free-energy by changing the internal states that encode the variational density makes it an approximate posterior or conditional density on the causes of sensations. This follows because free-energy is surprise plus a Kullback–Leibler divergence between the variational and conditional densities (the second expression for free-energy in Fig. 12.1). Because this difference is non-negative, minimizing free-energy makes the variational density an approximate posterior probability. This means the agent implicitly infers or represents the causes of its sensory samples in a Bayes-optimal fashion. At the same time, the free-energy becomes a tight bound on surprise that is minimized through action. The equation above perception simply states that internal states perform a gradient decent on (i.e., minimize) free-energy. This gradient decent is in a moving frame of reference for generalized states and accumulates gradients over time for the parameters. **Prediction error:** The equations show that the free-energy comprises a (Gibb's) energy, which is effectively the (precision weighted) sum of squared prediction error. This error contains the sensory prediction error and other differences that mediate empirical priors on the motion of hidden states. The predictions rest on a generative model of how sensations are caused. These models have to explain complicated dynamics on continuous states with hierarchical or deep causal structure. An example of one such generic model is shown on the right. **Generative model:** Here the model comprises continuous nonlinear functions of (hidden) causes and states, at the i-th level of a hierarchical dynamic model. The random fluctuations play the role of observation noise at the sensory level and state-noise at higher levels. In this model, hidden causes link hierarchical levels, where the output of one level provides input to the next. Conversely, hidden states link dynamics over time and lend the model memory. Gaussian assumptions about the random fluctuations specify the likelihood of the model and furnish empirical priors in terms of predicted motion. These assumptions are encoded by the precision or inverse variance of the random fluctuations on hidden causes and the motion of hidden states that depend on precision parameters. The associated message-passing scheme implementing perception is shown in the next figure. Please see Table 12.1 for a glossary of the variables in this figure

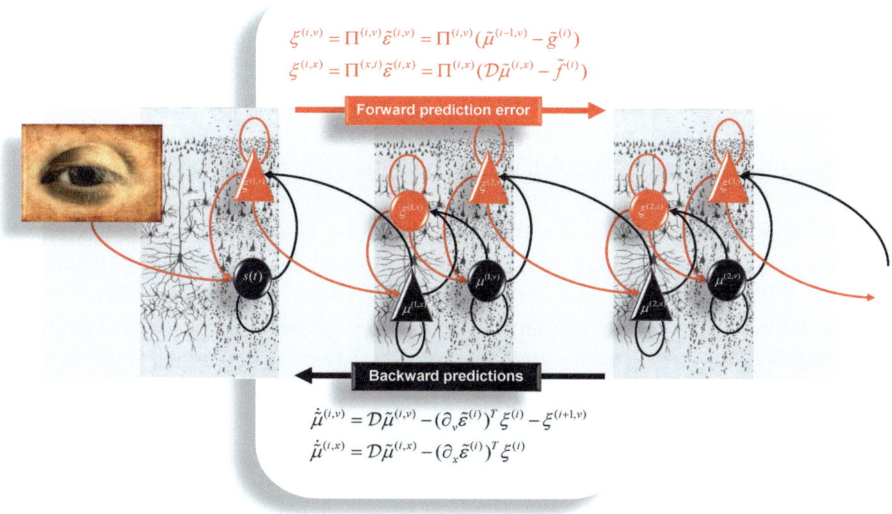

Fig. 12.3 The schematic details a neuronal architecture that optimizes the conditional expectations of causes in hierarchical models of sensory input of the sort illustrated in the previous figure. It shows the putative cells of origin of forward driving connections that convey prediction-error from a lower area to a higher area (*red arrows*) and nonlinear backward connections (*black arrows*) that construct predictions [22, 44]. These predictions try to explain away (inhibit) prediction-error in lower levels. In this scheme, the sources of forward and backward connections are superficial and deep pyramidal cells (triangles), respectively, where state-units are *black* and error-units are *red*. The equations represent a generalized gradient descent on free-energy using the generative model of the previous figure. **Predictions and prediction-error**: If we assume that synaptic activity encodes the conditional expectation of states, then recognition can be formulated as a gradient descent on free-energy. Under Gaussian assumptions, these recognition dynamics can be expressed compactly in terms of precision weighted prediction-errors on the causal states and motion of hidden states (at level i of the hierarchy). The ensuing equations suggest two neuronal populations that exchange messages; causal or hidden state-units encoding expected states and error-units encoding prediction-error. Under hierarchical models, error-units receive messages from the state-units in the same level and the level above whereas state-units are driven by error-units in the same level and the level below. These provide bottom-up messages that drive conditional expectations towards better predictions to explain away prediction-error. This scheme suggests the only connections that link levels are forward connections conveying prediction-error to state-units and reciprocal backward connections that mediate predictions. Note that the prediction errors that are passed forward are weighted by their precision. This tells us that precision may be encoded by the postsynaptic gain or sensitivity or error units, which also has to be optimized: see [23] for further details

In summary, surprise cannot be measured directly but we can induce a bound on surprise called free-energy and reduce this bound by optimizing the activity and connectivity in our brains. This renders free-energy approximately the same as surprise and obliges us to make Bayesian inferences about the state of our world. The implementation of this optimization may rest upon the minimization of prediction errors of the sort considered by predictive coding. In this context, the gradient ascent on free-energy (prediction errors) provides a plausible account of the dynamics that underlie synaptic activity (perceptual inference) and synaptic efficacy (perceptual learning). An important aspect of this optimization is the proper estimation of the precision (inverse variance or uncertainty) associated with prediction errors. In the generalized predictive coding scheme of Fig. 12.3, we consider this precision to be encoded by synaptic gain, which has to be optimized in exactly the same way as synaptic activity (encoding expected states of the world) and synaptic efficacy (encoding the coupling among these states). The role of precision or synaptic gain will become important later when we consider the difference between action and action-observation later. The scheme described in Fig. 12.3 has been used to explain many different aspects of perceptual learning and inference in psychophysics and psychology. However, perceptual inference and learning does not itself reduce surprise; it just reduces the difference between free-energy and surprise. To understand how surprise *per se* is reduced, we have to consider action and the active sampling of sensory data:

12.2.2 Active Inference

So far, we have seen that perception can be understood as furnishing a proxy for surprise, in the sense that perception reduces the divergence between the variational density and the true conditional density over hidden states causing sensations. In doing this, it makes free-energy a tighter bound or better approximation to surprise. Next, we consider how action can actually reduce surprise. In brief, we can minimize prediction error in one of two ways: We can either change our expectations or predictions (perception) or we can change the things that are predicted (action). This perspective suggests that we should selectively sample data (or place ourselves in relation to the world) so that we experience what we expect to experience. In other words, we will act upon the world to ensure that our predictions come true [23]. This is exactly the sort of behavior that we were trying to explain at the beginning; namely, how do biological systems avoid surprising exchanges with the environment?

It is fairly easy to show that the only part of free-energy that can be changed by action is sensory prediction error. This simple fact provides a nice explanation for how we interact with the world at a number of levels. First, in biological terms, it suggests that our muscles are wired to cancel sensory prediction errors. We are all familiar with this as a reflex: If I stretched the muscles in your leg by tapping the tendons below your knee, then they respond by contracting to cancel the unpredicted stretch-receptor signals. This reflects a basic functional architecture in movement (motor) control whereby movements are elicited by prediction errors about the position of limbs: at its simplest, this is the classical motor reflex. If we generalize this view of how the brain controls our bodies, then peripheral motor or muscle systems are enslaved to fulfill predictions. This means we only have to expect or predict an action and it will be executed automatically. The resulting perspective implies a curious and ambivalent relationship between action and perception: On the one hand, perception optimizes predictions so that action can minimize surprise, while, on the other hand, our motor behavior is prescribed entirely by perceptual predictions. If action and perception work in synergy, we will navigate our econiche, never straying from well trodden paths, eluding surprise (and potential danger).

12.2.3 Summary

In conclusion, we have reviewed the motivation for the free-energy principle in terms of explaining how self-organizing adaptive and biological systems manage to resist a tendency to disorder. When we unpack this principle, we see that it accommodates both perception and action, while embedding the action-perception cycle in an evolutionary context. We have seen that the underlying imperative of all biological systems is to minimize (a free-energy bound) on surprise; and that surprise, self-evidently, depends upon predictions. These predictions can be constrained by prior expectations, which allow our behavior to be optimized by evolution and neurodevelopment (learning). In the next section, we will apply these ideas to understand how agents emit sequences of movements or action. We will focus on handwriting, noting that the basic principles should apply to any structured and sequential pattern of behavior. This example has been chosen to highlight the central role of itinerant dynamics in furnishing prior expectations about action and concomitant perception.

12.3 Action and Its Observation

In this section, we describe a generative model of handwriting and then apply the fee-energy scheme of the previous section to simulate emergent neuronal dynamics and behavior. To create these simulations, all we have to do is specify a generative model. This model and (generalized) sensations define the free-energy, which determines the dynamics of action and neuronal states encoding the conditional expectations of hidden states in the world. Action and perception are prescribed by the equations in Fig. 12.2, which simulate neuronal and behavioral responses, respectively.

$$\begin{aligned} \dot{\tilde{\mu}} &= \mathcal{D}\tilde{\mu} - \partial_{\tilde{\mu}}\mathcal{F}(\tilde{s}(a), \mu) \\ \dot{a} &= -\partial_a \mathcal{F}(\tilde{s}(a), \mu) \end{aligned} \tag{12.1}$$

The first equation represents a generalized or instantaneous gradient descent on free-energy for the conditional expectations of hidden states causing sensory input (i.e., neuronal activity). The first term represents their expected generalized motion, while the second is simply the gradient of the free-energy with respect to the expectations. The reason that this is a generalized descent is that it is formulated in generalized coordinates of motion, such that the first term augments and anticipates the descent so that it becomes effectively instantaneous. The second equality is the equivalent gradient descent for action. Both of these equations rest upon the free-energy, which is a function of sensory information and current expectations. This function depends upon a generative model, which is specified completely by equations of motion of the hidden states and a function mapping hidden states to sensory signals (see Fig. 12.2). This means all we have to do to simulate action and perception is to specify the equations of the generative model and then solve or integrate Eq. (12.1) over time. In what follows, we describe the generative model that will be used for the remainder of this chapter. We have chosen this model because it embodies the sort of itinerant dynamics considered in previous chapters. In particular, we focus on stable heteroclinic channels and how they prescribed expectations about movements.

12.3.1 Itinerant Dynamics and Attractors

Our agent was equipped a simple hierarchical model of its sensorium based on a Lotka–Volterra system. The particular form of this model has been discussed previously as the basis of putative speech decoding [39]. Here, it is used to model a stable heteroclinic channel [48] encoding successive locations to which the agent expects its arm to move. The resulting trajectory was contrived to simulate synthetic handwriting.

A stable heteroclinic channel is a particular form of (stable) itinerant trajectory or orbit that revisits a sequence of (unstable) fixed points. In our model, there are two sets of hidden states, which we will associate with two levels of a hierarchical model. The first set $x^{(2)} \in \mathbb{R}^{6\times 1}$ corresponds to the state-space of a Lotka–Volterra system. This is an abstract (attractor) state-space, in which a series of attracting points are visited in succession. The second set $x^{(1)} = \{x_1, x_2, x'_1, x'_2\}$ corresponds to the (angular) positions and velocities of two joints in (two dimensional) physical space. The dynamics of hidden states at the first level embody the agent's prior expectation that the arm will be drawn to a particular location, $v^{(1)} = g(x^{(2)})$ specified by the attractor states of the second level. This is implemented simply by placing a (virtual) elastic band between the tip of the arm and the attracting location. The hidden states basically draw the arm's extremity (finger) to a succession of locations to produce an orbit or trajectory, under classical Newtonian mechanics. We chose the locations so that the resulting trajectory looked like handwriting. These hidden states generate both proprioceptive and visual (exteroceptive) sensory data: The proprioceptive data are the angular positions and velocities of the two joints

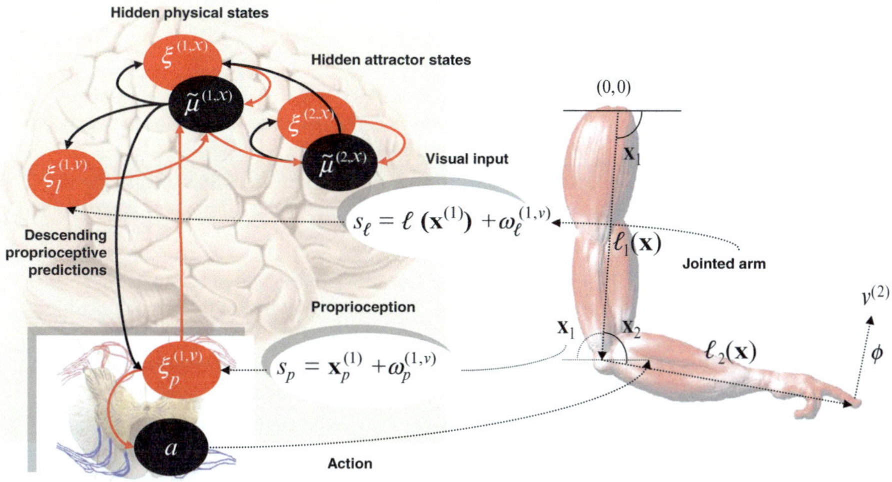

Fig. 12.4 This schematic details a simulated (mirror neuron) system and the motor plant that it controls (*left* and *right*, respectively). The right panel depicts the functional architecture of the supposed neural circuits underlying active inference. The red ellipses represent prediction error-units (neurons or populations), while the black ellipses denote state-units encoding conditional expectations about hidden states of the world (for simplicity, we have omitted hidden causes). The hidden states are split into two hierarchical levels: the higher abstract attractor states (that sup-ports stable heteroclinic orbits) and lower physical states of the arm (angular positions and velocities of the two joints). *Red* arrows are forward connections conveying prediction errors and *black* arrows are backward connections mediating predictions. Motor commands are emitted by the black units in the ventral horn of the spinal cord. Note that these just receive prediction errors about proprioceptive states. These, in turn, are the difference between sensed proprioceptive input from the two joints and descending predictions from optimized representations in the motor cortex. The two jointed arm has a state space that is characterized by two angles, which control the position of the finger that will be used for writing in subsequent figures

$x^{(1)}$, while the visual information was the location of the arm in physical (Cartesian) space $\{\ell_1, \ell_1 + \ell_2\}$, where $\ell_2(x^{(1)})$ is the displacement of the finger from the location of the second joint $\ell_1(x^{(1)})$ (see Fig. 12.4 and Table 12.2).

Crucially, because this generative model generates two (proprioceptive and visual) sensory modalities, the solutions to Eq. (12.1) implement Bayes-optimal multisensory integration. However, because action is also trying to reduce prediction errors, it will move the arm to reproduce the expected trajectory (under the constraints of the motor plant). In other words, the arm will trace out a trajectory prescribed by the itinerant priors (to cancel proprioceptive prediction errors). This closes the loop, producing autonomous self-generated sequences of behavior of the sort described below. Note that the real world does not contain any attracting locations or elastic bands: The only causes of observed movement are the self-fulfilling expectations encoded by the itinerant dynamics of the generative model. In short, hidden attractor states essentially prescribe the intended movement trajectory,

Table 12.2 Variables and quantities specific to the writing example of active inference (see main text for details)

Variable	Description
$x^{(2)} \in \mathbb{R}^{6\times 1}$	**Hidden attractor states:** A vector of hidden states that specify the current location towards which the agent expects its arm to be pulled
$x^{(1)} \in \mathbb{R}^{4\times 1}$	**Hidden effector states:** Hidden states that specify the angular position and velocity of the *i*-th joint of a two-jointed arm
$\ell_1(x^{(1)}) \in \mathbb{R}^{2\times 1}$ $\ell_2(x^{(1)}) \in \mathbb{R}^{2\times 1}$	**Joint locations:** Locations of the end of the two arm parts in Cartesian space. These are functions of the angular positions of the joints
$v^{(1)} = g(x^{(2)}) \in \mathbb{R}^{2\times 1}$	**Attracting location:** The location towards which the arm is drawn. This is specified by the hidden attractor states
$\phi(x^{(1)}, v^{(1)}) \in \mathbb{R}^{2\times 1}$	**Newtonian Force:** This is the angular force on the joints exerted by the attracting location
$A \in \mathbb{R}^{6\times 6} \subset \theta$	**Attractor parameters:** A matrix of parameters that govern the (sequential Lotka–Volterra) dynamics of the hidden attractor states
$L \in \mathbb{R}^{2\times 6} \subset \theta$	**Cartesian parameters:** A matrix of parameters that specify the attracting locations associated with each hidden attractor state

because they generate predictions that action fulfils. This means expected states encode conditional percepts (concepts) about latent abstract states (that do not exist in the absence of action), which play the role of intentions. We now describe the model formally.

12.3.2 *The Generative Model*

The model used in this section concerns the movements of a two-joint arm. When simulating active inference, it is important to distinguish between the agent's generative model and the actual dynamics generating sensory data. To make this distinction clear, we will use bold for true equations and states, while those of the generative model will be written in italics. Proprioceptive input corresponds to the angular position and velocity of both joints, while the visual input corresponds to the location of the extremities of both parts of the arm. This means the mapping from hidden states to sensory consequences is:

$$\mathbf{g}^{(1)} = g^{(1)} = \begin{bmatrix} x^{(1)} \\ \ell_1(x^{(1)}) \\ \ell_1(x^{(1)}) + \ell_2(x^{(1)}) \end{bmatrix} \tag{12.2}$$

We will ignore the complexities of inference on retinotopically mapped visual input and assume the agent has direct access to the locations of the arm in visual space. The kinetics of the arm conforms to Newtonian laws, under which action forces the angular position of each joint. Both joints have an equilibrium position at ninety degrees; with inertia $m_i \in (8,4)$ and viscosity $\kappa_i \in (4,2)$, giving the following equations of motion for the hidden states

$$\mathbf{x}^{(1)} = \begin{bmatrix} \mathbf{x}_1 \\ \mathbf{x}_2 \\ \mathbf{x}'_1 \\ \mathbf{x}'_2 \end{bmatrix} \quad \mathbf{f}^{(1)} = \begin{bmatrix} \mathbf{x}'_1 \\ \mathbf{x}'_2 \\ \left(a_1 + \mathbf{v}_1 - \frac{1}{4}(\mathbf{x}_1 - \frac{\pi}{2}) - \kappa_1 \mathbf{x}'_1\right)/m_1 \\ \left(a_2 + \mathbf{v}_2 - \frac{1}{4}(\mathbf{x}_2 - \frac{\pi}{2}) - \kappa_2 \mathbf{x}'_2\right)/m_2 \end{bmatrix} \tag{12.3}$$

However, the agent's empirical priors on this motion have a very different form. Its generative model assumes the finger is pulled to a (goal) location $v^{(1)}$ by a force $\phi(t)$, which implements the virtual elastic band above:

$$\begin{aligned} x^{(1)} &= \begin{bmatrix} x_1 \\ x_2 \\ x'_1 \\ x'_2 \end{bmatrix} \quad f^{(1)} = \begin{bmatrix} x'_1 \\ x'_2 \\ \left(\phi^T \ell_2 \ell_2^T O \ell_1 - \frac{1}{16}(x_1 - \frac{\pi}{2}) - \kappa_1 x'_1\right)/m_1 \\ \left(\phi^T O \ell_2 - \frac{1}{16}(x_2 - \frac{\pi}{2}) - \kappa_2 x'_2\right)/m_2 \end{bmatrix} \\ \ell_1 &= \begin{bmatrix} \cos(x_1) \\ \sin(x_1) \end{bmatrix} \quad \ell_2 = \begin{bmatrix} -\cos(-x_2 - x_1) \\ \sin(-x_2 - x_1) \end{bmatrix} \quad O = \begin{bmatrix} 0 & -1 \\ 1 & 0 \end{bmatrix} \\ \phi &= \tfrac{1}{2}(v^{(1)} - \ell_1 - \ell_2) \end{aligned} \tag{12.4}$$

The (moving) target location is specified by the second level of the hierarchy as a nonlinear (softmax) function of the hidden attractor states.

$$\begin{aligned} v^{(1)} &= g(x^{(2)}) = Ls(x^{(2)}) \\ f^{(2)} &= A\sigma(x^{(2)}) - \tfrac{1}{8}x^{(2)} + \begin{bmatrix} 1 \\ \vdots \\ 1 \end{bmatrix} \\ \sigma(x_i) &= \frac{1}{1+e^{2x_i}} \qquad s(x_i) = \frac{e^{2\alpha_i}}{\sum_j e^{2x_j}} \end{aligned} \tag{12.5}$$

Heuristically, these equations of motion mean that the agent thinks that changes in its world are caused by the dynamics of attractor states on an abstract (conceptual) space. The currently active state selects a location $v^{(1)}$ in the agent's physical (Cartesian) space, which exerts a force $\phi(t)$ on its finger. The equations of motion in Eq. (12.4) pertain to the resulting motion of the arm in Cartesian space, while Eq. (12.5) mediates the attractor dynamics driving these movements.

The (Lotka–Volterra) form of the equations of motion for the hidden attractor states ensures that only one has a high value at any one time and imposes a particular sequence on the underlying states. Lotka–Volterra dynamics basically induce competition among states that no state can win. The resulting winnerless competition rests on the (logistic) function $\sigma(x^{(2)})$, while the sequence order is determined by the elements of the matrix

$$A = \begin{bmatrix} 0 & -\frac{1}{2} & -1 & -1 & \cdots \\ -\frac{3}{2} & 0 & -\frac{1}{2} & -1 & \\ -1 & -\frac{3}{2} & 0 & -\frac{1}{2} & \ddots \\ -1 & -1 & -\frac{3}{2} & 0 & \\ \vdots & & \ddots & & \ddots \end{bmatrix} \tag{12.6}$$

Each attractor state has an associated location in Cartesian space, which draws the arm towards it. The attracting location is specified by a mapping from attractor space to Cartesian space, which weights different locations

$$L = \begin{bmatrix} 1 & 1.1 & 1.0 & 1 & 1.4 & 0.9 \\ 1 & 1.2 & 0.4 & 1 & 0.9 & 1.0 \end{bmatrix} \tag{12.7}$$

with a softmax function $s(x^{(2)})$ of the attractor states. The location parameters were specified by hand but could, in principle, be learnt as described in [25, 26]. The inertia and viscosity of the arm were chosen somewhat arbitrarily to reproduce realistic writing movements over 256 time bins, each corresponding to roughly eight milliseconds (i.e., a second). Unless stated otherwise, we used a log-precision of four for sensory noise and eight for random fluctuations in the motion of hidden states.

Figure 12.5 shows the results of integrating Eq. (12.1), using the generative model above. The top right panel shows the hidden states embodying Lotka-Volterra dynamics (the hidden joint states are smaller in amplitude). These generate predictions about the position of the joints (upper left panel) and consequent prediction errors that drive action. Action is shown on the lower right panel and displays intermittent forces that move the joint to produce a motor trajectory. This trajectory is shown on the lower left in visual space over time. This trajectory or orbit is translated as a function of time to reproduce handwriting. Although this is a pleasingly simple way of simulating an extremely complicated motor trajectory, it should be noted that this agent has a very limited repertoire of behaviors; it can only reproduce this sequence of graphemes, and will do so *ad infinitum*.

In summary, we have covered the functional architecture of a generative model whose autonomous (itinerant) expectations prescribe complicated motor sequences through active inference. This rests upon itinerant dynamics (stable heteroclinic channels) that can be regarded as a formal prior on abstract causes in the world. These are translated into physical movement through classical Newtonian mechanics, which correspond to the physical states of the model. Action tries to fulfill predictions about proprioceptive inputs and is enslaved by autonomous

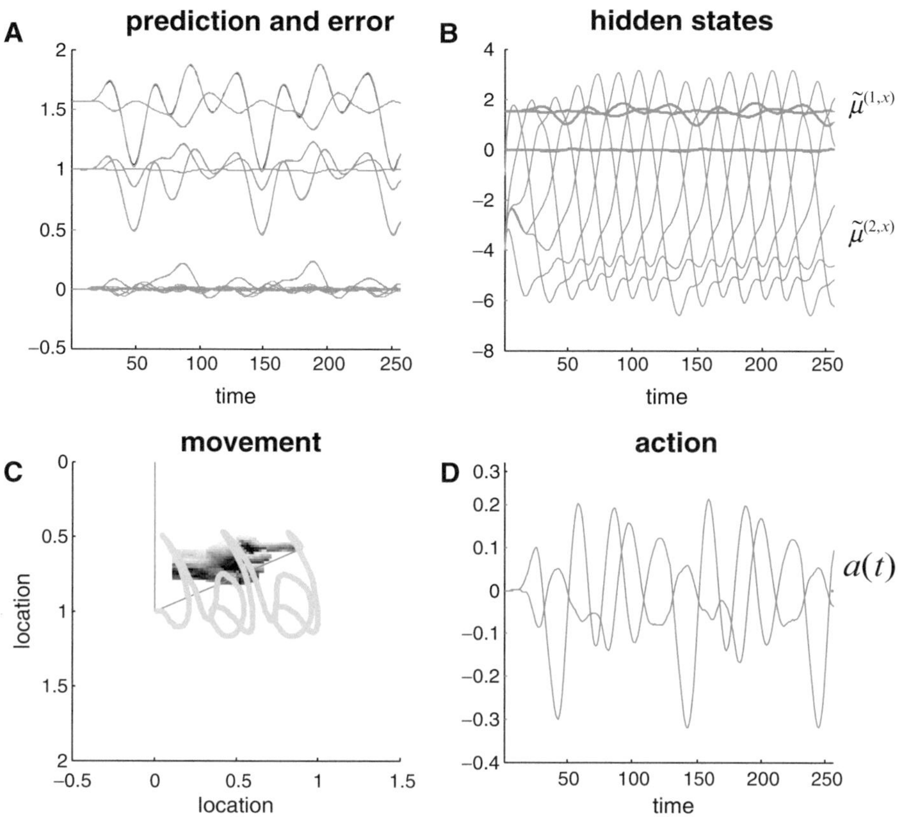

Fig. 12.5 This figure shows the results of simulated action (writing), under active inference, in terms of conditional expectations about hidden states of the world (*upper right*), consequent predictions about sensory input (*upper left*) and the ensuing behavior (*lower left*) that is caused by action (*lower right*). The autonomous dynamics that underlie this behavior rest upon the expected hidden states that follow Lotka–Volterra dynamics. These are the thinner lines in the *upper right* panel. The hidden physical states (thicker lines) have smaller amplitudes and map directly on to the predicted proprioceptive and visual signals (*shown on the left*). The visual locations of the two joints are shown above the predicted joint positions and angular velocities that fluctuate around zero. The *dotted lines* correspond to prediction error, which shows small fluctuations about the prediction. Action tries to suppress this error by "matching" expected changes in angular velocity through exerting forces on the joints. These forces are shown on the *lower right*. The subsequent movement of the arm is traced out on the *lower left*; this trajectory has been plotted in a moving frame of reference so that it looks like synthetic handwriting (e.g., a succession of 'j' and 'a' letters). The straight lines on the *lower left* denote the final position of the two jointed arm and the hand icon shows the final position of its finger

predictions, producing realistic behavior. These trajectories are both caused by neuronal representations of abstract (attractor) states and cause those states in the sense that they are conditional expectations. Closing the loop in this way ensures a synchrony between internal expectations and external outcomes.

In the next section, we will make a simple change which means that movements are no longer caused by the agent. However, we will see that the conditional expectations about attractor states are relatively unaffected, which means that they still anticipate observed movements. We conclude with this example because it illustrates nicely the potential role of itinerant dynamics in explaining some of the higher cognitive aspects of brain function. Our focus here is on emulating the electrophysiological phenomenology of the mirror neuron system; in particular, the fact that certain neurons in the ventral premotor cortex and inferior parietal cortex respond not only to the execution of particular movement primitives but also when these movements are observed in other agents [16, 21, 27, 53].

12.3.3 Action-Observation

The simulations above were repeated but with one small but important change. Basically, we reproduced the same movements but the proprioceptive consequences of action were removed, so that the agent could see but not feel the arm moving. From the agent's perspective, this is like seeing an arm that looks like its own arm but does not generate proprioceptive input (i.e., the arm of another agent). However, the agent still expects the arm to move with a particular itinerant structure and will try to predict the trajectory with its generative model. In this instance, the hidden states still represent itinerant dynamics (intentions) that govern the motor trajectory but these states do not produce any proprioceptive prediction errors and therefore do not result in action. Crucially, the perceptual representation still retains its anticipatory or prospective aspect and can therefore be taken as a perceptual representation of intention, not of self, but of another. We will see below that this representation is almost exactly the same under action-observation as it is during action.

Practically speaking, to perform these simulations, we simply recorded the forces produced by action in the previous simulation and replayed them as exogenous forces (real causes in Eq. (12.2)) to move the arm. This change in context (agency) was modeled by down-weighting the precision of proprioceptive signals. This is exactly the same mechanism that we have used previously to model attention [19]. In this setting, reducing the precision of proprioceptive prediction errors prevents them from having any influence on perceptual inference (i.e., the agent cannot feel changes in its joints). Furthermore, action is not compelled to reduce these prediction errors because they have no precision. In these simulations, we reduced the log-precision of proprioceptive prediction errors from eight to minus eight. To illustrate the key results of these simulations of action-observation, in relation to simulated action, we recorded the activity of units encoding hidden attractor states and examined and their relationship to observed movements:

12.3.4 Place-Cells, Itinerancy, and Oscillations

It is interesting to think about the attractor states as representing trajectories through abstract representational spaces (cf., the activity of place cells; [8, 47, 57]. Figure 12.5 illustrates the sensory or perceptual correlates of units representing expected attractor states. The left hand panels show the activity of one (the fourth) hidden state unit under action, while the right panels show exactly the same unit under action-observation. The top rows show the trajectories in visual space, in terms of horizontal and vertical displacements (grey lines). The black dots correspond to the time bins in which the activity of the hidden state unit exceeded an amplitude threshold of two arbitrary units. They key thing to take from these results is that the activity of this unit is very specific to a limited part of Cartesian space and, crucially, a particular trajectory through this space. The analogy here is between directionally selective place–cells of the sort studied in hippocampal recordings: In tasks involving goal-directed, stereotyped trajectories, the spatially selective activity of hippocampal cells depends on the animal's direction of motion [5]. A further interesting connection with hippocampal dynamics is the prevalence of theta rhythms during action: "Driven either by external landmarks or by internal dynamics, hippocampal neurons form sequences of cell assemblies. The coordinated firing of these active cells is organized by the prominent "theta" oscillations in the local field potential (LFP): place cells discharge at progressively earlier theta phases as the rat crosses the respective place field (phase precession)" [28]. Quantitatively, the dynamics of the hidden state-units in Fig. 12.5 (upper left panel) show quasiperiodic oscillations in the (low) theta range. The notion that quasiperiodic oscillations may reflect stable heteroclinic channels is implicit in many treatments of episodic memory and spatial navigation, which "require temporal encoding of the relationships between events or locations" [17], and may be usefully pursued in the context of active inference under itinerant priors.

Notice that the same "place" and "directional" selectivity is seen under action and observation (Fig. 12.6 right and left columns). The direction selectivity can be seen more clearly in the lower panels, in which the same data are displayed but in a moving frame of reference (to simulate writing). They key thing to note here is that this unit responds preferentially when, and only, when the motor trajectory produces a down-stroke, but not an up-stroke. There is an interesting dissociation in the firing of this unit under action and action-observation: during observation the unit only starts responding to down-strokes *after* it has been observed once. This reflects the finite amount of time required for visual information to entrain the perceptual dynamics and establish veridical predictions.

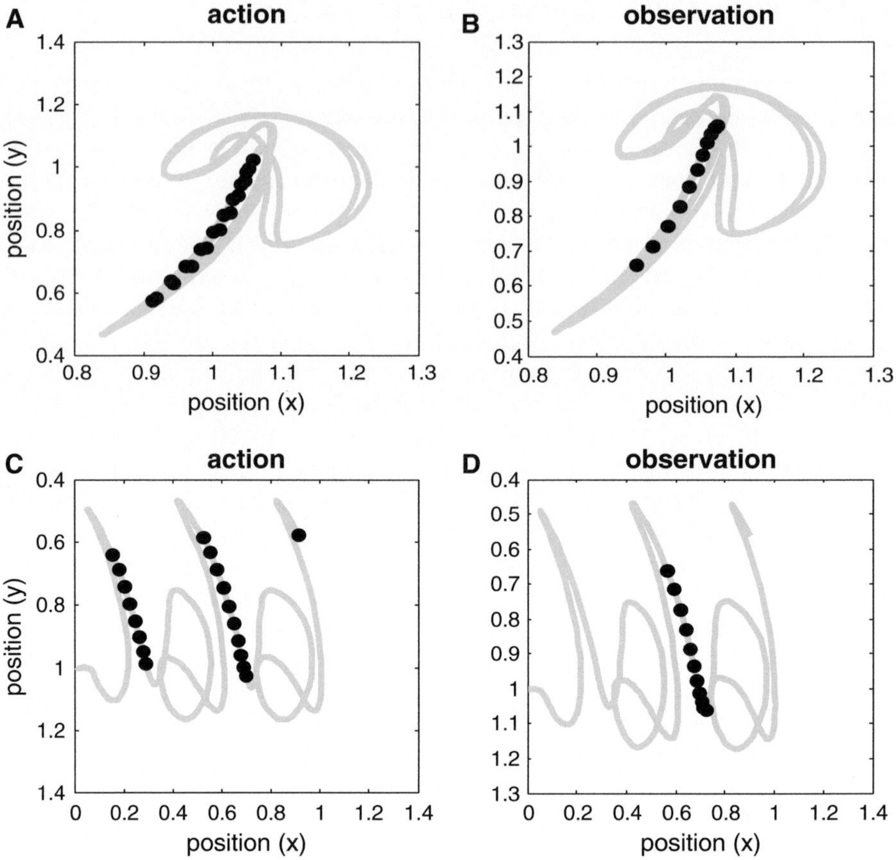

Fig. 12.6 These results illustrate the sensory or perceptual correlates of units encoding expected hidden (attractor) states. The left hand panels show the activity of one (the fourth attractor) hidden state-unit under action, while the *right panels* show exactly the same unit under action-observation. The *top rows* show the trajectory in visual space in terms of horizontal and vertical position (*grey lines*). The dots correspond to the time bins during which the activity of the state-unit exceeded an amplitude threshold of two arbitrary units. The key thing to take from these results is that the activity of this unit is very specific to a limited part of visual space and, crucially, a particular trajectory through this space. Notice that the same selectivity is seen under action and observation. The implicit direction selectivity can be seen more clearly in the lower panels, in which the same data are displayed but in a moving frame of reference to simulate writing. The key thing to note here is that this unit responds preferentially when, and only when, the motor trajectory produces a down-stroke, but not an up-stroke

12.4 Conclusion

In this chapter, we have tried to show that some aspects of action, perception and high-level (cognitive) inference are consistent with (Bayes-optimal) active inference under the free-energy principle. Put simply, the brain does not represent intended motor acts or the perceptual consequences of those acts separately. The constructs represented in the brain are both intentional and perceptual: They are amodal inferences about the states of the world generating sensory data that have both sensory and motor correlates, depending upon the context in which they are made.The predictions generated by these representations are modality-specific, prescribing both exteroceptive (e.g., visual) and interoceptive (e.g., proprioceptive) predictions, which action fulfils. The functional segregation of motor and sensory cortex could be regarded as a hierarchical decomposition, in the brain's model of its world, which provides predictions that are primarily sensory (e.g., visual cortex) or proprioceptive (motor cortex). If true, this means that high level representations can be used to furnish predictions in either visual or proprioceptive modalities, depending upon the context in which those predictions are called upon.

In one sense, this conclusion takes us back to very early ideas concerning the nature of movements and intentions. The notion of an ideomotor reflex or response was introduced in the 1840s by the Victorian physiologist and psychologist William Benjamin Carpenter. The ideomotor response (reflex) refers to the process whereby a thought or mental image induces reflexive or automatic movements, often very small and potentially outside awareness. Active inference formalizes this idea and suggests that all movements are prescribed by mental images that correspond to prior beliefs about what will happen next. These priors are inherently dynamic and itinerant. This suggests that our exchanges with our environment are constrained to an exquisite degree by local and global brain dynamics; and that these dynamics have been carefully crafted by evolution, neurodevelopment, and experience to optimize behavior.

Acknowledgements I am indebted to Mikhail Rabinovich for his guidance and insights into winnerless competition and its formulation in terms of generalized Lotka–Volterra systems that underly the work presented in this chapter. The Wellcome Trust funded this work.

References

1. Adams, R.A., Shipp, S., Friston, K.J.: Predictions not commands: active inference in the motor system. Brain Struct. Funct. **218**, 611–643 (2013)
2. Afraimovich, V., Tristan, I., Huerta, R., Rabinovich, M.I.: Winnerless competition principle and prediction of the transient dynamics in a Lotka-Volterra model. Chaos **18**, 043103 (2008)
3. Ashby, W.R.: Principles of the self-organizing dynamic system. J. Gen. Psychol. **37**, 125–128 (1947)
4. Bastos, A.M., Usrey, W.M., Adams, R.A., Mangun, G.R., Fries, P., Friston, K.J.: Canonical microcircuits for predictive coding. Neuron **76**, 695–711 (2012)
5. Battaglia, F.P., Sutherland, G.R., McNaughton, B.L.: Local sensory cues and place cell directionality: additional evidence of prospective coding in the hippocampus. J. Neurosci. **24**(19), 4541–50 (2004)
6. Bernard, C.: Lectures on the Phenomena Common to Animals and Plants, trans Hoff, H.E., Guillemin, R., Guillemin, L. Springfield, IL: Charles C Thomas (1974). ISBN 978–0398028572

7. Bick, C., Rabinovich, M.I.: Dynamical origin of the effective storage capacity in the brain's working memory. Phys. Rev. Lett. **103**, 218101 (2009)
8. Burgess, N., Barry, C., O'Keefe, J.: An oscillatory interference model of grid cell firing. Hippocampus **17**(9), 801–812 (2007)
9. Camerer, C.F.: Behavioural studies of strategic thinking in games. Trends Cogn. Sci. **7**(5), 225–231 (2003)
10. Carhart-Harris, R.L., Friston, K.J.: The default-mode, ego-functions and free-energy: a neurobiological account of Freudian ideas. Brain **133**(Pt 4), 1265–83 (2010)
11. Clark, A.: The many faces of precision. Front Psychol. **4**, 270 (2013)
12. Clark, A.: Whatever next? Predictive brains, situated agents, and the future of cognitive science. Behav. Brain Sci. **36**, 181–204 (2013)
13. Conant, R.C., Ashby, W.R.: Every good regulator of a system must be a model of that system. Int. J. Syst. Sci. **1**, 89–97 (1970)
14. Daw, N.D., Doya, K.: The computational neurobiology of learning and reward. Curr. Opin. Neurobiol. **16**(2), 199–204 (2006)
15. Dayan, P., Hinton, G.E., Neal, R.M.: The Helmholtz machine. Neural Comput. **7**, 889–904 (1995)
16. Di Pellegrino, G., Fadiga, L., Fogassi, L., Gallese, V., Rizzolatti, G.: Understanding motor events: a neurophysiological study. Exp. Brain Res. **91**, 176–80 (1992)
17. Dragoi, G., Buzsáki, G.: Temporal encoding of place sequences by hippocampal cell assemblies. Neuron **50**(1), 145–57 (2006)
18. Evans, D.J.: A non-equilibrium free-energy theorem for deterministic systems. Mol. Phys. **101**, 15551–1554 (2003)
19. Feldman, H., Friston, K.J.: Attention, uncertainty, and free-energy. Front. Hum. Neurosci. **4**, 215 (2010)
20. Feynman, R.P.: Statistical Mechanics. Benjamin, Reading, MA (1972)
21. Fogassi, L., Ferrari, P.F., Gesierich, B., Rozzi, S., Chersi, F., Rizzolatti, G.: Parietal lobe: from action organization to intention understanding. Science **308**, 662–667 (2005)
22. Friston, K.J.: A theory of cortical responses. Philos. Trans. R. Soc. Lond. B Biol. Sci. **360**, 815–36 (2005)
23. Friston, K.: The free-energy principle: a unified brain theory? Nat. Rev. Neurosci. **11**(2), 127–38 (2010)
24. Friston, K.: Life as we know it. J. R. Soc. Interface **10**, 20130475 (2013)
25. Friston, K., Daunizeau, J., Kiebel, S.: Active inference or reinforcement learning? PLoS One **4**(7), e6421 (2009)
26. Friston, K.J., Daunizeau, J., Kilner, J., Kiebel, S.J.: Action and behavior: a free-energy formulation. Biol Cybern. **102**(3), 227–60 (2010)
27. Gallese, V., Goldman, A.: Mirror-neurons and the simulation theory of mind reading. Trends Cogn. Sci. **2**, 493–501 (1998)
28. Geisler, C., Diba, K., Pastalkova, E., Mizuseki, K., Royer, S., Buzsáki, G.: Temporal delays among place cells determine the frequency of population theta oscillations in the hippocampus. Proc. Natl. Acad. Sci. USA **107**(17), 7957–62 (2010)
29. Gregory, R.L.: Perceptions as hypotheses. Phil. Trans. R. Soc. Lond. B **290**, 181–197 (1980)
30. Grist, M.: Changing the Subject. RSA. www.thesocialbrain.wordpress.com, pp. 74–80 (2010)
31. Haken, H.: Synergetics: an introduction. In: Non-equilibrium Phase Transition and Self-Organization in Physics, Chemistry and Biology. 3rd edn. Springer, New York (1983)
32. von Helmholtz, H.: Concerning the perceptions in general. In: Treatise on Physiological Optics, vol. III, 3rd edn. (translated by J. P. C. Southall 1925 Opt. Soc. Am. Section 26, reprinted New York: Dover, 1962) (1866)
33. Hinton, G.E., van Cramp, D.: Keeping neural networks simple by minimizing the description length of weights. In: Proceedings of COLT-93, pp. 5–13 (1993)
34. Hohwy, J.: The Predictive Mind. Oxford University Press, Oxford (2013)
35. Hohwy, J.: The self-evidencing brain. Noûs, n/a-n/a (2014)

36. Huang, G.: Is this a unified theory of the brain? New Scientist. Magazine issue 2658, 23 May 2008
37. Kauffman, S.: The Origins of Order: Self-Organization and Selection in Evolution. Oxford University Press, Oxford (1993)
38. Kersten, D., Mamassian, P., Yuille, A.: Object perception as Bayesian inference. Annu. Rev. Psychol. **55**, 271–304 (2004)
39. Kiebel, S.J., von Kriegstein, K., Daunizeau, J., Friston, K.J.: Recognizing sequences of sequences. PLoS Comput. Biol. **5**(8), e1000464 (2009)
40. Knill, D.C., Pouget, A.: The Bayesian brain: the role of uncertainty in neural coding and computation. Trends Neurosci. **27**(12), 712–9 (2004)
41. Kropotova, D., Vetrovb, D.: General Solutions for Information-Based and Bayesian Approaches to Model Selection in Linear Regression and Their Equivalence. Pattern Recognit Image Anal. **19**(3), 447–455 (2009)
42. Lifshitz, E.M., Pitaevskii, L.P.: Physical Kinetics. Course of Theoretical Physics, vol. 10, 3rd edn. Pergamon, London (1981). ISBN 0-08-026480-8 ISBN 0-7506-2635-6
43. MacKay, D.J.C.: Free-energy minimization algorithm for decoding and cryptoanalysis. Electron. Lett. **31**, 445–447 (1995)
44. Mumford, D.: On the computational architecture of the neocortex. II. The role of cortico-cortical loops. Biol. Cybern. **66**, 241–51 (1992)
45. Neal, R.M., Hinton, G.E.: A view of the EM algorithm that justifies incremental, sparse, and other variants'. In: Jordan, M.I. (ed.) Learning in Graphical Models, pp. 355–368. Kluwer Academic Publishers, Dordrecht (1998)
46. Nicolis, G., Prigogine, I.: Self-organization in non-equilibrium systems, p24. Wiley, New York (1977)
47. O'Keefe, J.: Do hippocampal pyramidal cells signal non-spatial as well as spatial information? Hippocampus **9**(4), 352–64 (1999)
48. Rabinovich, M., Huerta, R., Laurent, G.: Neuroscience. Transient dynamics for neural processing. Science **321**, 48–50 (2008)
49. Rabinovich, M.I., Afraimovich, V.S., Varona, P.: Heteroclinic binding. Dyn. Syst. Int. J. **25**, 433–442 (2010)
50. Rabinovich, M.I., Afraimovich, V.S., Bick, V., Varona, P.: Information flow dynamics in the brain. Phys. Life Rev. **9**(1), 51–73 (2012)
51. Rao, R.P., Ballard, D.H.: Predictive coding in the visual cortex: a functional interpretation of some extra-classical receptive field effects. Nature Neurosci. **2**, 79–87 (1998)
52. Rescorla, R.A., Wagner, A.R.: A theory of Pavlovian conditioning: variations in the effectiveness of reinforcement and nonreinforcement. In: Black, A.H., Prokasy, W.F. (eds.) Classical Conditioning II: Current Research and Theory, pp. 64–99. Appleton Century Crofts, New York (1972)
53. Rizzolatti, G., Craighero, L.: The mirror-neuron system. Annu. Rev. Neurosci. **27**, 169–92 (2004).
54. Sella, G., Hirsh, A.E.: The application of statistical physics to evolutionary biology. Proc. Natl. Acad. Sci. USA **102**(27), 9541–6 (2005)
55. Sutton, R.S., Barto, A.G.: Toward a modern theory of adaptive networks: expectation and prediction. Psychol. Rev. **88**(2), 135–70 (1981)
56. Thornton, C.: Some puzzles relating to the free-energy principle: comment on Friston. Trends Cogn. Sci. **14**(2), 53–4; author reply 54–5; (2010)
57. Tsodyks, M.: Attractor neural network models of spatial maps in hippocampus. Hippocampus **9**(4), 481–9 (1999)
58. Varona, P., Levi, R., Arshavsky, Y.I., Rabinovich, M.I., Selverston, A.I.: Competing sensory neurons and motor rhythm coordination. Neurocomputing **58**, 549–554 (2004)

Chapter 13
Olfactory Computation in Insects

M. Komarov, M. Stopfer, and M. Bazhenov

13.1 Synchronized Oscillations in Antennal Lobe

Olfactory neurons respond to static stimuli with oscillatory and temporally structured responses. The presence of odor-elicited oscillations in the olfactory systems of many animals including insects suggests oscillatory neural synchrony may play a fundamental role in odor encoding. Indeed, when the 20 Hz oscillations were specifically blocked by locally injecting picrotoxin into the antennal lobe (AL, Fig. 13.1), locust olfactory neurons in a brain area called the beta lobe underwent a breakdown in the specificity of odor responses [1], and honeybees, in a behavioral test, lost the ability to discriminate similar odors [2]. These results demonstrate oscillatory synchronization of AL neurons is needed for fine odor discrimination.

Computational models helped reveal mechanisms underlying synchronized network oscillations. Spikes in excitatory cells trigger spikes in widely branching postsynaptic inhibitory neurons, whose spikes, in turn, produce inhibitory potentials in excitatory cells. This inhibition systematically delays and synchronizes the generation of the next excitatory response, thereby creating widespread network oscillatory synchrony [3]. A model of spike synchronization in the honeybee AL, also dependent on local inhibition, has been proposed to explain the experimentally observed impairment of sensory discrimination when inhibition is blocked [4].

M. Komarov (✉) • M. Bazhenov
Department of Medicine, University of California, San Diego, CA 92093, USA
e-mail: mkomarov@ucsd.edu; mbazhenov@ucsd.edu

M. Stopfer
National Institutes of Health, NICHD, Bethesda, MD 20892, USA
e-mail: stopferm@mail.nih.gov

I.S. Aranson et al. (eds.), *Advances in Dynamics, Patterns, Cognition*,
Nonlinear Systems and Complexity 20, DOI 10.1007/978-3-319-53673-6_13

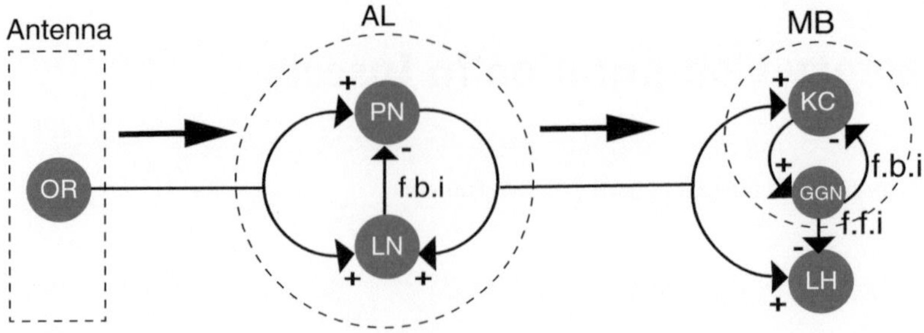

Fig. 13.1 Information flow in the insect olfactory system. Olfactory receptor (OR) neurons in the antenna provide simply structured input to the antennal lobe (AL). There, projection neurons (PNs) and local neurons (LNs) interact in part through feedback inhibition (f.b.i) to transiently synchronize spiking, and to generate complex temporal firing patterns in PNs. PNs then provide distributed and temporally structured output to the Mushroom body (MB) and lateral horn (LH). In the MB, intrinsic neurons called Kenyon cells (KCs) interact through feedback inhibition with the unique Giant GABAergic Neuron (GGN), resulting in very sparse and specific firing in KCs. GGN also provides feedforward inhibitions (f.f.i) to the LH

13.2 The Role of Plasticity in Dynamics of AL

Interestingly, AL oscillations are generally not evoked by the first presentation of an odorant. Oscillations, rather, emerge gradually over the course of multiple encounters with a given odorant, provided the encounters occur close together in time. Electrophysiological [5] and computational [6] evidence suggests the increase in oscillatory synchronization results from activity-dependent plasticity within the AL. In a computational test of this hypothesis, we prepared three versions of a realistic model of the AL [6]: one with fixed synaptic weights, one in which only fast $GABA_A$ receptors could facilitate given repeated stimulation, and one in which both fast and slow inhibitory receptors could facilitate (see complete model description in [6]). In facilitating models, the initial strengths of the inhibitory receptors were set to be too weak to maintain synchronous PN oscillations. Figure 13.2 shows the average network response (local field potential, LFP) and membrane potentials for one PN and one LN from the network during the first five trials with an odor stimulus. The model with fixed, strong synapses responded with relatively consistent patterns in all five trials (Fig. 13.2a). The model with initially weak, facilitating fast $GABA_A$ receptors displayed strong onset responses followed by reduced network activity, caused by the increasing activation of slow inhibitory receptors. Although oscillatory synchrony increased, as observed in vivo, the average number of PN spikes changed very little (less than 30%) during subsequent trials with the same odor, inconsistent with experimental results. Figure 13.2b illustrates the results obtained from the model in which both fast and slow inhibition could facilitate. This network started with intense PN responses only partially reduced by initially weak slow inhibition. PN firing rates were high during the first few trials and decreased over subsequent trials, a result of the facilitation of slow inhibition. Figure 13.2c

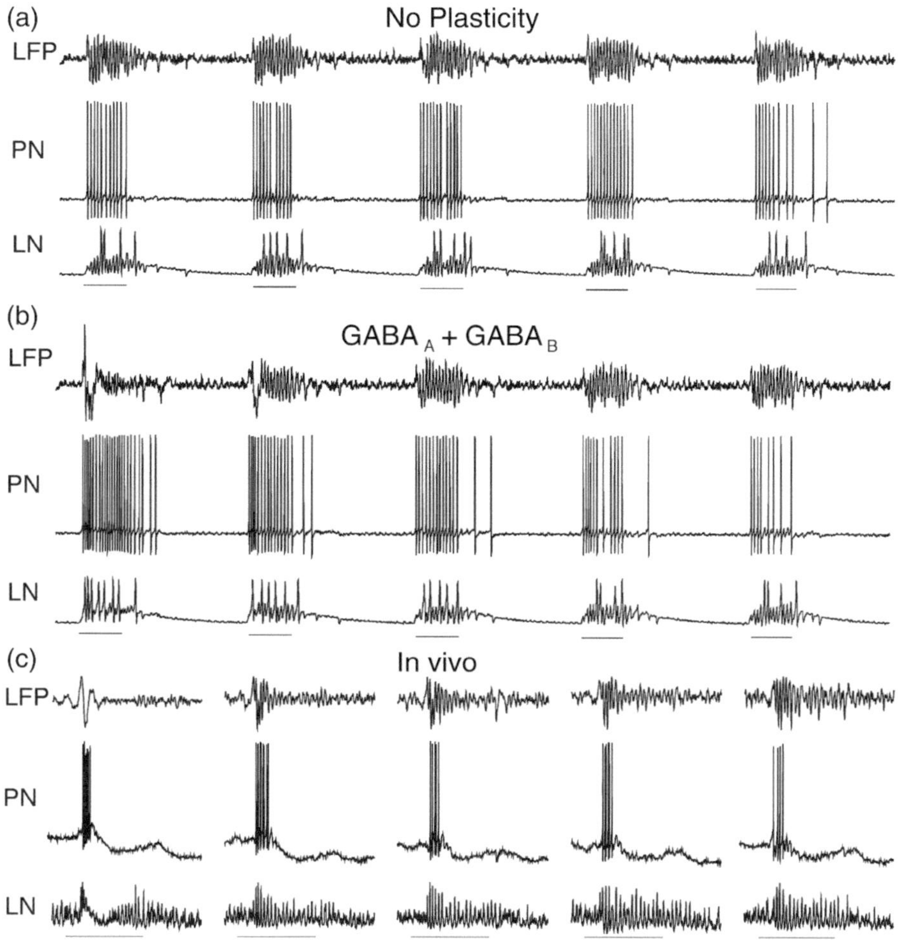

Fig. 13.2 Evolution of the AL responses over repeated stimulus presentations. (**a**) No plasticity. (**b**) Plasticity in both fast $GABA_A$-type and slow $GABA_B$-type synapses. Without facilitation of the slow inhibitory receptors, PN firing rates changed very little during training. Application of odor indicated by 500 ms line beneath responses in each panel. (**c**) Simultaneous recordings of LFP, PN, and LN from the locust reveal a similar response evolution. Calibration: stimulus bar, 1 s; vertical bar, 0.6 (LFP), 18 (PN), 5 (LN) mV

shows experimental results from locust illustrating both the increase in oscillatory power and the decrease in spike count over the first few trials.

Thus, our models suggest that facilitation of both fast and slow inhibition during repetitive trials is needed to account for our experimental results. The models indicate that this plasticity, when engaged by repeating odor stimuli (as often occurs in nature) achieves robust stability against noise.

13.3 Transient Synchronization and Role of Inhibition in Olfactory Computation

In the locust, $GABA_A$-dependent synchronization of a given pair of PNs is generally transient. Usually, after a few hundred milliseconds of firing spikes that are phase-locked to the oscillations, the synchrony between action potentials in a given PN and the field potential oscillations is lost; then, other subsets of PNs become transiently synchronized. Computer models of the locust AL have suggested that the transient nature of PN synchronization could be explained by variations in inhibitory drive from inhibitory LNs over the duration of the odor-elicited response (Fig. 13.3) [7]. To test these ideas we first simulated a small network that had intact inhibitory synapses between two LNs and from the LNs to PNs. The stimulus, delivered to both LNs and to two of the PNs (PN1 and PN2 in Fig. 13.3a), elicited an oscillatory response in the network. Both LNs started to fire synchronously, but the LN–LN inhibition quickly organized the fast periodic synchrony into a slower sequence of out-of-phase patterns (Fig. 13.3b): each LN produced periodic (20 Hz) Ca^{2+} spikes during 100–200 ms epochs, interrupted by epochs of subthreshold oscillations. These interruptions were caused by inhibitory input from the other LN, which displayed a complementary activity pattern. This slower patterning of LN oscillations and antagonistic activity depended on the rate at which LN responses adapted (see Fig. 13.3a) and also on the input from other neurons and external stimulation. The oscillations in one of the LNs always started at a higher frequency,

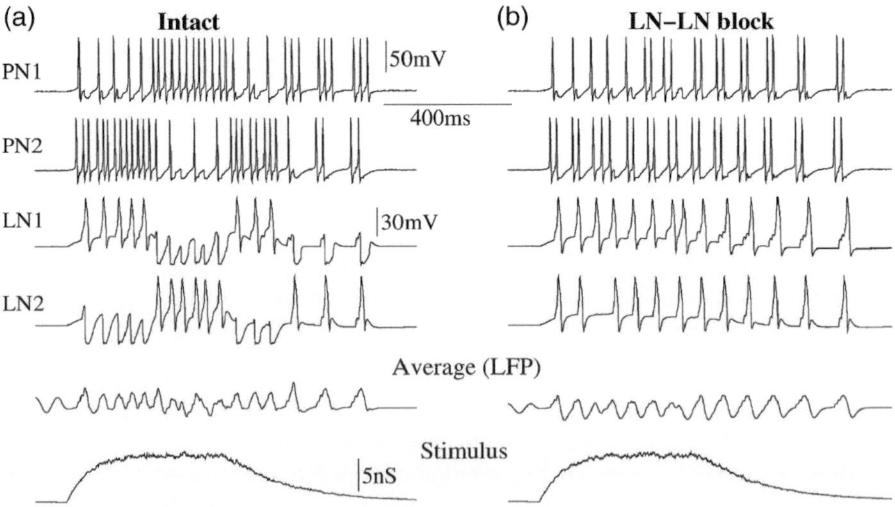

Fig. 13.3 (**a**) Lateral inhibition between LNs led to alternations in the temporal structure of LN oscillation. Variations in LN-evoked fast IPSPs changed the temporal patterns of PN activity. (**b**) Blocking LN–LN reciprocal connections made these cells oscillate regularly, thus decreasing the complexity of PN responses

which dampened the activation of the other LN. After a few cycles, activation of the Ca^{2+}-dependent potassium current in LN1 reduced the frequency of its Ca^{2+} spikes, thus diminishing the inhibitory input to LN2. LN2 then "escaped" from its "silent" state, producing periodic Ca^{2+} spikes, which, in turn, damped oscillations in LN1. The timescale of this slower temporal structure was determined by the rate of LN spike adaptation (controlled mainly by the $K^{+}(Ca^{2+})$ current in LNs) and the strength and time constant of LN–LN interactions. Blocking reciprocal inhibition between LNs eliminated this slow patterning; both neurons then displayed sustained synchronous periodic activity (Fig. 13.3b).

13.4 Computational Properties of Network of Inhibitory Interneurons in AL

Recent theoretical work led to a new strategy based on graph coloring theory to describe transient AL network dynamics [8]. In inhibitory neuronal networks, LNs that are not interconnected are labeled with a particular "color"; LNs sharing this color can be active at the same time and can collectively inhibit LNs associated with a different color. In a model of the AL, these dynamics lead to alternating patterns of firing among groups of LNs [9]. This model of inhibitory network dynamics mediated by lateral inhibition between LNs can account for transient synchronization patterns of olfactory neurons in the AL. The role LN–LN inhibition may play in promoting competition among olfactory glomeruli has been also explored [10, 11] and a model of olfactory memory based on altering lateral inhibition in the honeybee's AL has been proposed [12].

Another recent theoretical study [13] demonstrates that even a minimal inhibitory network can generate a great diversity of spatio-temporal patterning including complex regimes with non-trivial firing burst ratios. We show that the heterogeneity of the input profile leads to rich spatiotemporal bursting dynamics, including non-regular patterns and complex bursting regimes, when neurons are locked in non-trivial activation ratios (2:3, 1:2, 1:3 etc.). Figure 13.4 shows several distinctly different possible types of activity in small networks, which consist of two (panels a, b) and 3 (panels c–f) mutually connected inhibitory neurons. The pair of mutually inhibiting neurons demonstrated two types of dynamics: (1) the first type represents a complete inhibition of one neuron by another one (Fig. 13.4a); (2) the second type is an alternating bursting rhythm (Fig. 13.4b). The strength of internal adaptation current I_{AHP} (Ca^{2+}-dependent K^{+} afterhyperpolarization current) controlled the type of behavior in the neural circuit. In the first case (Fig. 13.4a) the spike-frequency adaptation was too weak (because of the small I_{ahp}), such that the presynaptic neuron oscillated at a high firing rate, sufficient to constantly suppress the postsynaptic neuron. In contrast, when I_{AHP} was large, the interplay between intrinsic spike-frequency adaptation and mutual inhibition led to the alternating bursting rhythm shown in Fig. 13.4b [14]. The strong adaptation current I_{AHP} caused

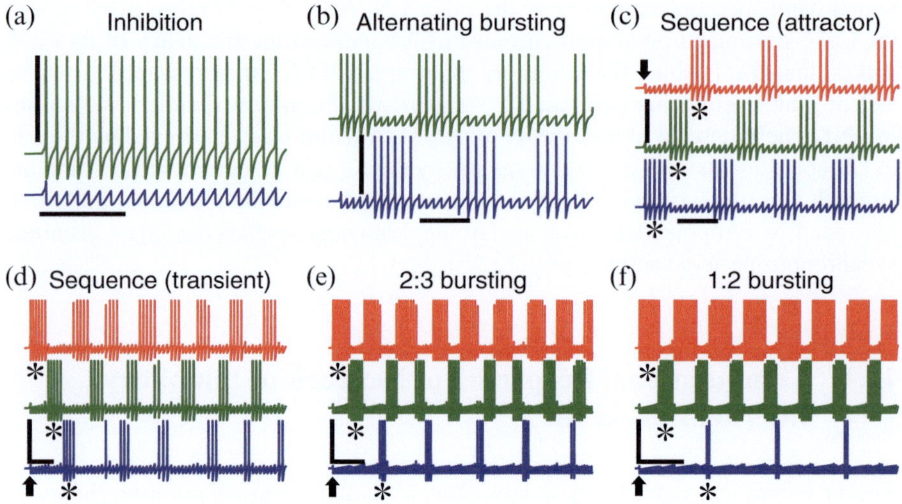

Fig. 13.4 Sequential bursting dynamics in a network of inhibitory neurons in a Hodgkin–Huxley-type model. (**a**) One neuron is active and completely inhibits activity in the other neurons. (**b**) Alternating bursting rhythm caused by an increase in the impact of adaptation current (I_{AHP}). (**c**) Simple bursting wave in the network of three mutually inhibiting neurons is shown. (**d**) Non-regular behavior induced by strong heterogeneity of input profile in the network. (**e**), (**f**) Strong difference in external inputs I_i^{ext} induced complex bursting, characterized by non-trivial burst activation ratios: 2:3 for panel (**e**) and 1:2 for panel (**e**). *Stars* denote bursts during initial transient period, which reflected ranking of external inputs: $I_3^{ext} > I_2^{ext} > I_1^{ext}$ for panel (**c**) and $I_1^{ext} > I_2^{ext} > I_3^{ext}$ for panels (**d**)–(**f**) (numeration from top to bottom). *Vertical arrows* indicate the stimuli onset. The network had been in a resting state before inputs were applied. Baseline current $I_0 = 2.2$

a significant decrease in the firing rate of the active presynaptic neuron (after a period of activity), which resulted in the escape of the postsynaptic neuron from inhibition. Hence, neither of the two neurons could suppress the activity of its postsynaptic target for very long, giving rise to the anti-phase bursting rhythm [14] shown in Fig. 13.4b.

Similarly, the intrinsic adaptation current determined the type of behavior generated by all-to-all connected networks that contain $N > 2$ inhibitory neurons with strong synaptic connections (Fig. 13.4c–f). As in the previous case, weak adaptation (small I_{AHP}) allowed one neuron to dominate with periodic spiking activity (not shown), but when I_{AHP} was strong, the interplay between intrinsic spike-frequency adaptation and mutual inhibition resulted in sequential bursting dynamics. However, in contrast to the relatively simple alternating bursting patterns observed in pairs of neurons, networks with more than two elements exhibited much more complex behavior [15], ranging from fairly simple bursting waves (for relatively small $|\Delta I|$, Fig. 13.4c) to the more complex bursting patterns shown in Fig. 13.4d–f for larger $|\Delta I|$. Here $|\Delta I|$ represents a key parameter in our study, which reflects the difference between external inputs I_i^{ext} (index i denotes number of the neuron), which were applied to inhibitory neurons in the network:

$$I_1^{ext} = I_0 + \Delta I,\ I_2^{ext} = I_0 + \Delta I/2,\ I_3^{ext} = I_0 \tag{13.1}$$

Large parameters of $|\Delta I|$ indicates large heterogeneity of the input profile. It is worth noting that in the simulation shown in Fig. 13.4c, the detuning parameter was negative ($\Delta I < 0$), meaning that the stimuli were ranked as $I_3^{ext} > I_2^{ext} > I_1^{ext}$ [see Eq. (13.1)]. This stimulus regime resulted in a counter-clockwise oriented bursting wave ($3 \rightarrow 2 \rightarrow 1$), with activity following the ranking of external stimuli (we refer to this as the "correct" sequence of activations). Notably, the counter-clockwise bursting wave was an attractor (for relatively small $|\Delta I|$), meaning that the network will go on to generate the same sequence repeatedly. For larger values of the heterogeneity parameter $|\Delta I|$ the resulting dynamics can be fairly complicated, as illustrated by several examples in Fig. 13.4c–f. Strongly heterogeneous input profiles led to complex dynamics when highly stimulated elements got activated more frequently (Fig. 13.4d–f). In some cases this stimulus regime led to the non-regular intermittent dynamics shown in Fig. 13.4d. However, high heterogeneity $|\Delta I|$ typically induced oscillations with fixed non-trivial ratios of bursting activation. Examples are given in Fig. 13.4e,f. For both cases ΔI was positive and sufficiently large, so the third element received the smallest stimulus. As a result, the ratio of activations of the third neuron to the first two neurons was 2:3 in panel h ($\Delta I = 1.3$) and 1:2 in panel i ($\Delta I = 1.75$).

To expand our hypothesis regarding the link between external inputs amplitudes I_i^{ext} and spatio-temporal bursting dynamics of the network, below we analyze sequences of bursts generated by the network of inhibitory $N = 3$ neurons given a generic form of the input.

Figure 13.5a presents a division of the plane of external inputs into regions of different activation sequences generated by an adaptive integrate-and-fire model (description of the model in [13]). Here, sequences are the orders in which neurons produced bursts during the initial period following stimuli onset (Fig. 13.5c–f). In areas $B_{1,2}$ and $C_{1,2}$ the activation order was strictly determined by the ranking of stimuli. In those regions of inputs we observe (a) high reliability of the network responses: different sets of external stimuli I_i^{ext} can elicit the same spatio-temporal patterns; and (b) sensitivity to the form of the input: a change in the ranking of I_i^{ext} immediately causes a change in the elicited spatio-temporal pattern. In the areas B_3 and C_3 the overall difference of external currents was large enough to cause the first two neurons to suppress the third one, at least during first two activation cycles. Hence, our analysis predicts that the knowledge of the sequence of bursting in the inhibitory network makes it possible to reconstruct the ranking of the amplitudes of external stimuli driving the neurons. We tested this hypothesis using the elementary network of $N = 3$ globally coupled inhibitory neurons. Below we will show that these principles are generic and hold for the larger networks.

In many systems, activation of sensory receptor neurons can be described by a sigmoidal function of the odor concentration (Fig. 13.6a) [16]. In the olfactory system, for example, different types of odor receptors have different sensitivities for a given odor that can be described as a relative shift in their activation curves

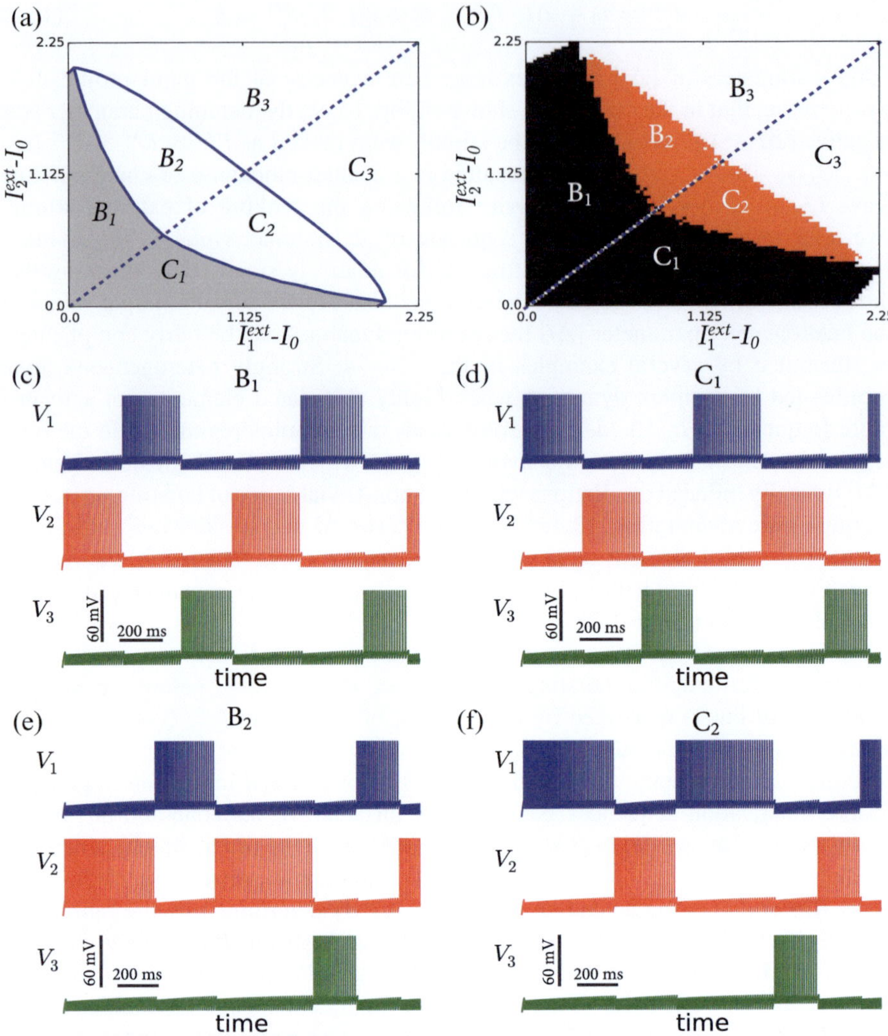

Fig. 13.5 Sequences of bursts in the network of $N = 3$ neurons predict the pattern of external inputs. (**a**) Different regions on plane of parameters $(I_1^{\text{ext}}, I_1^{\text{ext}})$ are shown for the integrate-and-fire model (see model description in [13]). Baseline current (which was the same for all three neurons) $I_0 = 2.0$. (**b**) The same as in the panel (**a**) but for the conductance-based model (see model description in [13]). (**c**)–(**f**) The dynamics of neurons modeled in different regions on the plane $(I_1^{\text{ext}}, I_2^{\text{ext}})$ [panel (**a**)]

(Fig. 13.6a). In insects, the axons of olfactory receptor neurons project to the antennal lobe (AL) where they make connections to populations of excitatory projection neurons (PNs) and local neurons (LNs) that are, for the most part, inhibitory. We showed previously that the inhibitory connections between LNs determine the pattern of sequential LN activation, and also that this pattern underlies

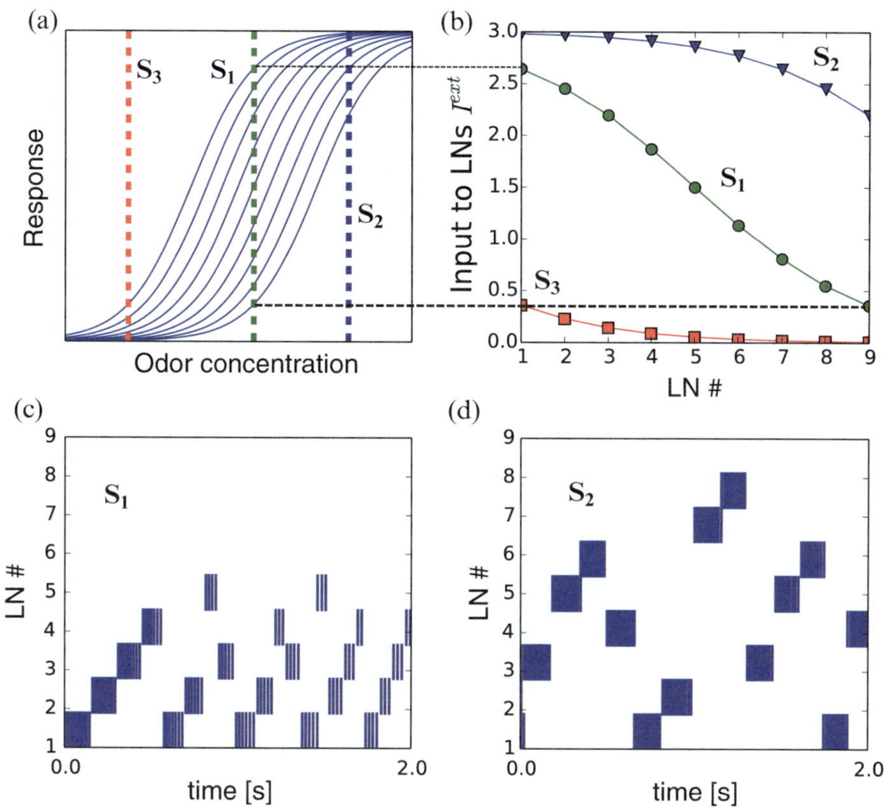

Fig. 13.6 Dynamics of the inhibitory network ($N = 9$) in the context of the information processing in the olfactory system. (**a**) Responses of different olfactory receptor neurons (different curves) to a given odor as a function of an odor concentration (x-axis). (**b**) Inputs I_i^{ext} from different olfactory receptor neurons to the inhibitory LNs of the AL network model are shown as a function of the LN index in the network (x-axis). The plot illustrates three different stimuli $S_{1,2,3}$. S_1 corresponds to the moderate odor concentrations and the large difference in the receptor neurons firing rates. S_2 and S_3 correspond to the strong and weak odor concentrations, respectively. (**c**), (**d**) Spatio-temporal network dynamics are plotted for the different stimuli S_1 [panel (**c**)], S_2 [panel (**d**)]

the synchronization properties of postsynaptic excitatory PNs [8, 14]. This earlier work, however, assumed all of the inhibitory neurons of the AL network received identical input. Below, we consider a more realistic network of $N = 9$ inhibitory AL neurons, each receiving different inputs from receptor neurons. Although this model was not specifically designed to simulate the detailed dynamics of the AL, it can nonetheless provide important insights about patterns of activity in the densely connected population of olfactory neurons. We tested a stimulus at three different concentrations, denoted $S_{1,2,3}$ (Fig. 13.6b). For the sake of simplicity, we assumed an equidistant profile of the inputs. For the stimulus (S1) with a "moderate" concentration, distinct olfactory receptor types responded with different and distinct firing rates without saturating, thus providing a representation of the input with a large dynamic range (differences in I_i^{ext}) to the inhibitory neurons (LNs of the AL) (Fig. 13.6a, b, green line).

The dynamics of the network simulated by the conductance-based model (see full description of the model in [13]) is shown in Fig. 13.6c. Relatively large differences of I_i^{ext} in the applied inputs resulted in a clear bursting pattern: the neurons were activated according to the ranking of external stimuli I_i^{ext}. Because of noise in the system, this precise dynamical pattern degraded over time, with subsequent sequences displaying less fidelity to the input ranking. Thus, in agreement with our theoretical analysis, the initial (upon stimuli onset) sequence of bursts generated by the network reflected the ranking of the inputs, and allowed reconstruction of the relative magnitudes of the external inputs from observed patterns of activity. When the odor intensity was high (S_2), olfactory receptors responded at a high firing rate, but within a narrow range of amplitudes across receptor types, providing a small dynamical range of inputs to the LNs of the AL (Fig. 13.6a, b, blue line). In this case the outputs of the receptors saturated ("upper" part of the sigmoid, see Fig. 13.6a, b, blue line). In a network given inputs containing only a small difference in the range of intensity, sequence generation is unstable against noise, so the network dynamics did not well reflect input ranking (Fig. 13.6d). As expected, a low odor concentration S_3 (Fig. 13.5a) resulted in a small and narrow distribution of inputs I_i^{ext} to LNs (Fig. 13.6a, b, red line), which failed to evoke any significant response from the LN network (data not shown).

We conclude that the relationship we propose between the dynamics of the inhibitory network and the structure of the input can degrade in noisy networks that are given only a narrow range of input amplitudes. However, the relationship held firm in the networks when the input pattern contains enough dynamic range to offset the effects of noise.

13.5 Conclusions

While the synchronization of PNs in the insects AL can be abolished by application of the GABA$_A$-blocker picrotoxin, experiments performed in vivo have shown that the slower firing pattern structures (alternating depolarizing and hyperpolarizing modes) are not affected by this treatment [1, 2]. In locusts, the origins of slow temporal patterning have been explored with electrophysiological studies and computational models. These patterns begin in olfactory receptor neurons, which respond to odors with a variety of simple firing patterns. The heterogeneity of these responses is an essential ingredient for the generation of more elaborate patterns downstream in the AL [17]. There, in the AL, the slow inhibitory receptors between LNs and PNs, which operate with a time constant of a few hundred milliseconds, are essential for translating the simple but varying input received from the receptor neurons into the stimulus-specific slow temporal patterning of PN responses [18].

The sequences in which periods of excitation, inhibition, and quiescence appear are different among PNs, and vary reliably with the stimulus. Relatively slow calcium and calcium-dependent potassium channels of PNs may further contribute to slow temporal patterning in the AL [19].

What is the functional significance of the temporal structure of olfactory neuron responses? As shown in zebrafish [20], the slow temporal patterning in mitral cells appears to play a major role in the decorrelation of odor representations. The responses of a population of mitral cells to a series of chemically similar odorants were highly correlated at the onset of the odor stimulation, but became progressively less correlated over the duration of the odor response. Recent computer modeling of the AL suggested that odor-specific evolution of PN activity can lead to an optimal encoding of the odor representation that decreases the overlap between odor-elicited PN spiking patterns [17, 21]. This increases the ability of downstream neurons to discriminate among odors. A balance between inhibitory and excitatory connections mediated by local AL interneurons enhances the decorrelation of similar odors while keeping the representation robust in the presence of noise.

In insects, output from relatively small populations of olfactory neurons in the AL fans out broadly to large interneuronal ensembles in the mushroom body (Fig. 13.1). Fully half of all PNs synapse directly upon each KC in the mushroom body, a connectivity scheme that maximizes the coding space available for odors [22]. Electrophysiological recordings indicate that KCs respond sparsely and specifically to odors: a given odor induces only a brief response in only a small subpopulation of KCs [23]. Mechanisms contributing to the specificity of this response include intrinsic nonlinear conductances in the KCs that respond best to coincident input from PNs, and inhibitory input from GGN [24] that regulates the amount of firing in KCs (Fig. 13.1). Physiological and modeling results show the precise timing established by AL circuitry is maintained not only within KCs, but also by their neural followers in the beta lobe [25]. Interestingly, odor specific timing precision is maintained in the beta lobe neurons, in part, by spike-timing dependent plasticity (STDP).

Computational models of the insect olfactory system have explored the proposal that AL circuitry reformats information about odors, distributing it into the identities of PNs responding to the odor, and into the relative timing of spikes in those PNs. Temporal features appear to become critical to the animal when discrimination tasks are more challenging, such as discriminating among PN ensembles that substantially overlap. Thus, AL actively sharpens information about the olfactory stimulus by employing time and transient synchrony as coding dimensions. Understanding and testing hypotheses about these complex biological processes has been greatly facilitated by the use of computational models.

Acknowledgements The work is supported by NIH (R01 DC012943) to M. Bazhenov and M. Stopfer, and an intramural grant from NIH-NICHD to M. Stopfer.

References

1. MacLeod, K., Becker, A., Laurent, G.: Who reads temporal information contained across synchronized and oscillatory spike trains? Nature **395**(6703), 693–698 (1998)
2. Stopfer, M., Bhagavan, S., Smith, B.H., Laurent, G.: Impaired odour discrimination on desynchronization of odour-encoding neural assemblies. Nature **390**(6655), 70–74 (1997)
3. Borgers, C., Kopell, N.: Synchronization in networks of excitatory and inhibitory neurons with sparse, random connectivity. Neural Comput. **15**, 509–538 (2003)
4. Linster, C., Cleland, T.: How spike synchronization among olfactory neurons can contribute to sensory discrimination. J. Comput. Neurosci. **10**, 187–193 (2001)
5. Stopfer, M., Laurent, G.: Short-term memory in olfactory network dynamics. Nature **402**, 664–668 (1999)
6. Bazhenov, M., Stopfer, M., Sejnowski, T.J., Laurent, G.: Fast odor learning improves reliability of odor responses in the locust antennal lobe. Neuron **46**, 483–492 (2005)
7. Bazhenov, M., Stopfer, M., Rabinovich, M., Huerta, R., Abarbanel, H.D.I., Sejnowski, T.J., Laurent, G.: Model of transient oscillatory synchronization in the locust antennal lobe. Neuron **30**(2), 553–567 (2001)
8. Assisi, C., Stopfer, M., Bazhenov, M.: Using the structure of inhibitory networks to unravel mechanisms of spatiotemporal patterning. Neuron **69**(2), 373–86 (2011)
9. Komarov, M.A., Osipov, G.V., Suykens, J.A.K.: Sequentially activated groups in neural networks. Europhys. Lett. **86**(6), 60006 (2009)
10. Linster, C., Marsan, D., Masson, C., Kerszberg, M.: Odor processing in the bee: a preliminary study of the role of central input to the antennal lobe. Adv. Neural Inf. Proces. Syst. **6**, 527–534–492 (1994)
11. Kerszberg, M., Masson, C.: Signal-induced selection among spontaneous oscillatory patterns in a model of honeybee olfactory glomeruli. Biol. Cybern. **72**, 487–495 (1995)
12. Linster, C., Masson, C.: A neural model of olfactory sensory memory in the honeybee's antennal lobe. Neural Comput. **8**, 94–114 (1996)
13. Komarov, M., Bazhenov, M.: Linking dynamics of the inhibitory network to the input structure. J. Comput. Neurosci. **41**, 367–391 (2016)
14. Assisi, C., Bazhenov, M.: Synaptic inhibition controls transient oscillatory synchronization in a model of the insect olfactory system. Front. Neuroeng. **5**, 7 (2012)
15. Wojcik, J., Schwabedal, J., Clewley, R., Shilnikov, A.L.: Key Bifurcations of bursting polyrhythms in 3-cell central pattern generators. PLoS One **9**(4), e92918 (2014)
16. Ito, I., Bazhenov, M., Ying Ong, R.C., Raman, B., Stopfer, M.: Frequency transitions in odor-evoked neural oscillations. Neuron **64**(5), 692–706 (2009)
17. Raman, B., Joseph, J., Tang, J., Stopfer, M.: Temporally diverse firing patterns in olfactory receptor neurons underlie spatiotemporal neural codes for odors. J. Neurosci. **30**, 1994–2006 (2010)
18. Bazhenov, M., Stopfer, M., Rabinovich, M., Abarbanel, H.D.I., Sejnowski, T. J., Laurent, G.: Model of cellular and network mechanisms for odor-evoked temporal patterning in the locust antennal lobe. Neuron **30**(2), 569–581 (2001)
19. Sivan, E., Kopell, N.: Oscillations and slow patterning in the antennal lobe. J. Comput. Neurosci. **20**, 85–96 (2006)
20. Friedrich, R.W., Laurent, G.: Dynamics of olfactory bulb input and output activity during odor stimulation in zebrafish. J. Neurophys. **91**(6), 2658–2669 (2004)
21. Assisi, C., Stopfer, M., Bazhenov, M.: Excitatory local interneurons enhance tuning of sensory information. PLoS Comput. Biol. **8**, e1002563 (2012)
22. Jortner, R.A., Farivar, S.S., Laurent, G.: A simple connectivity scheme for sparse coding in an olfactory system. J. Neurosci. **27**, 1659–1669 (2007)
23. Perez-Orive, J., Mazor, O., Turner, G.C., Cassenaer, S., Wilson, R.I., Laurent, G.: Oscillations and sparsening of odor representations in the mushroom body. Science **297**, 359–365 (2002)

24. Papadopoulou, M., Cassenaer, S., Nowotny, T., Laurent, G.: Normalization for sparse encoding of odors by a wide-field interneuron. Science **332**, 721–725 (2011)
25. Cassenaer, S., Laurent, G.: Hebbian STDP in mushroom bodies facilitates the synchronous flow of olfactory information in locusts. Nature **448**, 709–713 (2007)

Chapter 14
Respiratory Neural Network: Activity and Connectivity

Laurence Mangin and Maurice Courbage

14.1 Introduction

The human respiratory system displays several levels of complexity: the bronchial tree has a fractal structure with various degrees of self-similarity and the airflow dynamics inside exhibits chaos during rhythmic breathing (Fig. 14.1).

Why rhythmic breathing generates chaos in human airflow?

Misha Rabinovich, whom we are happy to dedicate this article on the occasion of his 75th anniversary, was the precursor to think this fundamental problem. Rabinovich and Abarbanel [20] consider that the utility of chaos for nervous system can come from the multitude of synchronization schemes forming regular and organized behaviors in the connected networks of the brain. In neural activity in general, synchronization is basic for maintaining organized mechanism. Its deregulation may lead to decease for many reasons among them respiratory failure.

Breathing involves a complex interplay between the automatic brainstem network and the voluntary cortical command [9, 22]. Both networks interact harmoniously to control respiratory muscles contraction, thereby ensuring normal blood

L. Mangin (✉)
Laboratoire Matière et Systèmes Complexes (MSC), UMR 7057 CNRS et Université Paris 7 - Denis Diderot, Case 7056, Bâtiment Condorcet, 10, rue Alice Domon et Léonie Duquet, 75205 Paris Cedex 13, France

Physiology Dpt, APHP, Bichat Hospital, Paris 7 University, Paris, France
e-mail: laurence.mangin@aphp.fr

M. Courbage
Laboratoire Matière et Systèmes Complexes (MSC), UMR 7057 CNRS et Université Paris 7 - Denis Diderot, Case 7056, Bâtiment Condorcet, 10, rue Alice Domon et Léonie Duquet, 75205 Paris Cedex 13, France
e-mail: maurice.courbage@univ-paris-diderot.fr

I.S. Aranson et al. (eds.), *Advances in Dynamics, Patterns, Cognition*,
Nonlinear Systems and Complexity 20, DOI 10.1007/978-3-319-53673-6_14

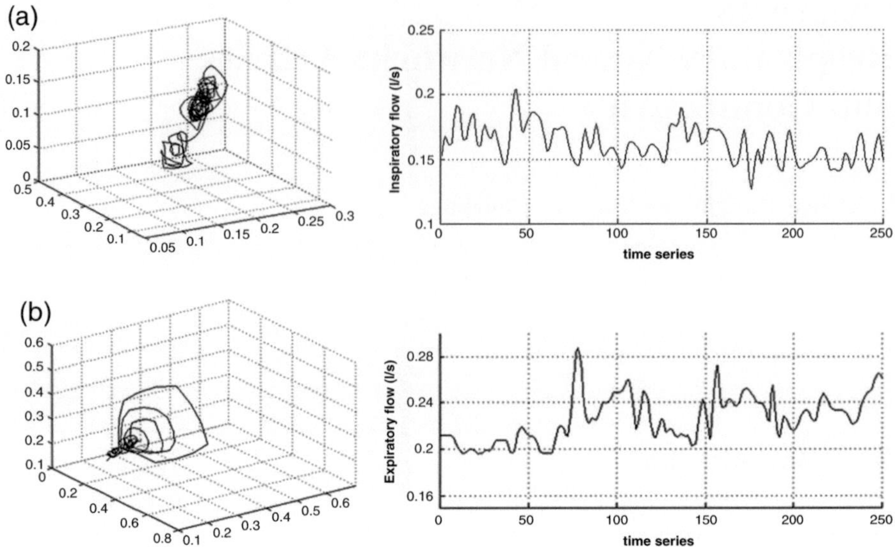

Fig. 14.1 The chaotic behaviors of the airflow in one healthy subject and one COPD patient are evidenced. The reconstructed attractors in the phase plane are shown on the *left panel* for one healthy subject during inspiration (**a**) and for one COPD patient during expiration (**b**). The corresponding time series are shown on the *right panel*

gas levels either during speech, volitional breathing, or ventilatory load increase. However until now the communication within the network is unknown.

In the first section we present a study of the link between chaos in the time series of successive amounts of airflow breath and the levels of activation of the neural circuits that control the automatic respiration. We will see how the activity of these circuits is affected by the disease of respiratory failure (COPD) and will study the role of chaos as a marker of the disease. A mathematical model, inspired by recent works on animals, was adapted to model these behaviors and compared with fMRI data.

In the second section we will see that when breathing at rest and during inspiratory loading, the motor cortex area in the patients had weaker connectivity, compared with the controls (healthy subjects), and no connectivity with the brainstem. These findings are crucial for patients. Causality analysis showed a reversal of the roles of the brainstem and motor cortex in the patients, as compared with the controls. This connectivity analysis highlights that the neural driver role of the network shifted in the patients from the medulla to the motor cortex. We shortly discuss the physiological impacts of these findings.

14.2 Chaos and Excitability in the Respiratory System

Breathing is maintained and controlled by a network of neurons in the brainstem that generates respiratory rhythm while receiving regulatory inputs. Pace-maker like neurons generating rhythmic breathing have been first identified in two brainstem regions in rodents, one located in the rostral ventro-lateral (VL) medulla, called the pre-Bötzinger complex [7], and the other close to this region, called the parafacial respiratory group located in the pons [15, 18, 24]. The pre-Bötzinger complex is in charge of inspiration and the parafacial group in charge of expiration. Recent evidence suggests that both groups of neurons are coupled oscillators that work in tandem to synchronize respiratory rhythm [15, 18, 25]. Moreover, these automatic neuronal groups have two important properties: they are capable of different synchronization regimes depending on the level of their excitabilities [25] and their dynamics exhibit chaotic spike-bursting oscillations in some circumstances [4]. Indeed, neural population activity recorded locally in the pre-Bötzinger complex of neonatal rat brainstem slices exhibits chaotic dynamics, when neuronal excitability is progressively elevated [4]. This is a strong argument to hypothesize that the chaos-like complexity of airflow in humans is also an intrinsic property of neural central respiratory generators. It is known that breathing is also modulated by the state of airways, by the chest wall, the lung, by chemical afferents sensitive to anomalies like hypercapnia, hypoxia, or acidosis and by mechanical afferents from the airway, lung, chest wall, respiratory muscles as well as by commands from the upper part of the pons. A study has shown that the structural and mechanical properties of the bronchial tree, lung, and chest wall in humans are not sufficient to generate chaos in airflow in the absence of a central neural drive [17]. Nevertheless, it was still unclear in humans to what extent the complex and chaotic dynamics of the respiratory center contributes to airflow complexity. Therefore, we used both experimental and theoretical approaches to decipher the brainstem neural substrates of ventilatory complexity in healthy humans and patients with chronic obstructive pulmonary disease (COPD).

COPD is the most frequent chronic lung disease in the general population and is mainly due to tobacco smoke. Patients with COPD have an impaired lung function with an increased respiratory load due to small airways obstruction by inflammation and remodeling. Lung parenchyma destruction is often associated with small airways obstruction and hyperinflation and respiratory muscles weakness. These are key factors contributing to load-capacity imbalance and hence increased respiratory drive [16]. At the end stage of the disease, the patients have respiratory insufficiency with home oxygen therapy while the neural respiratory drive is extremely high. We explored airflow data from a population of COPD patients and healthy subjects. We found that patients with COPD breathe with a higher level of complexity in airflow than healthy subjects (Fig. 14.2). These findings cast doubt on the traditional view that complexity systematically degrades in disease state [11, 19]. Inspiratory and expiratory complexity changes in parallel with the activity of the VL medulla and VL pons as assessed through fMRI (Fig. 14.2).

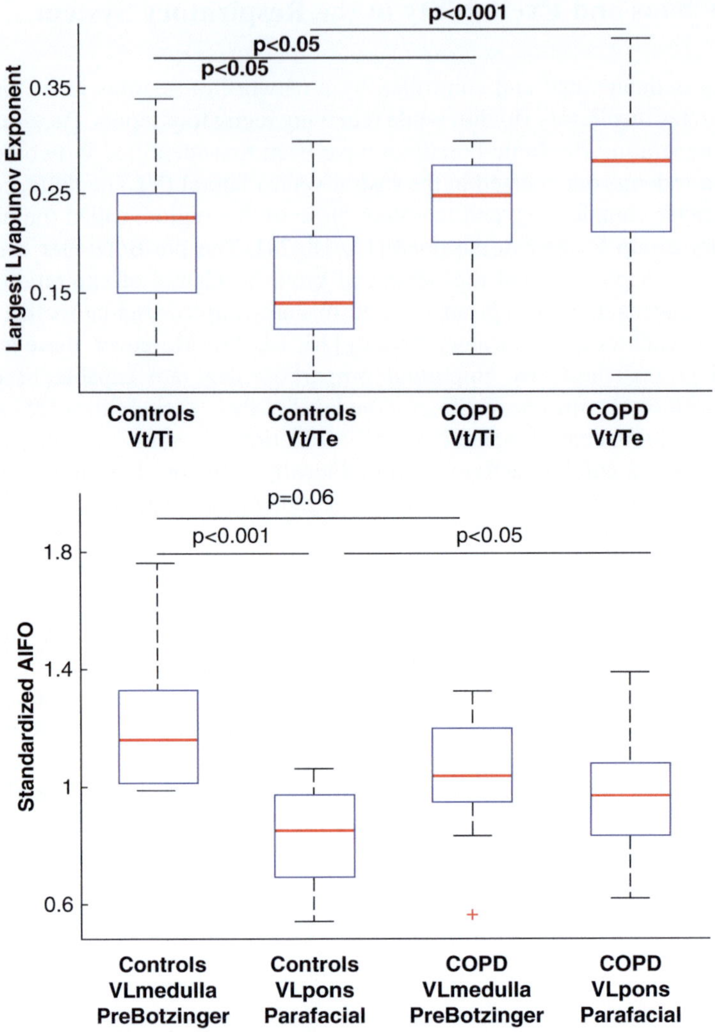

Fig. 14.2 *Top*. Largest Lyapunov exponent of airflow during inspiration (Vt/Ti) and expiration (Vt/Te) in the controls ($n = 25$) and COPD patients ($n = 25$). *Bottom*. fMRI results of the brainstem respiratory centers at rest. Amplitude of the low frequency oscillations (ALFO) of the resting state fMRI BOLD signal computed in controls and COPD patients. In healthy subjects the AlFO of the rostral ventro-lateral (VL) medulla that contains the pre-Bötzinger complex is higher than the VL medulla of the patients. Conversely, the ALFO of the caudal (VL) pons, which contains the parafacial respiratory group is higher in patients than the VL pons of healthy subjects. The boxes encompass the interquartile range with indication of the median, the whiskers delimit the 95th percentile of the data distribution. Paired and unpaired Ttest

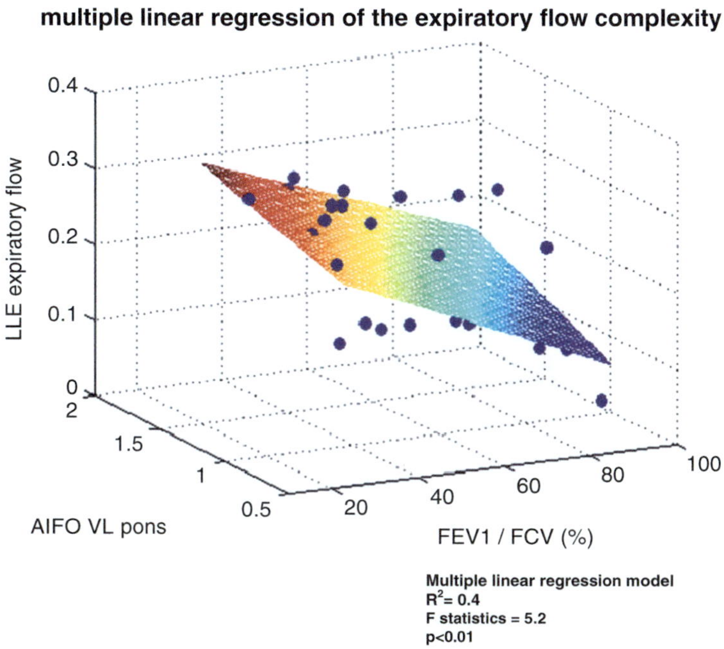

Fig. 14.3 Central neural correlates of airflow dynamics in the whole population using multiple linear regression. Both the amplitude of the low frequency oscillations (ALFO) and the pulmonary function index FEV1/FVC significantly predict expiratory flow complexity: the lower the pulmonary function, the higher the value of the AIFO of ventro-lateral pons and the higher the complexity of expiration

Moreover, we found in patients that chaos in the expiratory flow was higher than the inspiratory flow, and this was correlated with the high parafacial activity, the neural group responsible for expiration (Fig. 14.3). Remember that for healthy subjects important activation of the pre-Bötzinger complex is linked to a high active inspiration.

These results were reinforced by the mathematical model reproducing respiratory rhythmogenesis in healthy subjects and patients [13].

14.3 Mathematical Model of Respiratory Rhythmogenesis

Two pacemaker-like neurons have been identified in mammals in the ventro-lateral column of the brainstem, the pre-Bötzinger complex inspiratory group and parafacial expiratory group, respectively [7, 15, 18, 24]. Previous works showed that the parafacial group exhibits pre-inspiratory activity [8, 18] as well as a rebound bursting after inspiration [8] while the dynamics of both pacemakers display chaotic spike-bursting oscillations [4]. We therefore chose to develop a map-based model for respiratory rhythmogenesis for its relative simplicity compared with Hodgkin–Huxley formalism, and for its ability to generate several regimes of spontaneous bursting activity [1, 2]. The model is developed based on the discrete version of FitzHugh–Nagumo model by adding Heaviside step function $H(x)$. Each pacemaker is modeled by the two-dimensional original map which is further refined to incorporate post-inhibitory rebound bursting behavior typical in the automatic breathing involving the pre-Bötzinger complex inspiratory group and parafacial expiratory group:

$$\begin{cases} \bar{x} = x + F(x) - y - \beta H(x-d) \\ \bar{y} = y + \varepsilon(x_1 - (J + I_T + I\mathrm{syn})) \\ \bar{k} = k + G(k) \end{cases} \tag{14.1}$$

where x qualitatively defines the dynamics of the membrane potential of the neuron and y is the common variable specifying the dynamics of all outwards ionic currents (recovery variable). β and d control the threshold properties of the oscillation, ϵ is a positive parameter setting the time scale of the recovery variable y. J is associated with excitability properties of the neuron; $F(x)$ is a piece-wise linear version of the cubic function in the FitzHugh–Nagumo model and $H(x)$ as follows:

$$F(x) = \begin{cases} -m_0 x, & x < J_{\min}, \\ m_1(x-a), & J_{\min} \le x < J_{\max}, \\ m_0(x-1), & x \ge J_{\max}, \end{cases} \quad H(x) = \begin{cases} 1, & x \ge 0, \\ 0, & x < 0. \end{cases} \tag{14.2}$$

with

$$J_{\min} = \frac{am_1}{m_0 + m_1}, \quad J_{\max} = \frac{m_0 + am_1}{m_0 + m_1}, \quad m_0, m_1 > 0.$$

I_T is a low-threshold calcium Ca^{2+} current [23] defined as:

$$I_T = \delta k H(x - x_{\mathrm{th}}) \tag{14.3}$$

where k in Eq. (14.2) is a slow variable representing the inactivation of the low-threshold calcium conductance, which involves T-type Ca^{2+} calcium channels and produces a transmembrane current I_T. δ represents the maximum conductance associated with I_T. $G(k)$ represents the dynamics of I_T as follows:

$$G(k) = \begin{cases} -\frac{k}{\tau_1} & \text{if} \quad x \ge x_{\mathrm{th}} \\ \frac{1-k}{\tau_2} & \text{if} \quad x < x_{\mathrm{th}} \end{cases} \tag{14.4}$$

In this form the model is capable of post-inhibitory rebound bursting when x_{th} is below the resting values of x. In Eq. (14.4), τ_1 sets the duration of the burst and τ_2 sets the duration of the hyperpolarization necessary to recruit a maximal post-inhibitory rebound response.

In Eq. (14.1), *I*syn is the chemical synaptic coupling between the pre-Bötzinger complex and the parafacial group in the following form:

$$I_{syn} = \delta K \sum_{n^i<n} \text{rect}(n^i, n\tau) \tag{14.5}$$

where K is the coupling strength whose value is positive for excitatory synapse and negative for inhibitory synapse and rect is the rectangle function as described below:

$$\text{rect}(n^i, n, \tau) = \begin{cases} 0 & \text{if} \quad |n - n^i| > \tau \\ 1 & \text{if} \quad |n - n^i| \leq \tau \end{cases} \tag{14.6}$$

where n^i is the step of the ith spike in the presynaptic neuron and τ is the duration of the postsynaptic current. A post-inspiratory inhibitory feedback is introduced from the parafacial respiratory group with the same amplitude and duration of the rebound bursts for "inspiratory off-switch" to prevent the pre-Bötzinger from reactivation.

The present model is the first attempt to reproduce respiratory rhythmogenesis in healthy humans and COPD patients with experimental data. The model considers two chaotic pacemakers, the inspiratory (Pre-Bötzinger) and expiratory (parafacial) generators that work together via chemical synaptic connection, either activated or inhibited, to synchronize the respiratory cycle. Different dynamics are evidenced depending on the excitability level of the neurons. In the model, the parameters J_1 and J_2 represent the excitability level of the parafacial and pre-Bötzinger, respectively. Experimental results show that healthy subjects display more complexity during inspiration than expiration and that the low frequency oscillations of the fMRI BOLD ("blood oxygen level dependent") signal located in the rostral VL medulla have higher amplitude than oscillations of the caudal VL pons (Fig. 14.2 bottom). From this, we postulate that the pre-Bötzinger complex is highly likely more excitable than the parafacial group, and drives the respiratory rhythm (active inspiration). Simulation of this network scheme is shown in Fig. 14.4 with two possible regimes depending on the parameter values J_1 and J_2.

In the first regime (Fig. 14.4a), the parafacial has a very low excitability and is entirely depressed with no action potential. This network scheme is similar to the one described in adult rats, the "no-handshake process" [25]. The corresponding attractor of this scheme entirely relies on the pre-Bötzinger dynamics (Fig. 14.5a).

In the second regime (Fig. 14.4b), while the pre-Bötzinger is the dominant pacemaker still driving the respiratory cycle, the parafacial group is occasionally activated by specific physiological conditions (exercise for instance). Experimental results show in COPD patients that airflow complexity is higher during expiration than inspiration and that the low frequency oscillations of the BOLD signal located

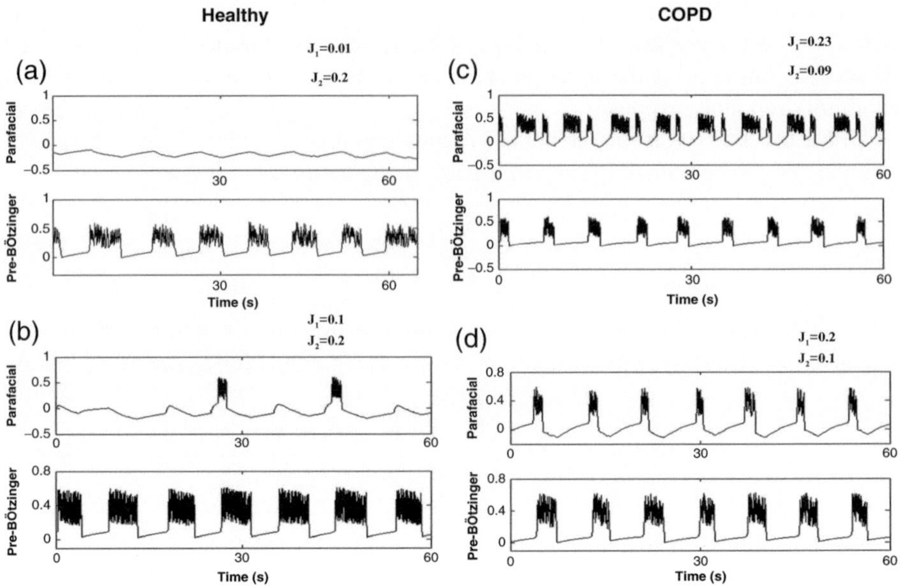

Fig. 14.4 Simulations of different synchronization regimes in healthy subjects (**a**, **b**) and COPD patients (**c**, **d**) are depending on the excitability level of the parafacial respiratory group (J_1) and the pre-Bötzinger complex (J_2). Other fixed parameter values of the model are: $\epsilon = 0.005$, $d = 0.4$, $\beta = 0.4$, $a = 0.2$, $m_0 = 0.864$, $m_1 = 0.65$, $\delta = 0.2$, $x_{th} = -0.02$ (threshold for calcium current), $\tau_1 = 10$ and $\tau_2 = 2$ for the parafacial while $\tau_1 = 5$ and $\tau_2 = 10$ for the pre-Bötzinger. In COPD patients, the parafacial respiratory group of the brainstem has a higher excitability level than healthy subjects and drives the pre-Bötzinger (active inspiration and expiration)

in the VL pons have higher amplitude than the oscillations of the VL pons of healthy subjects. In patients, we therefore hypothesize that the expiratory neurons located in the VL pons are more excitable than the pre-Bötzinger and drive the respiratory cycle. In this network, the more excitable parafacial group triggers the pre-Bötzinger which in turn inhibits the parafacial with a post-inhibitory rebound burst. Then, the parafacial switches-off the inspiration (Fig. 14.4c). This network scheme is similar to the "full-handshake process" described in neonatal rats [25]. The corresponding attractor of this synchronization process mainly relies on the parafacial neurons dynamics (Fig. 14.5c). Another synchronization regime may coexist in the disease state, when the excitability level of the expiratory group is slightly lower: the "half-handshake" process in which the parafacial still triggers the pre-Bötzinger which in turn induces a delayed post-inhibitory rebound burst that triggers a new pre-Bötzinger activation (Figs. 14.4d and 14.5d).

To conclude, the patients have an increased complexity of the airflow during expiration that correlates with the high activity of the VL pons. COPD patients reactivate the parafacial neuronal group, as shown with fMRI results and the mathematical model, to sustain ventilation.

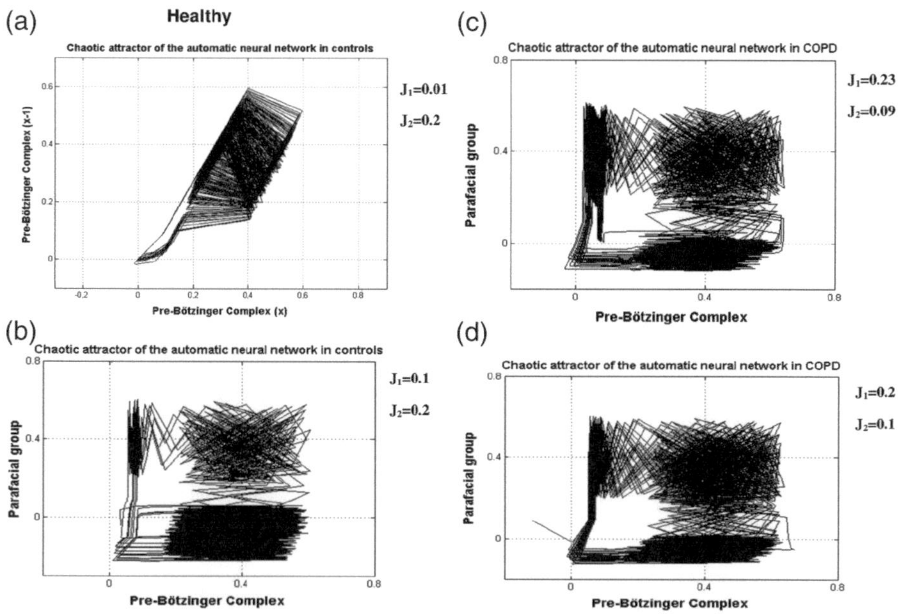

Fig. 14.5 Chaotic attractor of the two synchronized pacemakers for respiratory rhythmogenesis in healthy subjects (**a**, **b**) and COPD patients (**c**, **d**) after simulations. Each attractor is given according to the different network regime presented in Fig. 14.4. The figure reveals that the coupling between both neuronal pacemakers exhibits nonlinear deterministic chaos (**b**–**d**)

14.4 Connectivity and Causality of the Respiratory Neural Network

Breathing involves a complex interplay between the automatic brainstem network and the voluntary cortical command [9, 22]. Both networks interact harmoniously to control respiratory muscles contraction, thereby ensuring normal blood gas levels either during speech, volitional breathing or ventilatory load increase. However until now the communication within the network is unknown. This issue is crucial for many reasons in medicine, physiology and in physics: (1) increased respiratory loading is a major feature of several respiratory diseases, (2) failure of the voluntary motor and cortical sensory processing drives is among the mechanisms that precedes acute respiratory failure [3, 21], and (3) some of the cerebral structures involved in responding to inspiratory loading also participate in the perception of dyspnea, a common and often distressing symptom in many diseases. (4) this neural network vital for life would benefit from the building of a mathematical model able to simulate and analyze its dynamics in disease conditions and may serve as a paradigm of physiological and physical synchronization.

Transcranial magnetic stimulation (TMS) showed that the motor cortex driving the diaphragm had a low motor threshold, a high excitability level and a ceiling effect [14].

We hypothesized that the functional architecture and information flow of the brain network controlling the respiratory muscles exhibited specific impairments in patients with COPD, and that these impairments correlated with illness severity. The selected seed regions were identified during an inspiratory load design and involved seven key areas of the network (Fig. 14.6): two bilaterally in the motor cortex, two

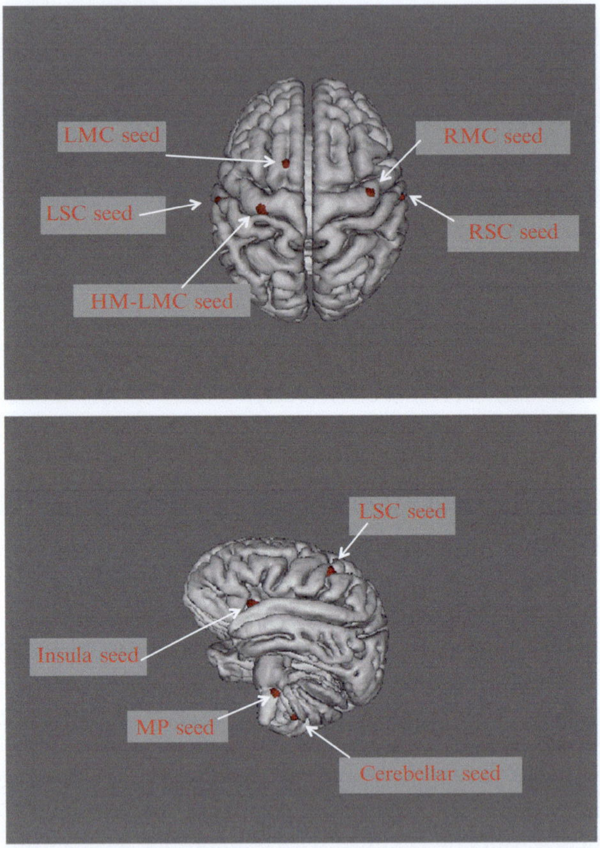

Fig. 14.6 Location of the seed regions (*red surface*) for functional connectivity analyses: the seeds were selected during an inspiratory loading block design paradigm by using a between groups voxel-wise contrast, controls > COPD. *Top panel*: the LMC seed was located in the left premotor cortex while the LSC seed belonged to the primary somatosensory cortex. The RMC (right motor and premotor cortex) seed was in the right primary and premotor cortex and the RSC seed in the right somatosensory cortex; the LMC-HM seed (left motor cortex related to hand movements) was identified by block-design imaging during repeated wrist extension. *Bottom panel*: The Insula seed was in the left anterior part of the Insula. The cerebellar seed was in the left cerebellar hemisphere. The MP (medulla and pons) seed was in the rostral medulla and caudal pons at the level of the Fissura Pontomedullaris. See [26] for details

bilaterally in the somatosensory cortex, one in the brainstem region including the medulla and pons, one in the cerebellum, and one in the insula. Correlations between the seed regions and the brain were evaluated and a Granger's predictability analysis was performed during resting-state fMRI and inspiratory loading to investigate full communication within the network. Let us first explain the Granger-causality method.

In 1969, Granger introduced the idea of G-causality in terms of linear regression modeling [12] that permits to determine the (fluctuating) variables of the system that are causes and those that are causal sinks, so introducing a causality direction. A variable X_2 "Granger causes" a variable X_1 if the inclusion of past observations of X_2 reduces the prediction error of X_1 in a linear regression model of X_1 and X_2, compared to a model including only previous observations of X_1. To illustrate G-causality, let us suppose that the temporal dynamics of two time series $X_1(t)$ and $X_2(t)$ (both of length T) can be described by a bivariate autoregressive model:

$$X_1(t) = \sum_{j=1}^{p} A_{11,j} X_1(t-j) + \sum_{j=1}^{p} A_{12,j} X_2(t-j) + \xi_1(t)$$

$$X_2(t) = \sum_{j=1}^{p} A_{21,j} X_1(t-j) + \sum_{j=1}^{p} A_{22,j} X_2(t-j) + \xi_2(t)$$

where p is the maximum number of lagged observations included in the model (the model order, $p < T$), A contains the coefficients of the model, and ξ_1, ξ_2 are the residuals (prediction errors) for each time series. If the variance of ξ_1 (or ξ_2) is reduced by including the X_2 (or X_1) terms in the first (or second) equation, then it is said that X_2 (or X_1) G-causes X_1 (or X_2). Assuming that X_1 and X_2 are covariance stationary (i.e., have unchanging mean and variance), the magnitude of this interaction can be measured by the log ratio of the prediction error variances for the restricted (R) and unrestricted (U) models:

$$F_{2\to 1} = \ln \frac{\mathrm{var}(\xi_{1R(12)})}{\mathrm{var}(\xi_{1U})} \tag{14.7}$$

where $\xi_{1R(12)}$ is derived from the model omitting the interaction coefficients $A_{12,j}$ (for all j) in the first equation and ξ_{1U} is derived from the full model. Importantly, it is possible to generalize to the multivariate (conditional) case in which the G-causality of X_2 on X_1 is tested in the context of multiple additional variables $X_3 \ldots X_n$ [10]. Our functional connectivity analysis results, based on correlations strength, showed that when breathing at rest and during inspiratory loading, the motor cortex area in the patients had weaker connectivity with their contralateral counterparts, compared to controls, and no connectivity with the brainstem (Fig. 14.7). These findings are crucial while it may explain why some patients with COPD are prone to acute respiratory failure, secondary to motor cortex failure [26]. Causality analyses confirmed this finding while the patients breathing at rest and

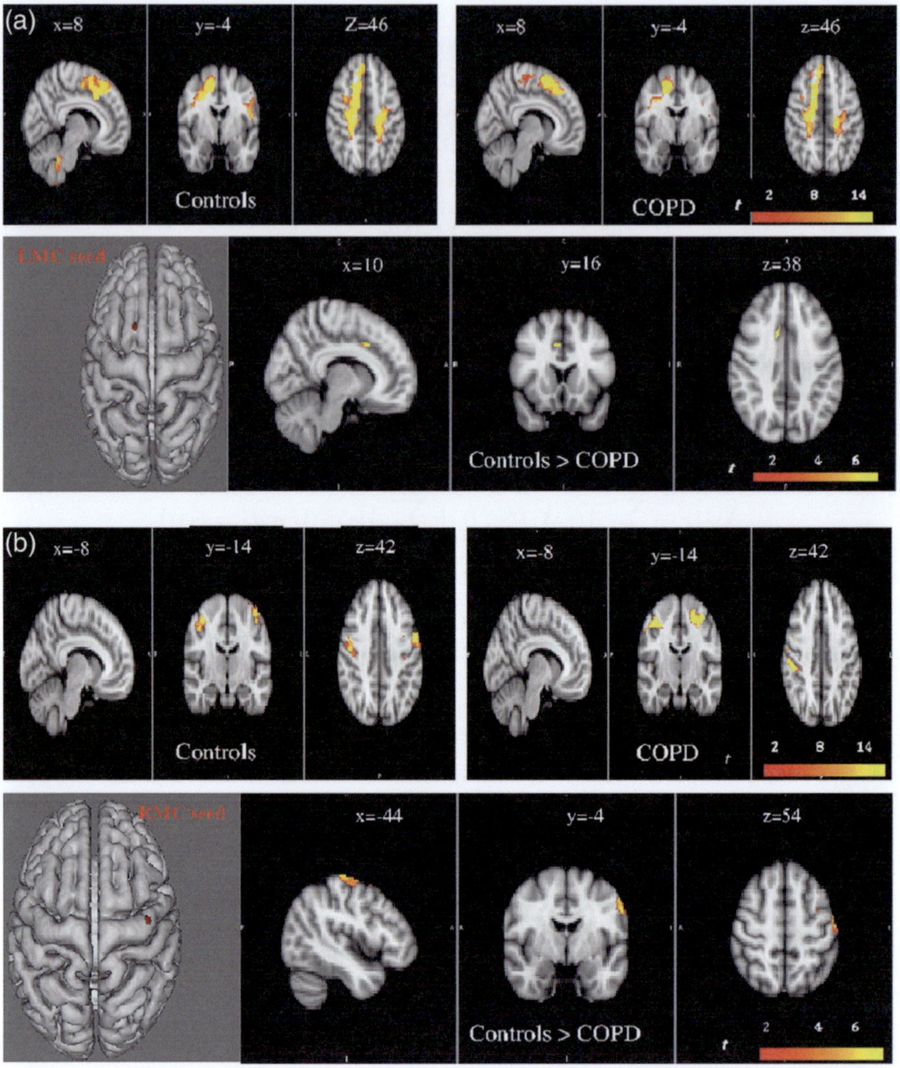

Fig. 14.7 Voxel-wise functional connectivity analyses at rest for the LMC (Left motor cortex, **a**) and RMC (right motor cortex, **b**) seeds. Sagittal, coronal, and axial slices are shown. Within ($P < 0.01$, FWE-corrected) and between-group ($P < 0.05$, FWE-corrected) differences are reported. A decreased connectivity of the networks linked with the right and left motor cortex seed regions is evident with contralateral motor area in the COPD group, as compared with controls. *S* superior, *I* inferior, *R* right, *L* left, *P* posterior. See [26] for details

during inspiratory loading showed a reversal of the roles of the MP (medulla pons) and motor cortex compared to controls (Fig. 14.8). As expected in the controls at rest, the MP, responsible for respiratory rhythmogenesis, exerted a significant causal

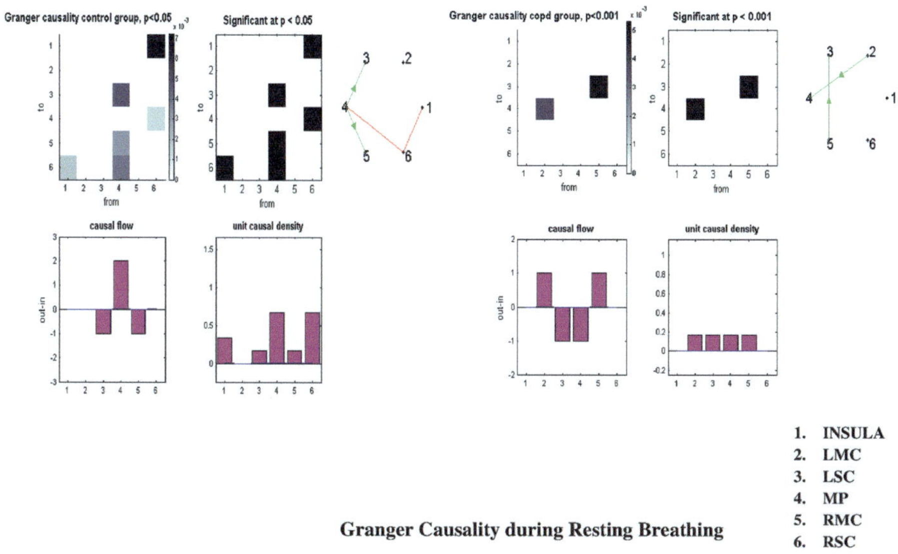

Fig. 14.8 Granger causality analysis at rest between six selected regions at rest. Control group results are reported on the *left*, COPD group results on the *right*. See text for comments. *LMC* left premotor cortex, *LSC* left primary somatosensory cortex, *MP* medulla and pons, *RMC* right motor and premotor cortex, *RSC* right primary somatosensory cortex

influence on the entire network, whereas the sensory-motor cortex was causal sink. Conversely in the patients, the motor cortex was a neural driver, whereas the MP was a causal sink. Thus, the neural driver of the network shifted in the patients from the medulla to the motor cortex.

The functional role of the motor cortex is disproportionate in patients during breathing at rest. During inspiratory loading, the right and left motor cortex were both neural drivers in the controls, and sinks in the patients (Fig. 14.9). The motor network, already overburdened at rest is overwhelmed during inspiratory loading. The results at rest and during inspiratory loading also illustrate the degeneracy of respiratory-network directionality during lung disease [5]. Degeneracy, the ability of structurally different elements to perform the same function or yield the same output, is a well-known physiological characteristic of neural networks.

We postulate that lung disease forces the brain network to adapt, and that the effects on the respiratory motor cortex are hyper-excitability, decreased functional connectivity, and loss of its major role in driving the network during inspiratory loading.

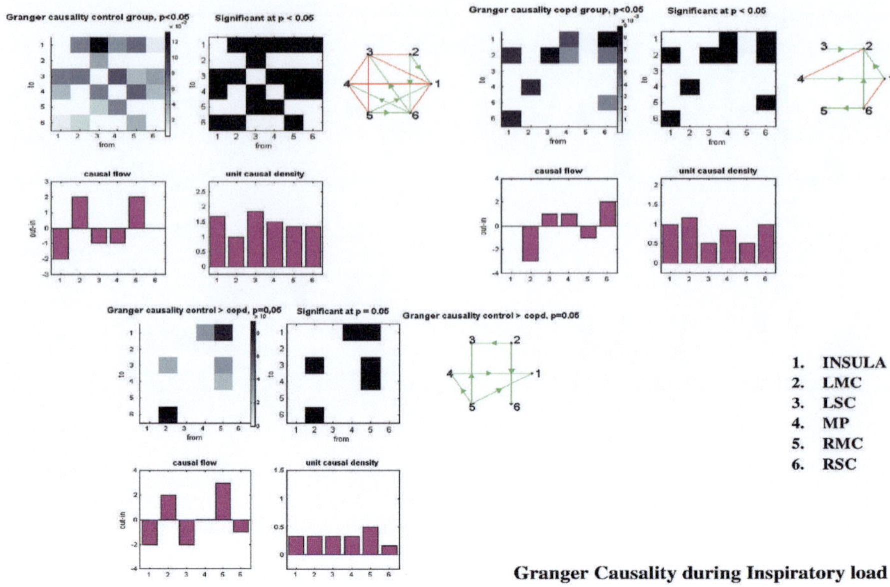

Fig. 14.9 Granger causality analysis between six selected regions during inspiratory loading. Control group results are reported on the *top left*, COPD group results on the *top right*, and controls > COPD at the *bottom*. See text for comments. Same abbreviations as in Fig. 14.8

14.5 Concluding Remarks

In this study, we shed new lights on the brainstem neural control of respiratory muscles in patients with COPD. These findings may be involved in the onset of respiratory failure when the neural network becomes overwhelmed by respiratory overload. Moreover, we show that COPD patients having a severe dyspnea unexplained by a worsening of their pulmonary function may exhibit an altered neuronal excitability of the VL pons, thereby reinforcing the central determinism of dyspnea. Identifying the activity of the respiratory pacemakers through both airflow complexity and functional imaging techniques opens new strategies to refine COPD patient phenotypes.

Our findings may open up new therapeutic avenues for COPD. Although brain networks are constrained by an anatomical structure, the strength of functional connectivity between network regions is a dynamic quantity measuring the degree of interaction between various subnetworks.

The intrinsic activity (motor driving, thresholds, etc.) of the brain can be modulated using TMS (transcranial magnetic stimulation) [6]. Future works should determine how TMS of the brain, at different frequencies, affects the cortical networks, and, more specifically, the connectivity of the sensorimotor and the insular cortical areas. The results may indicate ways to increase motor cortex functional connectivity and improve respiratory muscles performance in patients.

References

1. Courbage, M., Nekorkin, V.I. : Map based models in neurodynamics. Int. J. Bifurcation Chaos **20**, 1631–1651 (2010)
2. Courbage, M., Nekorkin, V.I., Vdovin, L.V.: Chaotic oscillations in a map-based model of neural activity. Chaos **17**, 043109 (2007)
3. Davenport, P.W., Cruz, M., Stecenko, A.A., Kifle, Y.: Respiratory-related potentials in children with life-threatening asthma. Am. J. Respir. Crit. Care Med. **161**, 1830–1835 (2000)
4. Del Negro, C.A., Wilson, C.G., Butera, R.J., Rigatto, H., Smith, J.C.: Periodicity, mixed mode oscillations, and quasi periodicity in a rhythm-generating neural network. Biophys. J. **82**, 206–214 (2002)
5. Edelman, G.M., Gally, J.A.: Degeneracy and complexity in biological systems. Proc. Natl. Acad. Sci. U. S. A. **98**, 13763–13768 (2001)
6. Eldaief, M.C., Halko, M.A., Buckner, R.L., Pascual-Leone, A.: Transcranial magnetic stimulation modulates the brain intrinsic activity in a frequency-dependent manner. Proc. Natl. Acad. Sci. U. S. A. **108**, 21229–21234 (2011)
7. Feldman, J.L., Del Negro, C.A.: Looking for inspiration: new perspectives on respiratory rhythm. Nat. Rev. Neurosci. **7**, 232–241 (2006)
8. Fortuna, M.G., West, G.H., Stornetta, R.L., Guyenet, P.G.: Bötzinger expiratory-augmenting neurons and the parafcial respiratory group. J. Neurosci. **28**, 2506–2515 (2008)
9. Gandevia, S.C., Rothwell, J.C.: Activation of the human diaphragm from the motor cortex. J. Physiol. **384**, 109–118 (1987)
10. Geweke, J.: Measurement of linear dependence and feedback between multiple time series. J. Am. Stat. Assoc. **77**, 304–313 (1982)
11. Goldberger, A.L., Amaral, L.A., Hausdorff, J., Ivanov, P.C., Peng, C.K., et al.: Fractal dynamics in physiology: alterations with disease and aging. Proc. Natl. Acad. Sci. U. S. A. **99**, 2466–2472 (2002)
12. Granger, C.W.J.: Investigating causal relations by econometric models and cross-spectral methods. Econometrica **37**, 424–438 (1969)
13. Hess, A., Yu, L., Klein, I., De Mazancourt, M., Jebrak, G., Mal, H., Brugiere, O., Fournier, M., Courbage, M., Dauriat, G., Schouman-Clayes, E., Clerici, C., Mangin, L.: Neural mechanisms underlying breathing complexity. PLoS One **8**, e75740 (2013)
14. Hopkinson, N.S., Sharshar, T., Ross, E.T., Nickol, A.H., Dayer, M.J., Porcher, R., Jonville, S., Moxham, J., Polkey, M.I.: Corticospinal control of respiratory muscles in chronic obstructive pulmonary disease. Respir. Physiol. Neurobiol. **141**, 1–12 (2004)
15. Janczewski, W.A., Feldman, J.L.: Distinct rhythm generators for inspiration and expiration in the juvenile rat. J. Physiol. **57**, 407–420 (2006)
16. Jolley, C.J., Luo, Y.M., Steier, J., Reilly, C., Seymour, J., Lunt, A., Ward, K., Rafferty, G.F., Polkey, M.I., Moxham, J.: Neural respiratory drive in healthy subjects and in COPD. Eur. Respir. J. **33**, 289–297 (2009)
17. Mangin, L., Fiamma, M.N., Straus, C., Derenne, J.P., Zelter, M., Similowski, T.: Source of human ventilatory chaos: lessons from switching controlled mechanical ventilation to inspiratory pressure support in critically ill patients. Respir. Physiol. Neurobiol. **161**, 189–196 (2008)
18. Onimaru, H., Homma, I.: A novel functional neuron group for respiratory rhythm generation in the ventral medulla. J. Neurosci. **23**, 1478–1486 (2003)
19. Poon, C.S., Merrill, C.K.: Decrease of cardiac chaos in congestive heart failure. Nature **389**, 492–495 (1997)
20. Rabinovich, M.I., Abarbanel, H.D.I.: The role of chaos in neural systems. Neuroscience **87**, 5–14 (1998)
21. Sassoon, C.S., Gruer, S.E., Sieck, G.C.: Temporal relationships of ventilator failure, pump fatigue, and diaphragm fatigue. J. Appl. Physiol. **81**, 238–245 (1996)

22. Sharshar, T., Hopkinson, N.S., Jonville, S., Prigent, H., Carlier, R., Dayer, M.J., Swallow, E.B., Lofaso, F., Moxham, J., Polkey, M.I.: Demonstration of a second rapidly conducting cortico-diaphragmatic pathway. J. Physiol. **560**, 897–908 (2004)
23. Smith, G.D., Cox, C.L., Sherman, M., Rinzel, J.: Fourier analysis of sinusoidally driven thlamocortical relay neurons and a minimal integrate-and-fire-or-burt model. J. Neurophysiol. **83**, 588–610 (2000)
24. Thoby-Brisson, M., Karlen, M., Charnay, P., Champagnat, J., Fortin, G.: Genetic identification of an embryonic parafacial oscillator coupling to the preBötzinger complex. Nature Neurosci. **12**, 1028–1036 (2009)
25. Wittmeier, S., Song, G., Duffin, J., Poon, C.S.: Pacemakers handshake synchronization mechanism of mammalian respiratory rhythmogenesis. Proc. Natl. Acad. Sci. U. S. A. **105**, 18000–18008 (2008)
26. Yu, L., De Mazancourt, M., Hess, A., Ashadi, F.R., Klein, I., Mal, H., Courbage, M., Mangin, L.: Functional connectivity and information flow of the respiratory neural network in chronic obstructive pulmonary disease. Hum. Brain Mapp. **37**, 2736–2754 (2016)

Chapter 15
Dynamics of Odor-Evoked Activity Patterns in the Olfactory System

Thomas Nowotny and Paul Szyszka

15.1 Introduction

Olfaction is being studied in a number of animal models, including a good number of invertebrates, in particular insects. Studying insects that have much smaller brains than typical mammalian models and less ethical encumbrances simplifies invasive experimentation, large throughput studies, and modeling. At the same time, olfactory systems have striking similarities across phyla [1] so that it is viable to study insects even if the final motivation is understanding the human system. In this chapter we will focus on the wealth of insect work accumulated over the last 3 decades. Insects' olfactory organs, the sensilla, are in direct contact to airborne odors. In contrast to the metabotropic mammalian receptors, insect olfactory receptors act as ligand-gated channels, with metabotropic auto-regulation [2–6]. Insects generally have dozens of different types of olfactory receptor neurons, each expressing a distinct receptor gene [7, 8]. Therefore, according to its type, each receptor neuron responds in a specific way to any given odorant, which generates a unique pattern of olfactory receptor neuron activity for every odorant [9]. These activity patterns are projected to the antennal lobe (AL), the first brain area that processes odor information in insects. The structure of the AL is similar to that of the mammalian olfactory bulb: olfactory receptor neurons of the same type converge onto the same glomerulus [7, 10]. As a consequence of this anatomical structure, the pattern of olfactory receptor neuron activity translates directly into an equivalent odor-specific pattern of activated glomeruli [11–14]. The simple

T. Nowotny (✉)
University of Sussex, Falmer, Brighton BN1 9QJ, UK
e-mail: t.nowotny@sussex.ac.uk

P. Szyszka
Department of Biology, University of Konstanz, 78457 Konstanz, Germany
e-mail: paul.szyszka@uni-konstanz.de

I.S. Aranson et al. (eds.), *Advances in Dynamics, Patterns, Cognition*,
Nonlinear Systems and Complexity 20, DOI 10.1007/978-3-319-53673-6_15

one-to-one connectivity between olfactory receptor neurons of the same type and corresponding glomeruli is complemented by a dense intra- and inter-glomerular network of inhibitory and excitatory local neurons (LN) that modify odor-evoked activity in the AL [15–23]. Projection neurons (PN) relay odor-evoked activity to various areas in the brain. The most prominent one is the mushroom body (MB), a brain region for odor learning and identification [24–27].

Natural stimulus dynamics and odor-driven behavior pose particular constraints on neural processing mechanisms for odor information. Here we discuss what aspects of odor-evoked receptor neuron activity patterns and what dynamical systems mechanisms the olfactory system could use for efficient encoding of chemical odorant identity, odorant concentration, and temporal stimulus structure.

15.2 Rate Coding

In a rate code, information is encoded in the average firing rate of neurons, or, equivalently, in their inter-spike-intervals. In the insect AL, a rate code for odor information would correspond to an odorant- and concentration-specific glomerular pattern of spike rates. Indeed, olfactory receptor neurons exhibit odorant specific spike rates, that increase with increasing concentration [28] and are sufficient for recognizing odors [29]. These odor- and concentration-specific spike-rate differences are relayed to antennal lobe PNs.

In first order approximation this can be observed by calcium imaging [30–32] assuming that the changes in calcium concentration can be mapped to the firing rate of PNs [33]. Calcium imaging experiments with back-filled PNs have provided first insights into the response profile of glomeruli, and hence indirectly of olfactory receptor neurons, in particular in bees where direct observation of olfactory receptor neuron activation is difficult. It has been found that glomerular rate patterns (measured as intra cellular calcium signals) are sufficient to distinguish odors [13, 32, 34] and assess their concentration [35].

In PNs the relative glomerular spike-rate differences remain stable over a wide range of concentrations [36], allowing to encode chemical odor-identity and concentration at the same time. Rate coding is further supported by the fact that the similarity between odor-evoked glomerular activity patterns correlates with the perceived similarity in the honeybee [37]. This suggests that glomerular rate differences contain behaviorally relevant information about odor identity. However, the mere fact that rate coding appears to be sufficient for recognizing the chemical identity of odor stimuli should not be taken as proof that the insect brain is indeed relying on a rate-coding scheme. Rate coding has its limitations, in particular with respect to rapid processing of stimuli: In order to read out a rate code, downstream areas from the AL would have to integrate over a considerable time span, at least on the order of the inter-spike interval of the least excited yet still relevant PN. This could easily reach 100s of milliseconds, in particular at low concentrations. However, the observed odor-information processing in insects is fast. Moths and honeybees, for instance, can detect millisecond short onset asynchrony between the arrivals of odorants [38, 39] and flies recognize odorants within less than 200 ms [40].

15.3 Temporal Coding Schemes

When stimulated with odor stimuli of long duration and constant concentration, PNs in the locust AL exhibit complex spiking patterns [41, 42]. The observed dynamics can be interpreted as fast, approximately 20 Hz oscillations super-imposed onto slower spatio-temporal patterns of activation of PNs and LNs. Oscillations are a familiar phenomenon observed in the olfactory system [43–46] but even though there are some indications that they may play a functional role for discrimination of similar odors [47, 48] (though see [18]), it is not clear yet whether they encode behaviorally relevant odorant information. The fact that oscillatory synchronization requires multiple odorant stimulations is not compatible with the proposed function for encoding odorant identity, as insects can form equally odor-specific associative memories during single- and multiple-trial conditioning [49]. The slower temporal patterns are a less widely reported phenomenon and their origin and function also remain disputed. Inspired by the complex dynamics observed in predator–prey models [50, 51] and the similarity of antagonistic (inhibitory) interactions between agents (neurons), Laurent and Rabinovich proposed that slow patterning of odor responses of PNs in the AL of insects might form a winnerless competition, mathematically realized as a heteroclinic orbit [52, 53] and physiologically implemented by $GABA_B$ transmission. In brief, the observed dynamics were described by generalized Lotka–Volterra equations,

$$\dot{a}_i = a_i\left(1 - \left(a_i + \sum_{i \neq j}^{N} \rho_{ij} a_j\right)\right), \tag{15.1}$$

where a_i is the firing rate of the ith neuron (group), ρ_{ij} are the strengths of inhibitory interaction of neuron (group) j onto neuron (group) i, and N is the number of relevant neurons (neuron groups). It was proven that if certain fairly mild conditions hold on ρ_{ij}, then a stable heteroclinic cycle exists and is structurally stable, lending this mathematical description credibility as a viable model. There were a number of strong arguments that supported the utility of heteroclinic dynamics. (1) The stability of the heteroclinic orbit provides robustness against intrinsic noise. (2) The interpretation of inputs modifying the synaptic inhibition between neurons, and the structural robustness of the heteroclinic structure with respect to changes in the strength of inhibition, make the model robust to noise in the odor input or the olfactory receptors. (3) On the other hand, the input setting the heteroclinic sequence made the responses sensitive to "meaningful differences" in the input signals.

One aspect of the original winnerless competition dynamical motif was that the patterning could continue essentially infinitely, only limited by the duration of the stimulus. This hypothesis was tested in locusts [54]. Giving long constant stimuli, the authors observed that the spatio-temporal dynamics in the AL appeared to stop evolving after approximately a second and the activity of the AL appeared to reach a stable odor-specific pattern of activation, equivalent to a fixed point. This led to the development of the stable heteroclinic channel (SHC), which, in contrast to the earlier models, can have a final fixed point [55].

The heteroclinic model was originally limited to the rate description described above. Ideally one would want to be re-assured that the same underlying dynamics could also arise in more detailed, e.g. Hodgkin–Huxley type, conductance based models. Working towards this goal, Nowotny and Rabinovich [56] developed a conductance based model that indeed displays the same phenomenology, in particular an apparently unlimited slowing-down of the cyclic bursting dynamics during strong constant input, which suggests the existence of a stable heteroclinic cycle. Later the quantitative reduction of the spiking model to a rate model developed in the same paper [56] was used to prove the structural robustness of the putative heteroclinic structure in the model under a large class of perturbations [57]. In independent work, the existence of heteroclinic structures in networks of phase oscillators was shown, where the saddle points correspond to the transient synchronization of clusters of neurons [58, 59]. This work can also be applied to intrinsically active Hodgkin–Huxley neurons [57].

Inspired by their work on the macro-glomerular complex in moths, Buckley and Nowotny [60] suggested an alternative dynamical systems model that was consistent with the observed odor patterns in the AL in response to long, constant stimuli. They proposed that spatio-patterning can arise from a dynamical system with a single, barely stable fixed point. The model can be formulated at multiple scales from detailed, Hodgkin–Huxley models to simplified mean field population descriptions. On a level of firing rate descriptions, the governing equations are

$$\dot{s}_i = -\beta s_i + \gamma \left(-\sum_{j \neq i} g_{ij} s_j + \theta_i + I_i \right) \tag{15.2}$$

where s_i are the activation variables of synapses, i.e. filtered versions of the presynaptic firing rate and β and γ define rates of decay in s_i and changes in s_i due to input, respectively. The input to neuron i is described by the conductance g_{ij} of (inhibitory) synapses from j to i, a neuron-specific bias term θ_i and an input current from ORNs, I_i. Note the similarity with the Lotka–Volterra equations (15.1) except that here the input is not multiplied by the current rate (or activation) variable. However, as it is this multiplicative interaction that generated the saddle points in (15.1), the phase space structure in this model is nevertheless very different. Parameters in the model, in particular the couplings g_{ij}, can be adjusted such that in absence of input there is a single fixed point that is stable but close to losing this stability if g_{ij} were further increased.

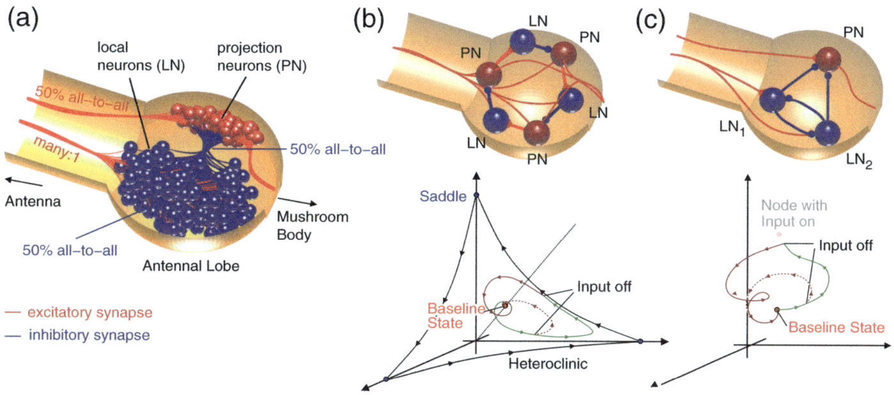

Fig. 15.1 Two dynamical systems interpretations for patterned activity in the AL. (**a**) Illustration of the AL depicting the densely inter-connected populations of LNs and PNs. (**b**) Interpretation of the AL network as a network motif giving rise to winnerless competition or a stable heteroclinic channel. The response of the AL is described as akin to a trajectory that approaches a stable heteroclinic cycle between a number of saddle points. The input to the system creates this heteroclinic structure by exciting LNs and PNs. (**c**) Interpretation of the AL network as a network motif giving rise to a single barely stable fixed point. Here, the response of the AL is interpreted as a trajectory that moves from the location of the baseline state to a barely stable fixed point. The input to the system in this view moves the single fixed point of the system from its baseline location to a new location by exciting LNs and PNs. (Modified from [60])

Odor input, i.e. a pattern of input currents I_i, in this model moves this fixed point in phase space, and the state of the system will follow it on a complex trajectory because of the weak stability. This gives rise to complex activity patterns, both for "stimulus on" and "stimulus off" conditions. Figure 15.1 illustrates the two competing views and their very similar phenomenology.

Unlike in Rabinovich's WLC model, the patterning of odor responses in this view arises more as a necessary evil than being a particular feature. The patterns are caused by the desire to be very sensitive to inputs and to have a large dynamical range of responses, which is achieved when the global fixed point is about to lose stability [61].

A lot of thought has been put into whether one could confirm experimentally which of the descriptions is more accurate. However, because the eventually observed dynamics of the system are always transients and the actual fixed points of the system dynamics are not visible under experimental stimulation, such an experimental proof is hard to come by. The field has also since made a different turn because of the observation that odor stimulation is of extremely transient nature in itself. Even the purported constant stimuli in the earlier experiments may well have been more dynamic than assumed [62].

In natural conditions odors occur in highly turbulent odor plumes (we will come back to this in detail below), which implies that long, constant odor stimuli are typically not encountered. Behavioral experiments in rats [63], mouse [64], bees

[65], and flies [40] also suggest that animals' odor recognition times are much faster than the time scales on which the spatio-temporal patterns discussed above emerge (see, however, [66], where longer sampling times of durations compatible with the use of spatio temporal dynamics appear to increase the accuracy in difficult odor discrimination problems). Taken together, the structure of odor input and the sufficiency of short sampling times, at least in a majority of conditions, seems to indicate that the full sequence of a stable heteroclinic channel or approach to a barely stable fixed point may not be necessary for odor recognition.

We also note that the marginally stable fixed point dynamics are advantageous regardless of whether long transient sequences are utilized. This dynamical motif offers high sensitivity and large dynamic range already for the very short stimuli which are typical in a natural environment. For the heteroclinic model one could argue that the position of the first saddle point on the SHC provides reproducible responses even without the benefit of observing the full sequence of the SHC. Coding capacity would be somewhat reduced when going from sequences to single saddle points but with realistic numbers for the number of independent units, typically glomeruli, even the number of individual saddles, with, e.g., k out of n active glomeruli, is very large. In the case of the fruit fly, there are about 42 glomeruli. If 10 out of 42 were activated for any given odor, there would be $\binom{42}{10} \approx 1.5$ billion patterns available to represent different odors. While this estimate incorrectly assumes that all combinations of glomerular activations would be available to represent existing odors, i.e. assumes some type of homogeneous distribution of odors in the space of glomerular activation patterns, it still suggests that coding capacity with a single glomerular activation pattern would not be a major capacity issue.

15.4 Processing of Natural odorant Stimuli

In natural conditions the structure of odor plumes is dominated by turbulent advection [67]. Turbulences disperse odor plumes into intermittent filaments which intermingle with filaments from different odor sources [67–70]. As illustrated in Fig. 15.2, this leads to a complex spatial pattern of high concentration filaments from one or several odor sources, interspersed with areas of relatively clean air. Wind and the movement of the animal itself transform the complex spatial pattern of odorant filaments into temporal patterns of changing odorant concentration and composition at the animals' olfactory organ. These temporal odorant patterns contain information about the distance of an odor source, as odorant intermittency decreases with decreasing distance [67]. They also contain information about the number of odor sources when different plumes mix, as odor compounds from the same source exhibit more correlated fluctuations than odor compounds from different sources [71]. Even at low wind speeds filaments in an odor plume have widths on the order of millimeters to centimeters. At a flight speed of 0.5 m/s a flying insect would encounter such filaments for only a few milliseconds at a time.

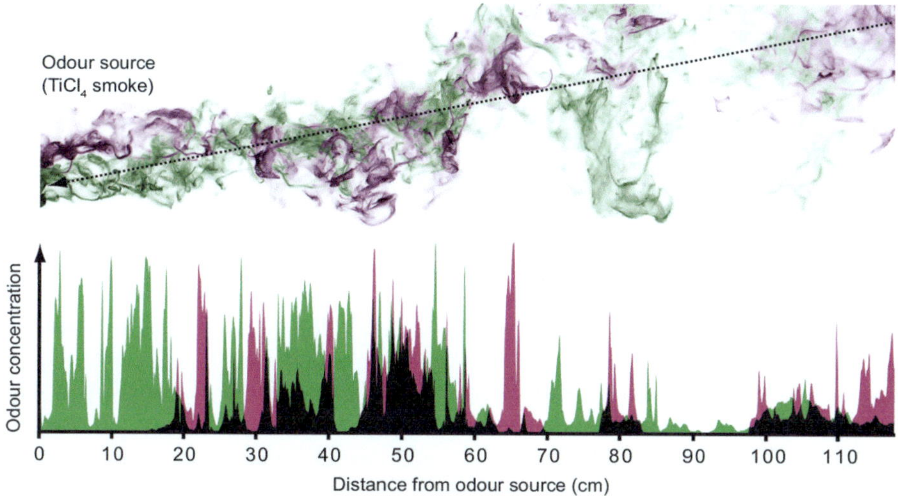

Fig. 15.2 Illustration of the spatial structure of odor plumes. *Top*: Smoke plumes emanating from two, sequentially imaged, close-by sources of $TiCl_4$ smoke. Note how smoke distribution in the plume is dominated by turbulence rather than diffusion so that the two different colored smoke filaments do not fully mix. *Bottom*: Computed smoke concentration profile a flying insect would encounter if it flew along the dotted arrow on the top. Note the small overlap of the two differently colored smoke plumes and the detailed spatial structure of the plumes

Insects can use short temporal asynchronies between stimulus onsets to segregate odorants from different sources. Moths and beetles can distinguish an attractive pheromone from an antagonistic pheromone based on millisecond short differences in the arrival of the pheromone components [38, 72, 73]. After learning that an initially neutral odorant A is associated with a food reward, honeybees show an increased probability to respond to an asynchronous mixture of A and a novel odor B in which A and B arrive with a delay of just 6 ms between them, as compared to the synchronous mixture AB [39]. Similarly, locusts can segregate a learned odorant A from a background odorant B, when A starts with a delay of 250 ms after B [74]. How is this temporal stimulus information represented in the olfactory system? In locusts, PN responses to synchronous mixtures cannot be predicted from the responses to single components. PN responses to asynchronous mixtures, however, partly match those evoked by the individual components [74, 75]. This sensitivity for odor onset asynchrony may underlie locusts' ability to determine whether two odors were presented simultaneously or with a particular time delay [74].

Consistent with honeybees' behavioral capability to use millisecond-short odor-onset asynchrony for odor-segregation, bees' PN responses are sensitive to millisecond stimulus asynchrony [76]. Stimulation with asynchronous mixtures results in more suppressive interactions, and compared to synchronous mixtures, PN responses to asynchronous mixtures contain more information about the components: the longer the delay between the components in an asynchronous mixture, the more do PN responses resemble the component response, and the less they resemble the mixture response [76].

These studies gave first evidence that the same odors can both be processed as a one-source mixture, or as a two-source mixture, on the basis of the temporal relationship between the components. But what are the neuronal mechanisms of odor segregation based on millisecond short stimulus onset asynchrony?

The few milliseconds of the minimal necessary odor onset delay for odor-object segregation observed in behavior [38, 39, 72, 73, 77] and physiology [76] is similar to the time scale of spikes, synaptic transmission, and minimal inter-spike intervals. Therefore, it seems unlikely that a head start of a few milliseconds is enough for a component to change the processing of the mixture such that it leads to a completely different perception and behavioral output. Baker and colleagues [38] suggested that the high-temporal resolution in the processing of pheromone mixtures relies on ephaptic mixture interactions between pairs of differently tuned olfactory receptor neurons within the same olfactory sensillum [78, 79]: the response of one receptor neuron to one component could suppress the response of the other receptor neuron to the other component. Indeed, in moths and beetles, olfactory receptor neurons, which are tuned to antagonistic pheromone components often are co-localized within the same sensilla [38, 73, 80]. Thus, on-site mixture processing in pairs of co-localized receptor neurons could provide the temporal accuracy needed for assessing synchronous versus asynchronous arrival of two odorants [38].

In addition or alternatively, the millisecond temporal resolution in the processing of odor mixtures could be achieved due to integration in the AL network. Nowotny et al. [81] offer a possible explanation for PNs' sensitivity to millisecond onset asynchrony between odorants using computational modeling. In their model (Fig. 15.3), LNs in the AL form a winner-takes-all competitive inhibitory network. Typically, the winning LN would be different for different odors, and hence different depending on which odor arrives first. As the inhibitory conductances of LNs onto PNs are likely not homogeneous [82], the resulting effect of the winner-take-all LN network is that the response pattern in PNs to an asynchronous mixture is different when different components arrive first and different from the response pattern to the corresponding synchronous mixture. This difference is preserved for an extended period of time beyond the initial odorant onset where the two mixture conditions actually differ, lasting as long as the winning LN remains active. The prolonged difference between response patterns to synchronous and asynchronous mixtures could then facilitate odor segregation in downstream circuits of the olfactory pathway. Thus, the dynamics of this network offer a possible explanation for the ability of insects to distinguish synchronous from asynchronous mixtures.

In order to capture millisecond short differences in stimulus onsets, insects need a fast and reliable odor transduction process, that is, the conversion of fluctuating odor stimuli into neuronal activity. Insects have a striking advantage over the mammalian olfactory system: they do not need to inhale in order to smell, which might delay odor transduction and decrease temporal precision. The morphology of insects' olfactory organs supports fast odor transduction, as olfactory sensilla have direct contact to the environment, with olfactory receptor neurons being less than 1 μm separated from the atmosphere [83]. Besides the morphology, biochemistry also is an important factor for the temporal resolution of odor transduction: perireceptor

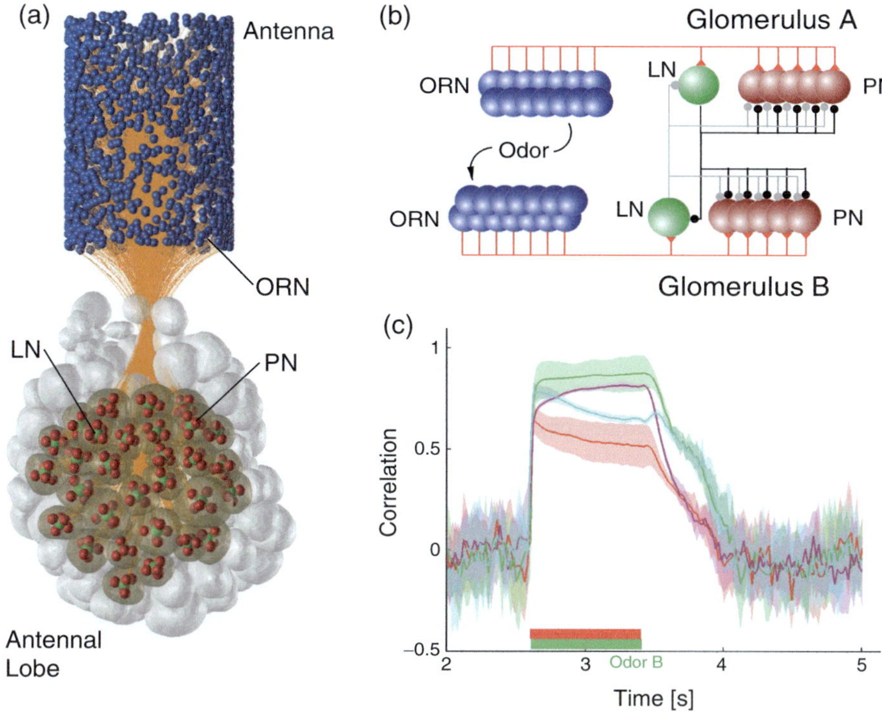

Fig. 15.3 Model of asynchronous mixture perception in the honeybee AL. (**a**) data driven network model of the bee antenna and AL (*ORN* olfactory receptor neuron, *PN* projection neuron, *LN* local interneuron). The *dark grey* glomeruli are routinely recorded and have been included in the model. (**b**) Circuit diagram of the hypothesized onset detection circuit. The LNs (*green*) form a winner-take-all circuit that helps preserving the memory of which odor was encountered first beyond the short time of actual onset difference. (**c**) results of the numerical simulations of the model. Correlations of the response to a delayed A-6 ms-B asynchronous stimulus is most similar to a B response template (*green*) and less to an A template (*red*). The similarity of the synchronous AB mixture response to the A response template (*magenta*) and to the B response template (*cyan*) is between

events [84, 85] are thought to be the time limiting processes in the speed and temporal resolution of odor transduction [86, 87]. They include the diffusion of the odorants in the sensillum lymph, the binding of the odorants to odor binding proteins [88], their release from odorant binding proteins and clearance of the odorant molecule from the receptor and sensillum lymph. The binding of odorant molecules to the receptors is followed by chemo-electric transduction processes at the olfactory receptor proteins. In contrast to the metabotropic olfactory receptors of mammals, insects have ionotropic receptors, that is, ligand-gated ion channels that are directly activated by the odorant molecule [3, 6, 89]. Ionotropic receptors generally have a shorter latency and a better temporal resolution than metabotropic receptors in which the production and degradation of second messengers takes time.

Indeed, measured transduction times in insect olfactory receptor neurons are fast and range from 2 to 30 ms [3, 28, 86, 90, 91]. Thus, insects' ionotropic olfactory receptors could be an adaptation to insects' need for a high-temporal olfactory resolution.

A second aspect of natural odor stimuli besides stimulus dynamics is that odors from natural sources are mixtures of numerous chemical compounds, whose composition gives rise to distinct odor percepts. For example, the aroma of brewed coffee contains in excess of 20 compounds well above odor detection threshold [92]. Similarly the number of compounds emitted by flowers is typically quite large [93] and each individual species may emit mixtures of easily dozens of compounds. This has two implications. First, olfactory systems need to recognize mixtures of chemical compounds, not necessarily any particular individual compound. Given the combinatorial response patterns of receptor neurons in which each receptor neuron responds to a number of compounds and each compound excites a number of receptor neurons a defined mixture of chemical compounds is just another combinatorial activation pattern. Gain control is in place in the AL [22, 36, 94–97] that prevents an increase in active glomeruli with increasing odorant concentration [98, 99]. Therefore, the responses to mixtures of many components is not necessarily qualitatively different to responses to a single compound. Nevertheless, the question of synthetic versus analytic odor processing remains open: When honeybees were trained to an odorant mixture, they show less responses to the individual components than to the trained mixture [100], demonstrating that in a mixture odorant information is processed synthetically and information about the components is partly lost. Further support for synthetic processing comes from biconditional discrimination learning and negative patterning: in biconditional discrimination, bees can learn to discriminate between two pairs of mixtures: one pair (AX and BY) is rewarded and the other pair (AY and BX) is not [101]. In negative patterning, bees learn that odors A and B are rewarded, while the mixture AB is not [102–104]. Because in these experiments each odor component is equally often paired with and without the reward, bees cannot discriminate between rewarded and non-rewarded odors based on a purely analytic strategy, that is, by identifying a particular component. The conclusion is that bees either learn the exact composition of the mixture or rely on a synthetic mixture representation. Whether the single components of a mixture can be perceived as individual odor objects can also be tested by blocking experiments [105]. In the blocking paradigm, animals are first conditioned to one stimulus A and then to a mixture of A and a novel odor B (AB). The learning of A typically reduces (or blocks) the learning of B during the subsequent conditioning of AB, as the reinforcement learning prediction error is reduced due to the previous learning of A, which means that B acquires less predictive value [106]. Thus, blocking requires analytic (elemental) processing of AB. Olfactory blocking experiments in honeybees gave controversial results. Blocking was demonstrated in some studies [107–109] but it could not be reproduced in two other studies [110, 111]. Taken together, these studies suggest that in most cases odor mixtures are processed in a synthetic rather than analytic manner. However, if we again refer back to a natural environment, the mixtures of compounds that define a given odor would never truly occur on their own. They would always be masked by background odors, e.g. bacon and eggs in case of coffee, and other flowers in the case of a flower in a field. This raises the question how animals can disambiguate the mixture of two odors from the mixture of all their constituent compounds?

In conjunction, the two aspects of complex plume structure and odorants being composed of many compounds offer an intriguing hypothesis: Animals use the fine temporal structure of odor plumes to distinguish the relevant odor sources in their environment. The underlying idea is simple. Unless the molecular weights differ too much, components of an odor should diffuse with comparable volatility and hence the filaments emanating from one odor source should have an approximate composition constancy across all filaments. Odors from other sources would form their own filaments. The filaments from all odor sources mix in the environment, but in a way such that they do not arrive at the same time at the receptors. Rather, there will be intermittent exposures to one or a small number of odors. Therefore, the correlated arrival of each odor signal allows distinguishing it from its background odors.

15.5 Pattern Recognition in Higher Brain Centers

Beyond the AL, odor information is processed in the MBs and in the lateral horn. Based on the extensive knowledge of the central role of MBs in learning and memory [24–27, 112] the generally accepted working hypothesis is that the MBs are involved in forming new associations between odor inputs and inputs from other modalities with rewards or punishments, while the lateral horn is involved in generating innate responses to odors, such as the natural attraction of flies to vinegar [113, 114]. Models of olfactory processing in the MBs are slowly converging onto a family of models with fundamentally similar features. The Kenyon cells (KCs) in the MB are seen to form a large "screen" of encoders [115] that translate the overlapping patterns of PN activity into a both temporally and spatially sparse representation as observed in experiments [116–118]. Whether this transformation is achieved by a purely random connectivity [115, 119, 120] or is refined by plasticity [121, 122] is still debated. Sparse KC activity patterns are then associated with output neurons in the MB lobes by plasticity in the KC to output neuron connections, of Hebbian nature [115, 119, 120] or by three factor rules [123–126] when actual rewards are considered. It is now also commonly assumed that ONs inhibit each other to force a decision on the identity or valence of a KC activity pattern [119, 125, 126].

The strong fan-out character of the PN to KC connectivity (100 to 5000 in fly, 800 to 50,000 in locust, 800 to 100,000 in bees) makes gain control mechanisms indispensable [127] to avoid instabilities in the MB activity. From a theoretical perspective this could take the form of feedforward gain control, e.g. mediated by lateral horn interneurons [119, 128], but recent experimental work [127, 129] and subsequent modelling [130] indicate that feedback gain control by MB extrinsic GABAergic neurons [127, 131] is the more likely scenario.

A series of recent works showed that within this general framework model, previously not considered phenomena such as peak-shift can be recapitulated [125, 132].

Based on the observation of sparse responses in the KC of the MB [116–118, 133] and an observed strong periodic inhibition at 20 Hz frequency [116], a common assumption has been that odor processing occurs in separate 50 ms time windows, so-called snapshot processing [119]. This view is appealing because it reduces the complexity of processing spatio-temporal patterns in the PNs. However, it is challenged by conceptual problems for both longer and very short time scales.

The view that odor discrimination of similar odors improves over time along the trajectory of odor responses in the AL networks is in conflict with strict snapshot processing in downstream brain areas. While temporal integration in higher brain areas could recover temporal sequences, it is not sensitive to the temporal order of patterns, leading to a potential loss of information.

Nowotny and colleagues [134] suggest in a model of the insect MB how sequences of PN activation patterns in the AL could be fully decoded in spite of snapshot processing. The model assumes random connections between PNs and KCs and slow excitatory connections between KC axons. It is these lateral connections that transform the temporal sequence of activation of PN patterns into a unique summed spatial pattern in the KCs. The mechanism is based on that lateral excitation primes "adjacent" KCs to respond preferentially to a subsequent pattern, which generates distinct summed KC activation patterns in response to pattern 1 followed by pattern 2, as opposed to pattern 2 followed by pattern 1.

On small time scales, snapshot processing is problematic as well. Here, one could think that the minute temporal differences of odor onsets could be lost in the relatively coarse 50 ms time window snapshot sampling. However, the model of a local winner-take-all LN network [135] discussed above offers a possible solution to this problem because the small temporal differences of stimulus arrival times are transformed into longer lasting differences in PN activation patterns that may well be read in snapshot sampling.

15.6 Conclusions

In this chapter we have given an overview over some aspects of olfactory information processing, focusing on the role of dynamics, both in the environment and intrinsic to the brain. In the environment, the nature of turbulent odor plumes suggests that sensing and encoding precise odor onsets is important to segregate intermingled odor plumes in natural odor scenes. On the other hand, the work on winnerless competition and barely stable fixed point dynamics suggests that within the brain, temporal dynamics is an essential ingredient to signal processing. Both discussed dynamical mechanisms have their own merits. In winnerless competition, the difficult balance between robustness to noise and sensitivity to informative differences in stimuli is implemented. The barely stable fixed point dynamics offers hypotheses how brains can balance extreme sensitivity to low concentration stimuli with large dynamic range of odor perception.

There are many more interesting questions in olfaction that we were not able to cover in this chapter, including the intriguing questions of how chemical structure maps to odor perception, how higher order learning and cognitive processing is achieved, what the principles behind the structure of odor receptors are, how odor transduction at receptors is accomplished, how olfaction is achieved in other animals outside the insect models discussed here and many more. While olfaction may not be the most relevant sense to humans and remains one of the less studied senses, it is a research area that keeps delivering interesting insights and holds the promise of important applications, e.g. in chemical sensing for food industries, security, and environmental monitoring.

References

1. Hildebrand, J.G., Shepherd, G.M.: Mechanisms of olfactory discrimination: converging evidence for common principles across phyla. Annu. Rev. Neurosci. **20**, 595–631 (1997)
2. Nakagawa, T., Vosshall, L.B.: Controversy and consensus: noncanonical signaling mechanisms in the insect olfactory system. Curr. Opin. Neurobiol. **19**(3), 284–292 (2009)
3. Sato, K., Pellegrino, M., Nakagawa, T., Nakagawa, T., Vosshall, L.B., Touhara, K.: Insect olfactory receptors are heteromeric ligand-gated ion channels. Nature **452**(7190), 1002–1006 (2008)
4. Silbering, A.F., Benton, R.: Ionotropic and metabotropic mechanisms in chemoreception: 'chance or design'? EMBO Rep. **11**(3), 173–179 (2010)
5. Vosshall, L.B., Hansson, B.S.: A unified nomenclature system for the insect olfactory coreceptor. Chem. Senses **36**(6), 497–498 (2011)
6. Wicher, D., Schäfer, R., Bauernfeind, R., Stensmyr, M.C., Heller, R., Heinemann, S.H., Hansson, B.S.: Drosophila odorant receptors are both ligand-gated and cyclic-nucleotide-activated cation channels. Nature **452**(7190), 1007–1011 (2008)
7. Clyne, P.J., Warr, C.G., Freeman, M.R., Lessing, D., Kim, J., Carlson, J.R.: A novel family of divergent seven-transmembrane proteins: candidate odorant receptors in Drosophila. Neuron **22**(2), 327–338 (1999)
8. Vosshall, L.B., Wong, A.M., Axel, R.: An olfactory sensory map in the fly brain. Cell **102**, 147–159 (2000)
9. Hallem, E.A., Ho, M.G., Carlson, J.R.: The molecular basis of odor coding in the Drosophila antenna. Cell **117**(7), 965–979 (2004)
10. Gao, Q., Yuan, B., Chess, A.: Convergent projections of Drosophila olfactory neurons to specific glomeruli in the antennal lobe. Nat. Neurosci. **3**, 780–785 (2000)
11. Galizia, C.G., Menzel, R.: Odour perception in honeybees: coding information in glomerular patterns. Curr. Opin. Neurobiol. **10**, 504–510 (2000)
12. Hansson, B.S., Carlsson, M.A., Kalinová, B.: Olfactory activation patterns in the antennal lobe of the sphinx moth, manduca sexta. J. Comp. Physiol. A Neuroethol. Sens. Neural Behav. Physiol. **189**(4), 301–308 (2003)
13. Joerges, J., Küttner, A., Galizia, C.G., Menzel, R.: Representations of odour mixtures visualized in the honeybee brain. Nature **387**, 285–288 (1997)

14. Wilson, R., Mainen, Z.F.: Early events in olfactory processing. Annu. Rev. Neurosci. **29**, 163–201 (2006)
15. Christensen, T.A., Waldrop, B.R., Harrow, I.D., Hildebrand, J.G.: Local interneurons and information processing in the olfactory glomeruli of the moth manduca sexta. J. Comp. Physiol. A **173**(4), 385–399 (1993)
16. Christensen, T.A., Waldrop, B.R., Hildebrand, J.G.: Gabaergic mechanisms that shape the temporal response to odors in moth olfactory projection neurons. Ann. N. Y. Acad. Sci. **855**, 475–481 (1998)
17. Christensen, T.A., Waldrop, B.R., Hildebrand, J.G.: Multitasking in the olfactory system: context-dependent responses to odors reveal dual gaba-regulated coding mechanisms in single olfactory projection neurons. J. Neurosci. **18**(15), 5999–6008 (1998)
18. Sachse, S., Galizia, C.G.: Role of inhibition for temporal and spatial odor representation in olfactory output neurons: a calcium imaging study. J. Neurophysiol. **87**, 1106–1117 (2002)
19. Wilson, R.I., Laurent, G.: Role of gabaergic inhibition in shaping odor-evoked spatiotemporal patterns in the Drosophila antennal lobe. J. Neurosci. **25**, 9069–79 (2005)
20. Olsen, S.R., Bhandawat, V., Wilson, R.: Excitatory interactions between olfactory processing channels in the Drosophila antennal lobe. Neuron **54**(1), 89–103 (2007)
21. Shang, Y., Claridge-Chang, A., Sjulson, L., Pypaert, M., Miesenböck, G.: Excitatory local circuits and their implications for olfactory processing in the fly antennal lobe. Cell **128**(3), 601–612 (2007)
22. Silbering, A.F., Galizia, C.G.: Processing of odor mixtures in the Drosophila antennal lobe reveals both global inhibition and glomerulus-specific interactions. J. Neurosci. **27**(44), 11966–11977 (2007)
23. Silbering, A.F., Okada, R., Ito, K., Galizia, C.G.: Olfactory information processing in the Drosophila antennal lobe: anything goes? J. Neurosci. **28**(49), 13075–13087 (2008)
24. Davis, R.L.: Mushroom bodies and Drosophila learning. Neuron **11**, 1–14 (1993)
25. Davis, R.L.: Traces of Drosophila memory. Neuron **70**(1), 8–19 (2011)
26. Heisenberg, M.: Mushroom body memoir: from maps to models. Nat. Rev. Neurosci. **4**(4), 266–275 (2003)
27. Menzel, R.: Searching for the memory trace in a mini-brain, the honeybee. Learn. Mem. **8**(2), 53–62 (2001)
28. de Bruyne, M., Clyne, P.J., Carlson, J.R.: Odor coding in a model olfactory organ: the Drosophila maxillary palp. J. Neurosci. **19**(11), 4520–4532 (1999)
29. Nowotny, T., de Bruyne, M., Berna, A.Z., Warr, C.G., Trowell, S.C.: Drosophila olfactory receptors as classifiers for volatiles from disparate real world applications. Bioinspir. Biomim. **9**(4), 046007 (2014)
30. Galizia, C.G., Joerges, J., Küttner, A., Faber, T., Menzel, R.: A semi-in-vivo preparation for optical recording of the insect brain. J. Neurosci. Methods **76**(1), 61–69 (1997)
31. Galizia, C.G., Sachse, S., Rappert, A., Menzel, R.: The glomerular code for odor representation is species specific in the honeybee Apis mellifera. Nat. Neurosci. **2**, 473–478 (1999)
32. Sachse, S., Rappert, A., Galizia, C.G.: The spatial representation of chemical structures in the antennal lobes of honeybees: steps toward the olfactory code. Eur. J. Neurosci. **11**, 3970–3982 (1999)
33. Galizia, C.G., Kimmerle, B.: Physiological and morphological characterization of honeybee olfactory neurons combining electrophysiology, calcium imaging and confocal microscopy. J. Comp. Physiol. A **190**, 21–38 (2004)
34. Galán, R.F., Sachse, S., Galizia, C.G., Herz, A.V.M.: Odor-driven attractor dynamics in the antennal lobe allow for simple and rapid olfactory pattern classification. Neural Comput. **16**, 999–1012 (2004)
35. Schmuker, M., Yamagata, N., Nawrot, M.P., Menzel, R.: Parallel representation of stimulus identity and intensity in a dual pathway model inspired by the olfactory system of the honeybee. Front. Neuroeng. **4**, 17 (2011)
36. Sachse, S., Galizia, C.G.: The coding of odour-intensity in the honeybee antennal lobe: local computation optimizes odour representation. Eur. J. Neurosci. **18**, 2119–2132 (2003)

37. Guerrieri, F., Schubert, M., Sandoz, J.-C., Giurfa, M.: Perceptual and neural olfactory similarity in honeybees. PLoS Biol. **3**(4), e60 (2005)
38. Baker, T.C., Fadamiro, H.Y., Cosse, A.A.: Moth uses fine tuning for odour resolution. Nature **393**(6685), 530–530 (1998)
39. Szyszka, P., Stierle, J.S., Biergans, S., Galizia, C.G.: The speed of smell: odor-object segregation within milliseconds. PLoS One **7**(4), e36096 (2012)
40. Badel, L., Ohta, K., Tsuchimoto, Y., Kazama, H.: Decoding of context-dependent olfactory behavior in Drosophila. Neuron **91**(1), 155–167 (2016)
41. Laurent, G., Wehr, M., Davidowitz, H.: Temporal representations of odors in an olfactory network. J. Neurosci. **16**, 3837–3847 (1996)
42. Laurent, G., MacLeod, K., Wehr, M.: Spatiotemporal structure of olfactory inputs to the mushroom bodies. Learn. Mem. **5**, 124–132 (1998)
43. Adrian, E.D.: Olfactory reactions in the brain of the hedgehog. J. Physiol. **100**(4), 459–473 (1942)
44. Freeman, W.J.: The physiology of perception. Sci. Am. **264**(2), 78–85 (1991)
45. Gelperin, A., Tank, D.W.: Odour-modulated collective network oscillations of olfactory interneurons in a terrestrial mollusc. Nature **345**(6274), 437–440 (1990)
46. Heinbockel, T., Kloppenburg, P., Hildebrand, J.G.: Pheromone-evoked potentials and oscillations in the antennal lobes of the sphinx moth manduca sexta. J. Comp. Physiol. A **182**(6), 703–714 (1998)
47. Stopfer, M., Bhagavan, S., Smith, B.H., Laurent, G.: Impaired odour discrimination on desynchronization of odour-encoding neural assemblies. Nature **390**(6655), 70–74 (1997)
48. Nusser, Z., Kay, L.M., Laurent, G., Homanics, G.E., Mody, I.: Disruption of gaba(a) receptors on gabaergic interneurons leads to increased oscillatory power in the olfactory bulb network. J. Neurophysiol. **86**(6), 2823–2833 (2001)
49. Biergans, S.D., Jones, J.C., Treiber, N., Galizia, C.G., Szyszka, P.: Dna methylation mediates the discriminatory power of associative long-term memory in honeybees. PLoS One **7**(6), e39349 (2012)
50. Lotka, A.J.: Contribution to the theory of periodic reaction. J. Phys. Chem. **14**(3), 271–274 (1910)
51. May, R.M., Leonard, W.J.: Nonlinear aspects of competition between three species. SIAM J. Appl. Math. **29**, 243–253 (1975)
52. Laurent, G., Stopfer, M., Friedrich, R.W., Rabinovich, M.I., Abarbanel, H.D.I.: Odor encoding as an active, dynamical process: Experiments, computation, and theory. Annu. Rev. Neurosci. **24**, 263–297 (2001)
53. Rabinovich, M. Volkovskii, A., Lecanda, P., Huerta, R., Abarbanel, H.D.I., Laurent, G.: Dynamical encoding by networks of competing neuron groups: winnerless competition. Phys. Rev. Lett. **87**(6), 068102 (2001)
54. Mazor, O., Laurent, G.: Transient dynamics versus fixed points in odor representations by locust antennal lobe projection neurons. Neuron **48**, 661 (2005)
55. Rabinovich, M.I., Huerta, R., Varona, P., Afraimovich, V.S.: Generation and reshaping of sequences in neural systems. Biol. Cybern. **95**(6), 519–536 (2006)
56. Nowotny, T., Rabinovich, M.I.: Dynamical origin of independent spiking and bursting activity in neural microcircuits. Phys. Rev. Lett. **98**, 128106 (2007)
57. Ashwin, P., Karabacak, O., Nowotny, T.: Criteria for robustness of heteroclinic cycles in neural microcircuits. J. Math. Neurosci. **1**(1), 13 (2011)
58. Ashwin, P., Swift, J.W.: The dynamics of n weakly coupled identical oscillators. J. Nonlinear Sci. **2**, 69–108 (1992)
59. Hansel, D., Mato, G., Meunier, C.: Clustering and slow switching in globally coupled phase oscillators. Phys. Rev. E **48**(5), 3470–3477 (1993)
60. Buckley, C.L., Nowotny, T.: Transient dynamics between displaced fixed points: an alternate nonlinear dynamical framework for olfaction. Brain Res. **1434**, 62–72 (2012)
61. Buckley, C.L., Nowotny, T.: Multiscale model of an inhibitory network shows optimal properties near bifurcation. Phys. Rev. Lett. **106**(23), 238109 (2011)

62. Vetter, R.S., Sage, A.E., Justus, K.A., Cardé, R.T., Galizia, C.G.: Temporal integrity of an airborne odor stimulus is greatly affected by physical aspects of the odor delivery system. Chem. Senses **31**(4), 359–369 (2006)
63. Uchida, N., Mainen, Z.F.: Speed and accuracy of olfactory discrimination in the rat. Nat. Neurosci. **6**, 1224–1229 (2003)
64. Resulaj, A., Rinberg, D.: Novel behavioral paradigm reveals lower temporal limits on mouse olfactory decisions. J. Neurosci. **35**(33), 11667–11673 (2015)
65. Ditzen, M., Evers, F., Galizia, C.G.: Odor similarity does not influence the time needed for odor processing. Chem. Senses **28**, 781–789 (2003)
66. Wright, G.A., Carlton, M., Smith, B.H.: A honeybee's ability to learn, recognize, and discriminate odors depends upon odor sampling time and concentration. Behav. Neurosci. **123**(1), 36–43 (2009)
67. Murlis, J., Elkinton, J.S., Cardé, R.T.: Odor plumes and how insects use them. Annu. Rev. Entomol. **37**, 505–532 (1992)
68. Celani, A., Villermaux, E., Vergassola, M.: Odor landscapes in turbulent environments. Phys. Rev. X **4**, 041015 (2014)
69. Riffell, J.A., Shlizerman, E., Sanders, E., Abrell, L., Medina, B., Hinterwirth, A.J., Kutz, J.N.: Sensory biology. flower discrimination by pollinators in a dynamic chemical environment. Science **344**(6191), 1515–1518 (2014)
70. Soltys, M.A., Crimaldi, J.P.: Joint probabilities and mixing of isolated scalars emitted from parallel jets. J. Fluid Mech. **769**, 130–153, 003 (2015)
71. Hopfield, J.J.: Olfactory Computation and Object Perception. Proc. Natl. Acad. Sci. U. S. A. **88**(15), 6462–6466 (1991)
72. Fadamiro, H.Y., Baker, T.C.: Reproductive performance and longevity of female European corn borer, Ostrinia nubilalis: effects of multiple mating, delay in mating, and adult feeding. J. Insect Physiol. **45**(4), 385–392 (1999)
73. Nikonov, A.A., Leal, W.S.: Peripheral coding of sex pheromone and a behavioral antagonist in the Japanese beetle, Popillia japonica. J. Chem. Ecol. **28**(5), 1075–1089 (2002)
74. Saha, D., Leong, K., Li, C., Peterson, S., Siegel, G., Raman, B.: A spatiotemporal coding mechanism for background-invariant odor recognition. Nat. Neurosci. **16**(12), 1830–1839 (2013)
75. Broome, B., Jayaraman, V., Laurent, G.: Encoding and decoding of overlapping odor sequences. Neuron **51**, 467–482 (2006)
76. Stierle, J., Galizia, C.G., Szyszka, P.: Millisecond stimulus onset-asynchrony enhances information about components in an odor mixture. J. Neurosci. **33**(14), 6060–6069 (2013)
77. Andersson, M.N., Binyameen, M., Sadek, M.M., Schlyter, F.: Attraction modulated by spacing of pheromone components and anti-attractants in a bark beetle and a moth. J. Chem. Ecol. **37**(8), 899–911 (2011)
78. Andersson, M.N., Larsson, M.C., Blazenec, M., Jakus, R., Zhang, Q.-H., Schlyter, F.: Peripheral modulation of pheromone response by inhibitory host compound in a beetle. J. Exp. Biol. **213**(Pt 19), 3332–3339 (2010)
79. Su, C.-Y., Menuz, K., Reisert, J., Carlson, J.R.: Non-synaptic inhibition between grouped neurons in an olfactory circuit. Nature **492**(7427), 66–71 (2012)
80. O'connell, R.J., Grant, A.J., Mayer, M.S., Mankin, R.W.: Morphological correlates of differences in pheromone sensitivity in insect sensilla. Science **220**(4604), 1408–1410 (1983)
81. Nowotny, T., Stierle, J.S., Galizia, C.G., Szyszka, P.: Data-driven honeybee antennal lobe model suggests how stimulus-onset asynchrony can aid odour segregation. Brain Res. **1536**, 119–134 (2013)
82. Linster, C., Sachse, S., Galizia, C.G.: Computational modeling suggests that response properties rather than spatial position determine connectivity between olfactory glomeruli. J. Neurophysiol. **93**(6), 3410–3417 (2005)
83. Shanbhag, S.R., Müller, B., Steinbrecht, R.A.: Atlas of olfactory organs of Drosophila melanogaster 2. Internal organization and cellular architecture of olfactory sensilla. Arthropod Struct. Dev. **29**(3), 211–229 (2000)

84. Getchell, T.V., Margolis, F.L., Getchell, M.L.: Perireceptor and receptor events in vertebrate olfaction. Prog. Neurobiol. **23**(4), 317–345 (1984)
85. Pelosi, P.: Perireceptor events in olfaction. J. Neurobiol. **30**(1), 3–19 (1996)
86. Kaissling, K.-E.: Kinetics of olfactory responses might largely depend on the odorant-receptor interaction and the odorant deactivation postulated for flux detectors. J. Comp. Physiol. A Neuroethol. Sens. Neural Behav. Physiol. **199**(11), 879–896 (2013)
87. Leal, W.S.: Odorant reception in insects: roles of receptors, binding proteins, and degrading enzymes. Annu. Rev. Entomol. **58**, 373–391 (2013)
88. Vogt, R.G., Riddiford, L.M.: Pheromone binding and inactivation by moth antennae. Nature **293**(5828), 161–163 (1981)
89. Benton, R., Sachse, S., Michnick, S.W., Vosshall, L.B.: Atypical membrane topology and heteromeric function of Drosophila odorant receptors in vivo. PLoS Biol. **4**(2), e20 (2006)
90. Schneider, D., Lacher, V., Kaissling, K.E.: Die Reaktionsweise und das Reaktionsspektrum von Riechzellen bei Antheraea pernyi (Lepidoptera, Saturniidae). Z. Vgl. Physiol. **48**, 632–662 (1964)
91. Szyszka, P., Gerkin, R.C., Galizia, C.G., Smith, B.H.: High-speed odor transduction and pulse tracking by insect olfactory receptor neurons. Proc. Natl. Acad. Sci. U. S. A. **111**(47), 16925–16930 (2014)
92. Semmelroch, P., Grosch, W.: Studies on character impact odorants of coffee brews. J. Agric. Food Chem. **44**, 537–543 (1996)
93. Knudsen, J.T., Tollsten, L., Bergström, L.G.: Floral scents - a checklist of volatile compounds isolated by head-space techniques. Phytochemistry **33**, 253–280 (1993)
94. Root, C.M., Masuyama, K., Green, D.S., Enell, L.E., Nässel, D.R., Lee, C.-H., Wang, J.W.: A presynaptic gain control mechanism fine-tunes olfactory behavior. Neuron **59**(2), 311–321 (2008)
95. Girardin, C.C., Kreissl, S., Galizia, C.G.: Inhibitory connections in the honeybee antennal lobe are spatially patchy. J. Neurophysiol. **109**(2), 332–343 (2013)
96. Hong, E.J., Wilson, R.I.: Olfactory neuroscience: normalization is the norm. Curr. Biol. **23**(24), R1091–R1093 (2013)
97. Hong, E.J., Wilson, R.I.: Simultaneous encoding of odors by channels with diverse sensitivity to inhibition. Neuron **85**(3), 573–589 (2015)
98. Deisig, N., Giurfa, M., Lachnit, H., Sandoz, J.-C.: Neural representation of olfactory mixtures in the honeybee antennal lobe. Eur. J. Neurosci. **24**(4), 1161–1174 (2006)
99. Deisig, N., Giurfa, M., Sandoz, J.-C.: Antennal lobe processing increases separability of odor mixture representations in the honeybee. J. Neurophysiol. **103**(4), 2185–2194 (2010)
100. Smith, B.H.: Analysis of interaction in binary odorant mixtures. Physiol. Behav. **65**(3), 397–407 (1998)
101. Chandra, S., Smith, B.H.: An analysis of synthetic processing of odor mixtures in the honeybee (Apis mellifera). J. Exp. Biol. **201**(Pt 22), 3113–3121 (1998)
102. Deisig, N., Lachnit, H.,Giurfa, M., Hellstern, F.: Configural olfactory learning in honeybees: negative and positive patterning discrimination. Learn. Mem. **8**(2), 70–78 (2001)
103. Deisig, N., Lachnit, H., Giurfa, M.: The effect of similarity between elemental stimuli and compounds in olfactory patterning discriminations. Learn. Mem. **9**(3), 112–121 (2002)
104. Deisig, N., Lachnit, H., Sandoz, J.-C., Lober, K., Giurfa, M.: A modified version of the unique cue theory accounts for olfactory compound processing in honeybees. Learn. Mem. **10**(3), 199–208 (2003)
105. Kamin, L.J.: Predictability, surprise, attention, and conditioning. Punishment and Aversive Behavior, pp. 279–296. Appleton-Century-Crofts, New York (1969)
106. Rescorla, R.A., Wagner, A.R., et al.: A theory of pavlovian conditioning: variations in the effectiveness of reinforcement and nonreinforcement. Classical Conditioning II: Current Research and Theory, vol. 2, pp. 64–99. Appleton-Century Crofts, New York (1972)
107. Hosler, J.S., Smith, B.H.: Blocking and the detection of odor components in blends. J. Exp. Biol. **203**(Pt 18), 2797–2806 (2000)

108. Smith, B.H.: An analysis of blocking in odorant mixtures: an increase but not a decrease in intensity of reinforcement produces unblocking. Behav. Neurosci. **111**(1), 57–69 (1997)
109. Smith, B.H., Cobey, S.: The olfactory memory of the honeybee Apis mellifera. II. Blocking between odorants in binary mixtures. J. Exp. Biol. **195**, 91–108 (1994)
110. Gerber and Ullrich: No evidence for olfactory blocking in honeybee classical conditioning. J. Exp. Biol. **202**(Pt 13), 1839–1854 (1999)
111. Guerrieri, F., Lachnit, H., Gerber, B., Giurfa, M.: Olfactory blocking and odorant similarity in the honeybee. Learn. Mem. **12**(2), 86–95 (2005)
112. Zars, T., Fischer, M., Schulz, R., Heisenberg, M.: Localization of a short-term memory in Drosophila. Science **288**(5466), 672–675 (2000)
113. Becher, P.G., Bengtsson, M., Hansson, B.S., Witzgall, P.: Flying the fly: long-range flight behavior of Drosophila melanogaster to attractive odors. J. Chem. Ecol. **36**(6), 599–607 (2010)
114. Strutz, A., Soelter, J., Baschwitz, A., Farhan, A., Grabe, V., Rybak, J., Knaden, M., Schmuker, M., Hansson, B.S., Sachse, S.: Decoding odor quality and intensity in the Drosophila brain. Elife **3**, e04147 (2014)
115. Huerta, R., Nowotny, T., Garcia-Sanchez, M., Abarbanel, H.D.I., Rabinovich, M.I.: Learning classification in the olfactory system of insects. Neural Comput. **16**, 1601–1640 (2004)
116. Perez-Orive, J., Mazor, O., Turner, G.C., Cassenaer, S., Wilson, R., Laurent, G.: Oscillations and sparsening of odor representations in the mushroom body. Science **297**(5580), 359–365 (2002)
117. Szyszka, P., Ditzen, M., Galkin, A., Galizia, C.G., Menzel, R.: Sparsening and temporal sharpening of olfactory representations in the honeybee mushroom bodies. J. Neurophysiol. **94**(5), 3303–3313 (2005)
118. Lin, A.C., Bygrave, A.M., de Calignon, A., Lee, T., Miesenböck, G.: Sparse, decorrelated odor coding in the mushroom body enhances learned odor discrimination. Nat. Neurosci. **17**(4), 559–568 (2014)
119. Nowotny, T., Huerta, R., Abarbanel, H.D.I., Rabinovich, M.I.: Self-organization in the olfactory system: rapid odor recognition in insects. Biol Cybern. **93**, 436–446 (2005)
120. Huerta, R., Nowotny, T.: Fast and robust learning by reinforcement signals: explorations in the insect brain. Neural Comput. **21**, 2123–2151 (2009)
121. Finelli, L.A., Haney, S., Bazhenov, M., Stopfer, M., Sejnowski, T.J.: Synaptic learning rules and sparse coding in a model sensory system. PLoS Comput. Biol. **4**(4), e1000062 (2008)
122. Szyszka, P., Galkin, A., Menzel, R.: Associative and non-associative plasticity in Kenyon cells of the honeybee mushroom body. Front. Syst. Neurosci. **2**, 3 (2008)
123. Smith, D., Wessnitzer, J., Webb, B.: A model of associative learning in the mushroom body. Biol. Cybern. **99**(2), 89–103 (2008)
124. Wessnitzer, J., Young, J.M., Armstrong, J.D., Webb, B.: A model of non-elemental olfactory learning in Drosophila. J. Comput. Neurosci. **32**(2), 197–212 (2012)
125. Bazhenov, M., Huerta, R., Smith, B.H.: A computational framework for understanding decision making through integration of basic learning rules. J. Neurosci. **33**(13), 5686–5697 (2013)
126. Perry, C.J., Barron, A.B., Cheng, K.: Invertebrate learning and cognition: relating phenomena to neural substrate. Wiley Interdiscip. Rev. Cogn. Sci. **4**(5), 561–582 (2013)
127. Papadopoulou, M., Cassenaer, S., Nowotny, T., Laurent, G.: Normalization for sparse encoding of odors by a wide-field interneuron. Science **332**(6030), 721–725 (2011)
128. Assisi, C., Stopfer, M., Laurent, G., Bazhenov, M.: Adaptive regulation of sparseness by feedforward inhibition. Nat. Neurosci. **10**(9), 1176–1184 (2007)
129. Gupta, N., Stopfer, M.: Functional analysis of a higher olfactory center, the lateral horn. J. Neurosci. **32**(24), 8138–8148 (2012)
130. Kee, T., Sanda, P., Gupta, N., Stopfer, M., Bazhenov, M.: Feed-forward versus feedback inhibition in a basic olfactory circuit. PLoS Comput. Biol. **11**(10), e1004531 (2015)
131. Grünewald, B.: Morphology of feedback neurons in the mushroom body of the honeybee, Apis mellifera. J. Comp. Neurol. **404**, 114–126 (1999)

132. Andrew, S.C., Perry, C.J., Barron, A.B., Berthon, K., Peralta, V., Cheng, K.: Peak shift in honeybee olfactory learning. Anim. Cogn. **17**(5), 1177–1186 (2014)
133. Campbell, R.A.A., Honegger, K.S., Qin, H., Li, W., Demir, E., Turner, G.C.: Imaging a population code for odor identity in the Drosophila mushroom body. J. Neurosci. **33**(25), 10568–10581 (2013)
134. Nowotny, T., Rabinovich, M.I., Huerta, R., Abarbanel, H.D.I.: Decoding temporal information through slow lateral excitation in the olfactory system of insects. J. Comput. Neurosci. **15**, 271–281 (2003)
135. Nowotny, T., Berna, A.Z., Binions, R., Trowell, S.: Optimal feature selection for classifying a large set of chemicals using metal oxide sensors. Sens. Actuators B **187**, 471–480 (2013); Selected Papers from the 14th International Meeting on Chemical Sensors.

Chapter 16
Dynamics of Intermittent Synchronization of Neural Activity

Leonid L. Rubchinsky, Choongseok Park, and Sungwoo Ahn

16.1 Synchronized Neural Activity

Electrical activity of neurons and neural populations in the brain frequently exhibits some degree of synchrony (defined in some specific ways). Multiple experimental studies indicate that neural synchronization is important for various functions of the brain. For example, neural synchrony has been observed in relation to perception [11], memory [12], other cognitive functions [13], and motor functions [16, 31]. Different mechanisms underlying the involvement of oscillatory neural activity in neural function have been considered [8]. Abnormally high (and sometimes abnormally low) strength of neural synchrony has been implied to be critical for the symptoms of several neurological and psychiatric disorders [32, 34], including Parkinson's disease (e.g., [29]) and schizophrenia (e.g., [35]).

Synchronization is a widely observed phenomenon and has been traditionally studied with the methods of physics and nonlinear dynamics [23]. One of the scholars, who early recognized the importance and relevance of physical studies of synchronization to neuroscience was Misha (Mikhail Izrailevich, as one would

L.L. Rubchinsky (✉)
Department of Mathematical Sciences, Indiana University Purdue University Indianapolis, Indianapolis, IN 46202, USA

Stark Neurosciences Research Institute, Indiana University School of Medicine, Indianapolis, IN 46202, USA
e-mail: lrubchin@iupui.edu

C. Park
Department of Mathematics, North Carolina A&T State University, Greensboro, NC, USA
e-mail: cpark@ncat.edu

S. Ahn
Department of Mathematics, East Carolina University, Greenville, NC, USA
e-mail: ahnmath@gmail.com

I.S. Aranson et al. (eds.), *Advances in Dynamics, Patterns, Cognition*, Nonlinear Systems and Complexity 20, DOI 10.1007/978-3-319-53673-6_16

politely call him in his native Russian) Rabinovich. He inspired his students and engaged his collaborators [2], and authored many important and elegant studies of the mechanisms and functions of neural synchrony (for example, [1, 7, 10, 17, 33]). Misha pointed out that neural synchronization may have some peculiar properties, which are not frequently observed in other synchronized systems in nature, because of the very specific functions of neural systems [18].

Interestingly, Misha Rabinovich discussed the potential generality of transient dynamics in neuroscience [24–26]. The transient (intermittent) synchrony appears to be the norm in the synchronized dynamics of neural circuits of the brain. Even at the rest state, without any transient stimuli, perfect synchrony in the brain has not been reported. This is probably not very surprising. At the rest state these circuits should be ready to respond to different stimuli. If they were in a completely synchronized stable state, it would probably require more time and/or efforts to respond to stimuli.

16.2 Fine Temporal Structure of Intermittent Synchronization

Neuroscience traditionally operates with observables. Usually, in the context of neural synchrony, these are intra- or extracellularly recorded electric potentials as a function of time. However, a consideration of the dynamics of synchronization and desynchronization in terms of the phase space helps to understand the nature of these phenomena.

16.2.1 Phase Space-Based View and Time-Series-Based View of Synchronized and Desynchronized Episodes

Let's think of coupled neurons or neural populations where each individual unit exhibits oscillatory dynamics. Coupling is strong enough and dynamic is synchronous so that there is a stable synchronization manifold in the phase space. As we discussed above, this is a straightforward, but not an experimentally realistic scenario. Experimentally relevant, but relatively weak intermittent synchrony may correspond to the following dynamics: the system moves into vicinity of an unstable synchronization manifold, but eventually leaves it because this manifold is unstable. Since the synchrony is relatively weak, the system will spend substantial amount of time away from the synchronization manifold. In this case a study of the properties of synchronization manifold (e.g., its loss of stability and associated types of intermittency) informs of what happens in the system of interest only for a small fraction of time.

Unlike the stability of the synchronization manifold, however, the properties of the periphery of the phase space are not universal in general. But it does not mean they are irrelevant to the dynamics of the neuronal networks. To connect these ideas to experimental data we will look at these issues from the time-series analysis perspective.

16.2.2 Dynamics of Desynchronization Episodes: Time-Series Analysis

The phase can be extracted from a "good" oscillatory data (the data with relatively narrow and prominent peak in the spectrum) in different ways, we will use Hilbert phase [23]. Using Hilbert transform one obtains an analytic signal $\xi(t)$ from a real time series $x(t)$ as follows:

$$\xi(t) = x(t) + i\bar{x}(t) \tag{16.1}$$

$$\bar{x}(t) = \frac{1}{\pi}\,\mathrm{PV}\int_{-\infty}^{\infty} \frac{x(\tau)}{t-\tau} d\tau. \tag{16.2}$$

Let the phase of the analytic signal $\xi(t)$, say $\varphi(t)$, be the Hilbert phase of the time series. Then it is given by

$$z(t) = \frac{\xi(t)}{\|\xi(t)\|} = e^{i\varphi(t)}. \tag{16.3}$$

If the phase difference between two oscillators tends to be close (in some specific sense) to some constant value, then we can consider this as a synchronized dynamics.

One can compute a fairly standard phase locking index for two phases $\varphi_1(t)$ and $\varphi_2(t)$:

$$\gamma = \|\frac{1}{N}\sum_{j=1}^{N} e^{i\Phi_j}\|, \tag{16.4}$$

where $\Phi_j = \varphi_1(t_j) - \varphi_2(t_j)$ and N is the number of data points (for the case of discrete time-series). This index varies between zero (no phase locking) and one (perfect phase locking) [23].

For further analysis (originally developed in [3, 21]), we consider a first-return map for the phase difference (see also [30]). In other words, we are considering whether the phase difference is close to its preferred (locked) state or not once per cycle of oscillations (we assume this preferred state exists and can be extracted from the data, otherwise the described procedure does not make sense). How close it should be depends on a particular problem under consideration. We will consider the case where we require the phase difference to be within $\pi/2$ of the preferred phase difference.

Whenever the phase of one signal crosses a check point (which may be assumed to be zero) from negative to positive values, the phase of the other signal is recorded, resulting in a set of consecutive phase values $\{\phi_i\}_{i=1}^{N}$. Since the phase of one signal is zero, this is actually the phase difference between two oscillators measured once per cycle of oscillations. Now consider (ϕ_i, ϕ_{i+1}) space. The predominantly synchronous dynamics will appear as a cluster of points on the diagonal $\phi_{i+1} = \phi_i$ (note that this phase space is actually a torus). For the uniformity of analysis, all values of the phases may be shifted in such a way that the center of the cluster lies at the center of the first quadrant. The phase space is then partitioned into four equally spaced regions. Figure 16.1 shows a diagram for this first-return map. The first region is considered to be a synchronous state while other regions (II, III, and IV) are considered to be desynchronized states. If the phase difference is required to be not within $\pi/2$ of the preferred phase difference, but within different tolerance limit, different partition will be required.

One can define the transition rates $r_{1,2,3,4}$ for transitions between four regions of the phase space as a ratio of the number of points leaving a region to the total number of points in that region (see [3, 21, 30]). For example, r_1 is the number of points leaving the region I for the region II divided by the total number of points in the region I. One can also define the duration of desynchronization events to explore how long do the desynchronization events last. Here, the duration

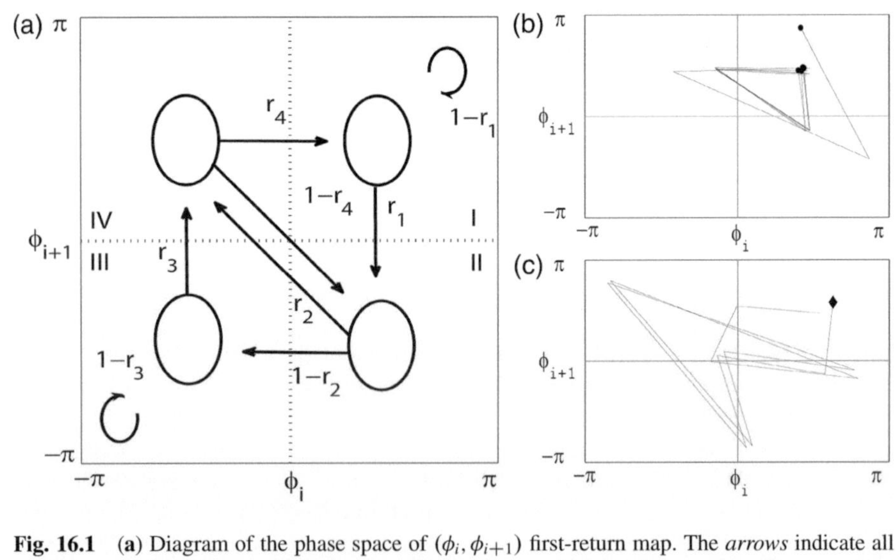

Fig. 16.1 (**a**) Diagram of the phase space of (ϕ_i, ϕ_{i+1}) first-return map. The *arrows* indicate all possible transitions from one region to another and $r_{1,2,3,4}$ indicate the corresponding transition rates. The synchronized state is placed at the center of the region I and three other regions are desynchronized states. (**b**) presents an example of dynamics with numerous short desynchronizations. (**c**) presents an example of dynamics with a very long desynchronization event

of desynchronization events is defined as the number of steps that the system spends away from the region I minus one. For example, the shortest duration of a desynchronization event corresponds to the shortest path II → IV → I (Fig. 16.1b). This corresponds to the length of one cycle of desynchronization events. Length of two cycles corresponds to the path II → III → IV → I and longer lengths of desynchronization events will have many different paths.

16.2.3 An Analysis of a Simple Model System

To illustrate some of the ideas discussed above, following [3] we will consider an example of a very simple coupled system: two coupled skewed tent maps. While this example may be ill-suited to study phase synchronization [28], it helps to illustrate the major ideas in a very simple setting. Consider a skew tent map

$$f(a,x) = \begin{cases} \frac{x}{a}, & \text{if } 0 \le x \le a, \\ \frac{1-x}{1-a}, & \text{if } a < x \le 1, \end{cases} \tag{16.5}$$

where $0 < a < 1$. Two such maps, described by variables x and y, are coupled in the following way:

$$\begin{aligned} x(t+1) &= (1-\varepsilon)f(a,x(t)) + \varepsilon f(a,y(t)), \\ y(t+1) &= \varepsilon f(a,x(t)) + (1-\varepsilon)f(a,y(t)), \end{aligned} \tag{16.6}$$

where ε is the coupling strength. The difference of the variables of two maps $\phi(t) = y(t) - x(t)$ may serve as a proxy for the phase difference. The synchronous state is $x = y$. It becomes stable for ε larger than a critical value ε_c. Two Lyapunov exponents ($\lambda(a)$ and $\lambda_{\perp}(a,\varepsilon)$) can be computed analytically [23] and are not changed if a is changed into $(1-a)$, i.e. they are symmetrical about $a = 1/2$.

Therefore, two different pairs of maps with symmetrical values of a have the same values of Lyapunov exponents (in particular, the same value of $\lambda_{\perp}(a,\varepsilon)$, which characterizes the stability of the synchronous state). Thus they have the same expansive/contractive properties on the average. But the two systems are different. In one case, the map is strongly expansive in a small area of the phase space, while in the other case the map is less expansive, but the corresponding area is larger. The properties of the desynchronized dynamics are different between the two systems in the intermittently synchronous dynamics (the coupling value ε is less than ε_c) [3]. The transition rates r_i and the distributions of desynchronization events durations are markedly different between the dynamics of $a = 0.1$ and $a = 0.9$ (Fig. 16.2). This example shows that there may be different temporal patterns of synchronized and desynchronized dynamics in the coupled systems, which have the same stability properties of synchronized state.

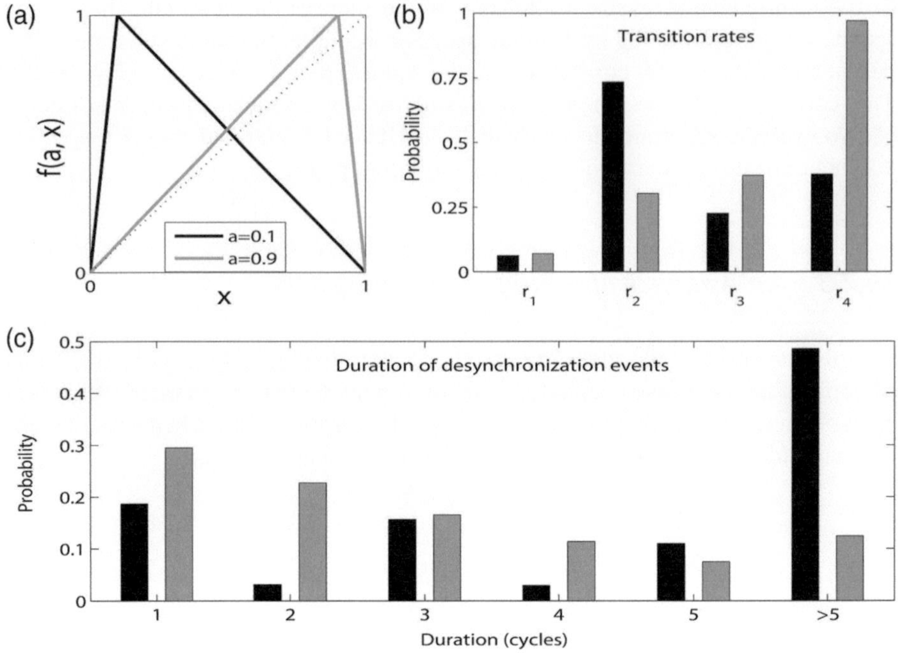

Fig. 16.2 (**a**) Two kinds of skew tent maps, with $a = 0.1$ (*black*) and $a = 0.9$ (*gray*). The coupled black maps and gray maps have identical Lyapunov exponents but different expansive/contractive properties in different areas of the phase space. (**b**) transition rates and (**c**) distribution of durations of desynchronization events for both types of coupled maps [30]. Here, ">5" is a sum of the relative frequencies of all desynchronizations longer than five

16.3 Applications to Experimental Neuroscience Data

In this section we will discuss the analysis of the temporal dynamics of synchronization in several different neuroscience experiments. There are different species involved (rodents and humans), different types of recorded data (spikes, local field potentials (LFP), and electroencephalogram (EEG)), different brain areas, and different brain states. What is general here is that in all these cases we are dealing with the synchronized oscillations in the neural activity of mammalian brains.

16.3.1 EEG Recordings in Healthy Human Subjects

In a recent study [4], EEGs recorded in a group of a hundred of healthy subjects were subjected to the analysis described above. EEGs were recorded from scalp electrodes and beta-band oscillations were extracted from the recordings. The data were acquired while subjects were at rest or executed a simple motor task (open

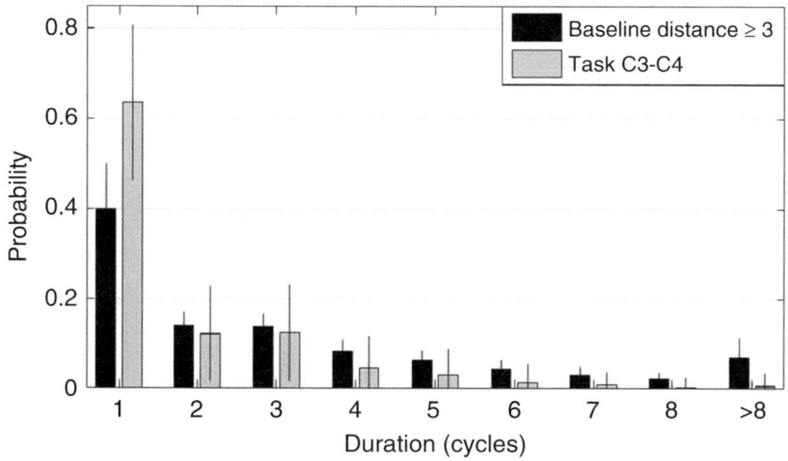

Fig. 16.3 Distribution of desynchronization durations for "Baseline distance $\geq$ 3" (*black*) and "Task C3-C4" (*gray*). Here, ">8" is a sum of the relative frequencies of all desynchronizations longer than eight. Mean $\pm$ SD is presented

and close fists). The resulting distribution of the durations of desynchronizations is presented in Fig. 16.3. In the rest state the pairs of distant electrodes, "Baseline distance $\geq$ 3", were considered to minimize a cross-talk between nearby electrodes. During a motor task, synchronization between a pair of electrodes over motor cortices was analyzed,"Task C3-C4."

The results do depend quantitatively on many factors, but qualitatively the shortest possible desynchronization is always the most frequent: for different brain rhythms, at rest and during a motor task execution, and for different arrangements of considered electrodes.

16.3.2 Spikes and Local Field Potentials in the Basal Ganglia of the Parkinson's Disease Patients

A study by Park et al. [21] analyzed the fine temporal structure of the phase-locking of neural oscillations in Parkinson's disease in the subthalamic nucleus (STN, a part of the basal ganglia, subcortical brain nuclei related to Parkinson's disease). Spiking units and LFPs were recorded with microelectrodes in a group of patients during a neurosurgery to implant deep brain stimulator electrode. LFPs are usually believed to be formed by synaptic currents and STN apparently lacks intranuclear connections. Thus, unlike cortex, spikes and LFPs in STN are formed by different processes. Episodes of beta-band activity were extracted from the data because this activity is associated with hypokinetic motor symptoms of the disease. The segments of data recordings with statistically significant synchrony strength were further selected.

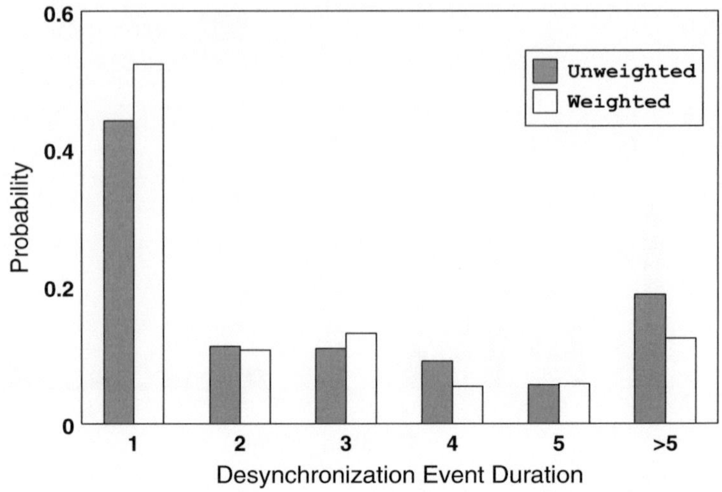

Fig. 16.4 The distribution of the durations of desynchronization events in the data from patients with Parkinson's disease. The histograms of desynchronization event durations were computed in two ways, unweighted (*gray bars*) and weighted (*white bars*) proportionally to the length of the analyzed data segments. All durations that are greater than or equal to six cycles of oscillations are pooled together in ">5" group

The distributions of the durations of desynchronization events in the resulting data are presented in Fig. 16.4. The most frequent duration of desynchronization events is the shortest one and the probability to observe a desynchronization event of a duration decreases as the event duration increases. Two different ways of computing the frequencies, unweighted and weighted proportionally to the length of the analyzed data segments, yield qualitatively the same results.

Similar results were obtained after the analysis of the recordings from a different brain nuclei, internal Globus Pallidus [27]. Modeling studies of the beta-band oscillations in parkinsonian basal ganglia reproduce short desynchronization dynamics even in very small networks [19, 20] and in more realistic, larger networks [22] reach quantitative agreement with experiments.

16.3.3 Local Field Potentials in the Cortex and Hippocampus in the Rodents Undergoing Behavioral Sensitization

Another application of the analysis of desynchronization durations to neurophysiology was done at [5]. The data were recorded from hippocampus and prefrontal cortex of rats undergoing the protocol of behavioral sensitization. This protocol is used to study drug addiction in experimental animals and aforementioned brain areas are known to be directly involved in the addiction phenomena. The animals received an injection of amphetamine every other days for several days. The dynamics of theta-band oscillation were analyzed. In all animals (including the ones from the control group) and at every stage of the protocol short desynchronizations were observed most frequently. The distribution of the durations of desynchronization events was different for different stages of the protocol, but short desynchronizations always prevailed.

Interestingly, at the very beginning of the protocol, initial drug injection does not alter average synchrony strength of theta-band oscillation. But the distribution of desynchronization durations is changed in response to the action of the drug [5]. Eventually, after more injections, average synchrony started to exhibit differences as well. There are two (not necessarily mutually exclusive) explanations for this. The first one is that the distribution of desynchronization durations is altered independently of average synchrony strength in an experiment. The second one is that as desynchronization durations are reorganized, so the average synchrony is, but it changes so weakly that the change of average synchrony has not been detected statistically. Either way the distribution of desynchronization durations turns to be more sensitive to the changes in a neuronal system, rather than average synchrony.

16.3.4 Cardiac and Respiratory Rhythms in Healthy Subjects and Subjects with Coronary Artery Disease

Final experimental example considered here is the analysis of phase-locking between respiration and heartbeat rhythms [6]. These rhythms are not usually thought of as brain rhythms. However, respiratory rhythm is generated in the brain and cardiac rhythm is generated by the electrically active cells in the heart's sinoatrial node, which bear some similarity to neurons (both cardiac cells and neurons have active membrane channels and produce relaxational oscillations). It is interesting to mention that cardiorespiratory phase-locking is almost never $1:1$, unlike the neurophysiological phenomena described above. In general, this is $n:m$ phase-locking, n and m are small integers, and respiration is a few times slower than heartbeat. In the study [6] the methods described in the Sect. 16.2.2 were generalized to $1:m$ case. This generalization is relatively straightforward: the value of the phase of the fast rhythm is recorded when the phase of the slow rhythm is going through the checkpoint.

The results are qualitatively similar to the experiments with the neural signals described above. The distributions of durations of desynchronization events do depend on many factors such as age, coronary artery disease status, ratio of frequencies of the oscillations. However, short desynchronizations are the most prevalent (although their relative frequencies in cardiorespiratory synchronization generally tend to be a bit lower than that in the neural systems described above).

16.4 Discussion

Since neural synchronization is usually intermittent even in the rest states, the analysis of distributions of synchronization and desynchronization intervals provides some interesting information about dynamics of neural systems. Properties of the synchronization intervals in neural systems have been studied earlier in the context of study of different types of intermittencies (e.g., [14, 15, 36]). But if neural synchrony is weak (which is usually the case), substantial fraction of time is spent in the desynchronized events. The studies reviewed in this paper indicate that quite different neural systems share similar properties of desynchronization events, which implies certain universality among these systems. We will discuss some potential ramifications of these observations.

16.4.1 Measuring Fine Temporal Structure of Synchronized Dynamics in Neuroscience

Measuring properties of desynchronization durations opens a way for two interesting possibilities in neuroscience. The first one is the use of the desynchronization durations as a diagnostic tool. The experiments with behavioral sensitization discussed above indicate that the distribution of desynchronization durations may be altered more easily than the synchrony strength and may serve as an early predictor of less subtle changes in a neuronal circuit and its dynamics in response to the drug injection. Perhaps properties of the distribution of desynchronization durations may serve as a clinical tool to diagnose otherwise undetectable changes in the neurophysiology.

The second one is the use of the desynchronization properties to match dynamical models to the real systems. If a system spends a substantial fraction of time away from synchronous state, then properties of the desynchronized states are important in order to have an adequate model of this system. One possibility is to use the transition rates, that describe the transitions between different parts of the phase space, and to develop a model in such a way that these transition rates in the model are matched to the ones derived from experimental data (like it was done in [9, 22]). This kind of matching may be important in the modeling studies of the modulation of synchrony in neuronal networks. If only the synchronized states in the model and real phase space are similar, but peripheries of the phase spaces are organized in different ways, then a modulation or stimulation during desynchronized episodes may lead to different consequences in the model and real systems (see [9]).

16.4.2 Potential Functional Significance of Observed Short Desynchronizations Dynamics in Neural Data

An interesting similarity among several experimental results described in the previous section is that in all these cases short desynchronizations dominate in the time-series (the mode of distribution of desynchronization durations equals to 1). For generic coupled oscillators, the mode of this distribution may have a higher value. This distribution is defined by the properties of a phase space periphery away from the synchronization state. Thus, for generic coupled oscillators, one may not necessarily expect much of universality here. Unlike the loss of stability of synchronization manifold, the reinjection mechanism is not necessarily universal. However, neural desynchronizations in the brain under very much different conditions and in different species appear to be universal: short desynchronizations prevail. There may be something in the very common properties of neurons (perhaps properties of kinetics of membrane channels), that universally facilitates short desynchronization dynamics.

The observed prevalence of short desynchronizations naturally brings a question of whether this property may have any significant functional advantage. We suppose that short desynchronization dynamics may make neuronal circuits more amenable to control by other brain parts or sensory inputs. Neural circuits need to be able to transiently create some synchronized states in response to external inputs. Short desynchronization dynamics means that although average synchrony is low, the system moves into vicinity of a synchronized state fairly frequently. This also implies that (for average synchrony been equal) synchronized state without inputs should be relatively strongly unstable. This arrangement may be more conducive to quick and efficient formation and break-up of transiently synchronized states in response to the external inputs to the network.

References

1. Abarbanel, H.D.I., Huerta, R., Rabinovich, M.I., Rulkov N.F., Rowat P.F., Selverston A.I.: Synchronized action of synaptically coupled chaotic model neurons. Neural Comput. **8**, 1567–1602 (1996)
2. Abarbanel, H.D.I., Rabinovich, M.I., Selverston, A., Bazhenov, M.V., Huerta, R., Sushchik, M.M., Rubchinskii, L.L.: Synchronization in neural networks. Phys. Usp. **39**, 337–362 (1996)
3. Ahn, S., Park, C., Rubchinsky, L.L.: Detecting the temporal structure of intermittent phase locking. Phys. Rev. E **84**, 016201 (2011)
4. Ahn, S., Rubchinsky, L.L.: Short desynchronization episodes prevail in synchronous dynamics of human brain rhythms. Chaos **23**, 013138 (2013)
5. Ahn, S., Rubchinsky, L.L., Lapish, C.C.: Dynamical reorganization of synchronous activity patterns in prefrontal cortex-hippocampus networks during behavioral sensitization. Cereb. Cortex **24**(10), 2556–2561 (2014)
6. Ahn, S., Solfest, J., Rubchinsky, L.L.: Fine temporal structure of cardiorespiratory synchronization. Am. J. Physiol. Heart Circ. Physiol. **306**, 755–763 (2014)

7. Bazhenov, M., Stopfer, M., Rabinovich, M., Huerta, R., Abarbanel, H.D.I., Sejnowski, T.J., Laurent, G.: Model of transient oscillatory synchronization in the locust antennal lobe. Neuron **30**, 553–567 (2001)
8. Buzsáki, G., Draguhn, A.: Neuronal oscillations in cortical networks. Science **304**, 1926–1929 (2004)
9. Dovzhenok, A., Park, C., Worth R.M., Rubchinsky L.L.: Failure of delayed feedback deep brain stimulation for intermittent pathological synchronization in Parkinson's disease. PLoS One **8**, e58264 (2013)
10. Elson, R.C., Selverston, A.I., Huerta, R., Rulkov, N.F., Rabinovich, M.I., Abarbanel, H.D.I.: Synchronous behavior of two coupled biological neurons. Phys. Rev. Lett. **81**, 5692–5695 (1998)
11. Engel, A.K., Fries, P., Singer, W.: Dynamic predictions: oscillations and synchrony in top-down processing. Nat. Rev. Neurosci. **2**, 704–716 (2001)
12. Fell, J., Axmacher, N.: The role of phase synchronization in memory processes. Nat. Rev. Neurosci. **12**, 105–118 (2011)
13. Fries, P.: Rhythms for cognition: communication through coherence. Neuron **88**(1), 220–235 (2015)
14. Gong, P., Nikolaev, A.R., Leeuwen, C.V.: Intermittent dynamics underlying the intrinsic fluctuations of the collective synchronization patterns in electrocortical activity. Phys. Rev. E **76**, 011904 (2007)
15. Hramov, A.E., Koronovskii, A.A., Midzyanovskaya, I.S., Sitnikova, E., Rijn, C.M.V.: On-off intermittency in time series of spontaneous paroxysmal activity in rats with genetic absence epilepsy. Chaos **16**, 043111 (2006)
16. Murthy, V.N., Fetz, E.E.: Oscillatory activity in sensorimotor cortex of awake monkeys: synchronization of local field potentials and relation to behavior. J. Neurophysiol. **76**(6), 3949–3967 (1996)
17. Nowotny, T., Zhigulin, V.P., Selverston, A.I., Abarbanel, H.D.I., Rabinovich, M.I.: Enhancement of synchronization in a hybrid neural circuit by spike-timing dependent plasticity. J. Neurosci. **23**(30), 9776–9785 (2003)
18. Nowotny, T., Huerta, R., Rabinovich, M.I.: Neuronal synchrony: peculiarity and generality. Chaos **18**, 037119 (2008)
19. Park, C., Rubchinsky, L.L.: Intermittent synchronization in a network of bursting neurons. Chaos **21**, 033125 (2011)
20. Park, C., Rubchinsky, L.L.: Potential mechanisms for imperfect synchronization in parkinsonian basal ganglia. PLoS One **7**(12), e51530 (2012)
21. Park, C., Worth, R.M., Rubchinsky, L.L.: Fine temporal structure of beta oscillations synchronization in subthalamic nucleus in Parkinson's disease. J. Neurophysiol. **103**, 2707–2716 (2010)
22. Park, C., Worth, R.M., Rubchinsky, L.L.: Neural dynamics in Parkinsonian brain: the boundary between synchronized and nonsynchronized dynamics. Phys. Rev. E **83**, 042901 (2011)
23. Pikovsky, A., Rosenblum, M., Kurths, J.: Synchronization: A Universal Concept in Nonlinear Sciences. Cambridge University Press, Cambridge (2001)
24. Rabinovich, M.I., Huerta R., Laurent G: Transient dynamics for neural processing. Science **321**, 48–50 (2008)
25. Rabinovich, M.I., Varona, P., Selverston, A.I., Abarbanel, H.D.I.: Dynamical principles in neuroscience. Rev. Mod. Phys. **78**, 1213–1265 (2006)
26. Rabinovich, M.I., Simmons, A.N., Varona, P.: Dynamical bridge between brain and mind. Trends Cogn. Sci. **19**, 453–461 (2015)
27. Ratnadurai-Giridharan, S., Zauber, S.E., Worth, R.M., Witt, T., Ahn, S., Rubchinsky, L.L.: Temporal patterning of neural synchrony in the basal ganglia in Parkinson's disease. Clin. Neurophysiol. **127**, 1743–1745 (2016)
28. Rosenblum, M.G., Pikovsky, A.S., Kurth, J.: Comment on phase synchronization in discrete chaotic systems. Phys. Rev. E **63**, 058201 (2001)

29. Rubchinsky, L.L., Park, C., Worth R.M.: Intermittent neural synchronization in Parkinson's disease. Nonlinear Dyn. **68**, 329–346 (2012)
30. Rubchinsky, L.L., Ahn, S., Park, C.: Dynamics of desynchronized episodes in intermittent synchronization. Front. Phys. **2**, 38 (2014)
31. Sanes, J.N., Donoghue, J.P.: Oscillations in local field potentials of the primate motor cortex during voluntary movement. Proc. Natl. Acad. Sci. U. S. A. **90**, 4470–4474 (1993)
32. Schnitzler, A., Gross, J.: Normal and pathological oscillatory communication in the brain. Nat. Rev. Neurosci. **6**, 285–296 (2005)
33. Szücs, A., Huerta, R., Rabinovich, M.I., Selverston, A.I.: Robust microcircuit synchronization by inhibitory connections. Neuron **61**, 439–453 (2009)
34. Uhlhaas, P.J., Singer, W.: Neural synchrony in brain disorders: relevance for cognitive dysfunctions and pathophysiology. Neuron **52**, 155–168 (2006)
35. Uhlhaas, P.J., Singer, W.: Abnormal neural oscillations and synchrony in schizophrenia. Nat. Rev. Neurosci. **11**, 100–113 (2010)
36. Velazquez, J.L., Khosravani, H., Lozano, A., Bardakjian, B.L., Carlen, P.L., Wennberg, R.: Type III intermittency in human partial epilepsy. Eur. J. Neurosci. **11**(7), 2571–2576 (1999)

Part IV
Waves

Chapter 17
Vortices Termination in the Cardiac Muscle

Valentin I. Krinsky, Vadim N. Biktashev, Niels F. Otani, and Stefan Luther

17.1 Background

Control of chaos in the cardiac muscle for a long time was in hands of cardiologists only. They found a rotating wave in the heart, and called it "re-entry" since it enters to the same place again and again.

A seminal publication in 1946 by Wiener and Rosenblueth [1] induced interest of mathematicians and physicists to this problem. It became known to Soviet physicists due to M.I. Gelfand's seminars in Moscow and schools on Nonlinear Dynamics in Nizhny Novgorod organized by A. Gaponov-Grekhov and M. Rabinovich. The interest much increased when the rotating spiral waves were found in an active chemical medium based on the Belousov oscillating chemical reaction. The Belousov chemical reaction became known to scientists due to Simon Shnol (Moscow University and Biological Research Center Puschino of the USSR Academy of Science). Shnol found general Belousov whose attempts to publish

V.I. Krinsky (✉)
Max Planck Institute DS, BMPG, Göttingen, Germany

INLN, CNRS, Valbonne, France
e-mail: Valentin.Krinsky@ds.mpg.de

V.N. Biktashev
University of Exeter, Exeter, UK
e-mail: V.N.Biktashev@exeter.ac.uk

N.F. Otani
Rochester Institute of Technology, Rochester, NY, USA
e-mail: nfosma@rit.edu

S. Luther
Max Planck Institute DS, BMPG, Göttingen, Germany
e-mail: Stefan.Luther@ds.mpg.de

I.S. Aranson et al. (eds.), *Advances in Dynamics, Patterns, Cognition*,
Nonlinear Systems and Complexity 20, DOI 10.1007/978-3-319-53673-6_17

his discovery of the oscillatory chemical reaction were rejected by referees as contradicting to thermodynamics. Shnol convinced Belousov to publish his results in the Collection of Abstracts on Radiation Medicine [2]. Shnol requested one of his students, A. Zhabotinsky, to reproduce Belousov's result. I (V.K.) met A. Zhabotinsky in Puschino, he was analysing the chemical mechanism of the Belousov reaction. I told him about results of the Gelfand's seminar how to induce a rotating spiral wave in any excitable medium, and proposed to induce a rotating wave in the Belousov reaction together. A rotating wave did not arrive during weeks of experiments. The main difficulty was that the measured wave length was larger than the size of the Petri dish. This difficulty was overcome by chance, and the spiral rotating waves were obtained.

Experiments with rotating waves in this reaction became a fashionable field of research, and stimulated the theory. During long time of successful development, no decent applications arrived from this field. We discussed it with Misha Rabinovich during his visits to me in Nice, and with H. Abarbanel during my visits to Misha.

Then, after a meeting on excitable media and heart in the Kavli Institute of theoretical physics, together with Robert Gilmour, E. Bodenschatz and S. Luther we started cardiac experiments to implement the developed understanding [3–5] for creating low energy methods for termination chaos in the heart. Results of these experiments were published in [6, 7]. Physical mechanisms underlying termination of free and pinned vortices are described in this short review.

Rotating electrical waves (vortices) and their instabilities underlie cardiac chaos (fibrillation) [8–10]. Physics of the vortices is well understood, e.g. [11–15]. But contemporary method of terminating the life-threatening cardiac fibrillation is still aimed at termination of not vortices, but all waves in the heart.

Over a century ago, it was found that a single vortex (rotating wave or anatomical reentry) in a heart can be terminated with an electric pulse [16]. An electrode was placed close to the anatomical obstacle around which the wave rotates and a small energy electric pulse was delivered within a certain time interval, called critical window, or vulnerable window, VW (note that for a rotating wave, such intervals repeat within each lap).

This approach alone cannot terminate fibrillation since there are multiple rotating waves with unknown and changing geometric locations and phases [10]. That is, we have two main problems: (1) the geometric positions of their cores and (2) the positions of their critical time windows are not known during fibrillation.

An approach to overcome the problem (1) was developed [4, 17]. Due to the bi-domain electric nature of cardiac muscle [18], every defect in it that can serve as a pinning centre for a vortex is at the same time an electric inhomogeneity. This allows an electric field pulse (E-pulse) to excite the cores of all pinned vortices simultaneously, regardless of the geometric positions of their cores.

Approaches to resolve the problem (2) are being developed. They are aimed to deliver a pulse into VWs of all vortices without knowing their relative phases ("positions in time"). One of them is the phase scanning by E-pulses, with a phase step that is shorter than the VW, for all vortices in parallel. It was tested in experiment to terminate a vortex in a rabbit heart preparation [19]. Scanning with periodic E-pulses was used to terminate fibrillation [6, 7]. Termination of one vortex with periodic E-pulses was numerically investigated in [20–22].

17.2 Termination of Multiple 2 Dim Vortices

With multiple vortices, difficulties arise due to their interaction. We investigate the excitation dynamics in the vicinity of the cores of pinned vortices. This allows to draw conclusions about the overall dynamics. When the VW of a vortex is hit by the E-pulse, this vortex is displaced to a new position. If the vortex was situated close to the tissue boundary, it is terminated. Our aim is to hit the VW of every vortex by an E-pulse ("all vortices are terminated").

Wave patterns were calculated using the Barkley model

$$u_t = \varepsilon^{-1}u(1-u)[u-(v+b)/a] + \nabla^2 u, v_t = u - v \quad (17.1)$$

in a rectangular domain with circular holes, with no-flux boundary conditions at the outer boundaries.

Pulses of electric field **E** are implemented as in [23] using the boundary conditions $\mathbf{n}\cdot(\nabla u - \mathbf{E}) = 0$ at the boundaries of the holes. The numerical integration used an explicit Euler scheme with a time step of 1.6×10^{-3} and central-difference approximation of Laplacian with a space step of $\frac{1}{6}$. The Barkley model is formulated in non-dimensional units; for presentation purposes, we postulate that the time unit of the Barkley model is 20 ms and the space unit of the Barkley model is 0.5 mm; this gives physiologically reasonable time and space scales.

Figure 17.1 shows termination of two pinned vortices by E-pacing. This can be achieved generically, for any parameters of the vortices, without knowing their geometric location and time positions of the VWs.

To hit the VW with an E-pulse, the phase scanning (Fig. 17.2b) should be performed with steps $0 < s <$ VW. Thus, the VW length (at the chosen **E**, see Fig. 17.4h) determines suitable values of s. Then, the number of pulses N to cover the whole phase of a vortex is $N \geqslant T_v/s$, where T_v is the period of the vortex, $s = T - T_v$ is the scanning step, and T is the period of E-pacing. This gives the E-pacing period $T = s + T_v$. Thus, all parameters of E-pacing (E, N, T) can be set following equations

$$0 < s < \mathrm{VW}(E), N \geqslant T_v/s, T = s + T_v \quad (17.2)$$

to guarantee that at least one E-pulse hits the VW.

What does interaction of vortices change here? In cardiac muscle, the fastest vortex entrains (or "enslaves") slower vortices if there is normal wave propagation between them. Then, only one frequency remains; this facilitates vortices termination.

But entrainment ceases if the fastest vortex is terminated before the slower vortices, and then the frequency of the system changes (period increases). Here, two wave scenarios are possible, which we describe for the case of just two vortices with periods T_{v1} and T_{v2}, such that $T_{v1} > T_{v2}$:

1. If the periods of the two vortices are not much different, so that $T_{v2} < T_{v1} < T$, then the pacing is still under-driving, and the slower vortex (T_{v1}) can be terminated by E-pacing with same period T (see Figs. 17.1 and 17.2b), provided that termination conditions (17.2) are met for the slower vortex.

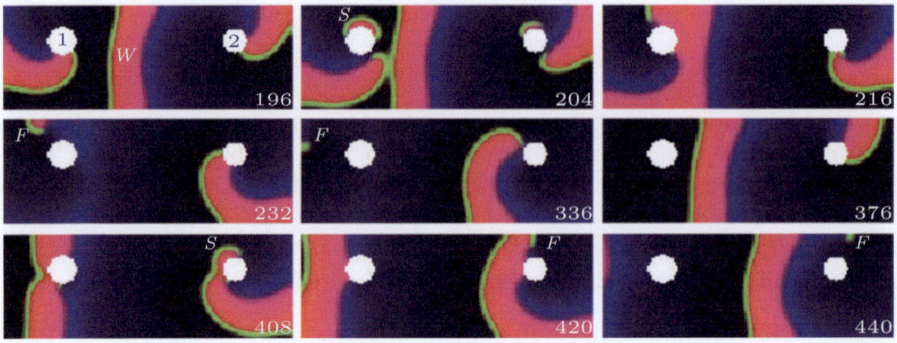

Fig. 17.1 Parallel termination of two pinned vortices with unknown both geometrical locations and time positions of the critical (vulnerable) windows (VW). The slow vortex 1 (period T_{v1} = 87 ms, pinned to the 1.2 mm defect 1) is entrained by the fast vortex 2 (period T_{v2} = 83 ms, 1.0 mm defect 2). They are paced with electric field directed from top to bottom, $|\mathbf{E}| = 1.3\,\mathrm{V/cm}$, pulses 2 ms duration, period 100 ms. This induces the phase scanning with the time step $s = 17$ ms. Colour code: *red* is a wave, *green* is the wave front. Time is measured from the start of pacing at $t = 0$ ms. **196** ms: a wave W emitted by vortex 2 enslaves vortex 1. **204** ms: an E-pulse delivered at $t = 200$ ms induces a wave S. **216** ms: the right wavebreak of wave S annihilates with the tip of vortex 1 (they have opposite topological charges). **232** ms: vortex 1 is unpinned and terminated. The left wavebreak of S created a free vortex F. **336**...**376** ms: F disappears on the boundary. **408**...**440** ms: Next E-pulse similarly terminates vortex 2. Barkley model, parameters $a = 0.8$, $b = 0.09$, $\epsilon = 0.02$

2. If however the periods of the two vortices are so different that

$$T_{v2} < T < T_{v1} \tag{17.3}$$

then the pacing with the same period is no longer under-driving, but over-driving. And overdrive pacing will typically entrain the remaining vortex rather than eliminate it.

For successful termination of fibrillation, the E-pacing period should be increased to a higher value T_2, such that $T_{v1} < T_2$. Thus, vortices can be terminated in any case. Experiments [7] underestimated the potential of the method since this mechanism was not known yet.

A similar mechanism can also terminate a free (not pinned) vortex when its moving core passes not very far (at distance $L < \lambda$, where λ is the wave length) from a defect in the medium, serving as a virtual electrode, Fig. 17.3. The success rate is the higher, the smaller is the distance L.

A mechanism reliably terminating a free rotating wave was found in 1983 [25]: waves with a frequency higher than the frequency of a rotating wave, induce its drift and termination on the boundary. Cardiologists used a high frequency pacing (anti-tachycardia pacing, ATP) well before the mechanism was understood. But ATP cannot terminate high frequency rotating waves. The mechanism found here uses electric field induced wave emission; electric field penetrates everywhere, there is no frequency limitations that exist for waves propagating in the excitable cardiac tissue.

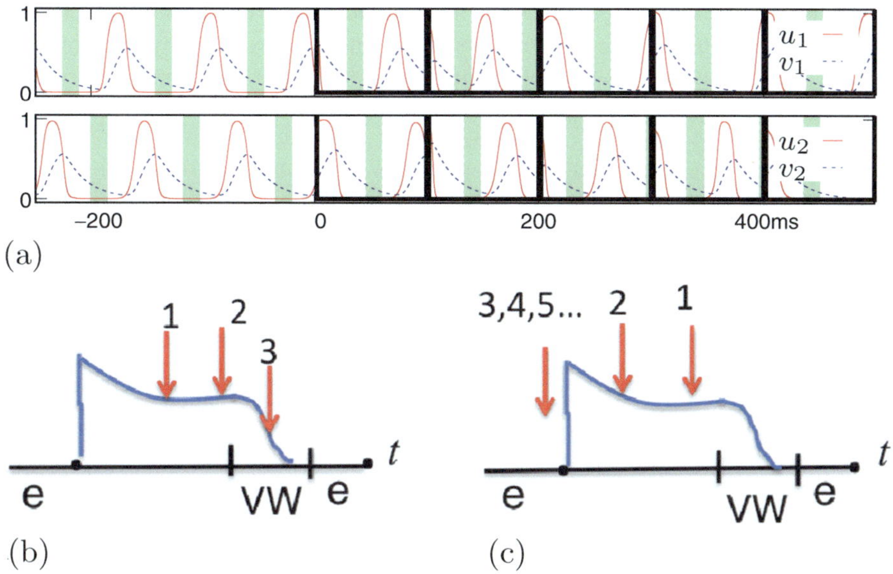

Fig. 17.2 Phase scanning. **(a)**: $u_1(t)$, $v_1(t)$ are recordings from the point just above defect 1, Fig. 17.1, and $u_2(t)$, $v_2(t)$ are same for defect 2. The *bold black lines* indicate timing of the delivered E-pulses. *Shaded areas* are vulnerable windows, defined as time intervals where $v \in (0.0871, 0.18)$, $u < b/a$. Seen that in spite of small phase disturbances produced by E-pulses, the topological features of the scanning are not disturbed, scanning successfully terminates the vortices. E-pulse 3 ($t = 200$ ms) reaches VW of vortex 1 and terminates it, E-pulse 5 ($t = 400$ ms) reaches VW of vortex 2 and terminates it, compare with Fig. 17.1. **(b, c)**: schematic. Superimposed action potentials (AP) are shown. *Red arrows* indicate timing of the delivered E-pulses; "e" is an excitable gap, s is the scanning step, $s = T - T_v$. **(b)** $s > 0$ for $T > T_v$, ("under-driving"), scanning reaches the VW. **(c)** $s < 0$, ("over-driving"), for faster pacing $T < T_v$, the scanning moves in the opposite direction. E-pulse reaches the excitable gap "e", excites an AP thus resetting the rotation phase, and all subsequent pulses get into the same phase [22, 24]. It does not reach the VW

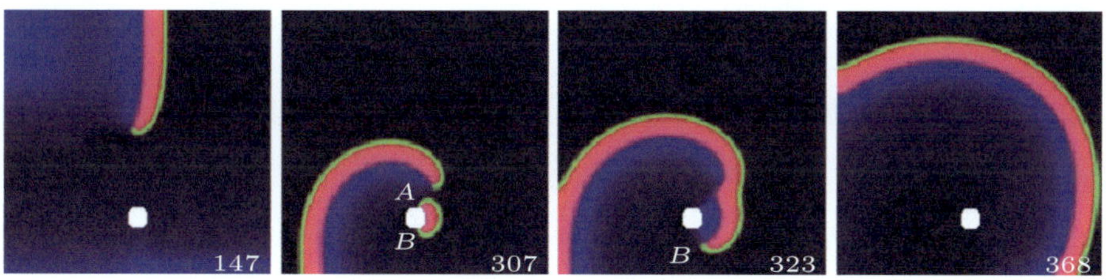

Fig. 17.3 Termination of a free vortex by an E-pulse. **147** ms: A free vortex and a defect (*white*). **307** ms: A semi-circular wave (with wavebreaks *A* and *B*) emitted from the defect by an E-pulse, electric field directed from right to left. **323** ms: Wavebreak *A* fused with the vortex tip. **368** ms: After annihilation of wavebreak *B* with the border, only a wave without wavebreaks is left in the medium. Barkley model, parameters $a = 0.6$, $b = 0.075$, $\epsilon = 0.02$

An increased amplitude of electric field |**E**| results in defibrillation. The only known mechanism was: the wave emission is induced from a larger number of defects [17]. We describe here another mechanism: the duration of the VW increases with the electric field, Fig. 17.4g.

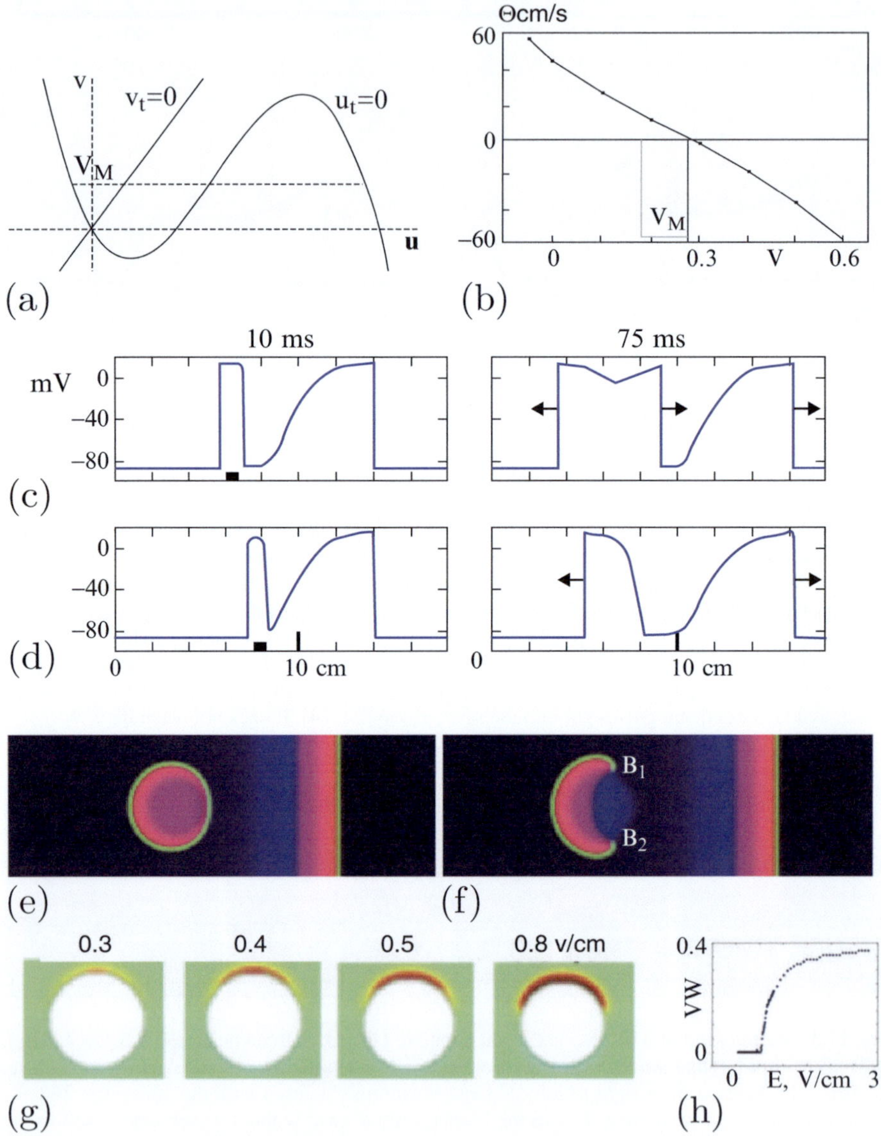

Fig. 17.4 Change of topological charge, creation of phase singularities, and vulnerable window VW. **(a–d)** *1 dim mechanism*. **(a)** Nullclines of FHN equations. *M* is the Maxwell point. The topological charge of a wave pattern is changed by an E-pulse only when an image of a nucleated

The mechanism of the VW is related to change of topological charge in 1D and creation of new topological singularities in 2D. We illustrate it using time-separation analysis for the FitzHugh–Nagumo (FHN) equations:

$$u_t = f(u) - v + Du_{xx}, \tag{17.4}$$

$$v_t = \varepsilon(u - kv). \tag{17.5}$$

Here $f(u) = Au(1-u)(u-\alpha)$, and $\epsilon \ll 1$ is a small parameter permitting the time scales separation (for details of relevant formalisms see review [27]). The wavefront propagation velocity θ can be estimated by assuming that the slow variable v is approximately constant across the wavefront. The propagation of the front is then described by Eq. (17.4) alone, where v is a constant parameter. Transforming the independent variables such that $\xi = x - \theta t$ makes Eq. (17.4) an ordinary differential equation

$$-\theta u_\xi = f(u) - v + Du_{\xi\xi}$$

which together with boundary conditions $u(\infty) = u_1$, $u(-\infty) = u_3$, where $u_1 = u_1(v)$ and $u_3 = u_3(v)$ are, respectively, the lowest and highest roots of $f(u) = v$, define θ as a function of v, see Fig. 17.4b. Here, velocity $\theta(v)$ is negative for $v > v_M$, where v_M is the Maxwell point, $\int_{u_1(v_M)}^{u_3(v_M)} (f(u) - v_M)\, \mathrm{d}u = 0$, $\theta(v_M) = 0$ [4].

Vulnerability is a cardiological term coined for initiation of fibrillation by an electric pulse. In the physical language, vulnerability in 1 dim can be related to a change of the topological charge, and in 2 and 3 dim to creation of new phase singularities. In 1 dim, the topological charge changes when the current injection nucleates a wave propagating in only one direction, Fig. 17.4d. This is in contrast to the generic case, where the topological charge is conserved, when the new wave propagates in two directions, Fig. 17.4c, or new wave is not nucleated at all (not shown). For one-directional propagation to happen, the nucleated wave should cover the points which have $v = v_M$ corresponding to the Maxwell point $\theta = 0$. Then, a part of the nucleated wave has positive velocity (becoming the front of the wave)

Fig. 17.4 (continued) wave contains the Maxwell point. **(b)** Wave front velocity θ vs the slow variable v. The value of v corresponding to velocity $\theta = 0$ is the ordinate v_M of the Maxwell point on panel **(a)**. **(c)** The topological charge conservation in 1 dim. Generic case: an electric pulse 3 ms duration is delivered far from the tail of an action potential (AP). $t = 10$ ms after pacing: a nucleated wave, very narrow, and the electrode (*black square*) below it. $t = 75$ ms: the nucleated wave developed into two counter propagating APs. Their total topological charge is zero. **(d)** Violation of the topological charge conservation. $t = 10$ ms: an electric pulse is delivered closer to the tail of the AP, inside VW. $t = 75$ ms: only one AP is induced. It propagates to the left only. The topological charge is changed. Cardiac ionic model [26]. **(e, f)** *2 dim mechanism.* **(e)** No phase singularities are created. An electric pulse is delivered as in **(c)**. **(f)** Creation of two phase singularities, B1 and B2. An electric pulse is delivered as in **(d)**. **(g, h)** VW increases with electric field in 2 dim. **(g)** Mechanism: the larger E, the larger is the depolarized region. **(h)** Graph VW(E)

and another part has a negative velocity (becoming the tail of the wave), as in Fig. 17.4d, f. Otherwise, all parts of the nucleated wave have velocity of the same sign. When velocity $\theta < 0$, the nucleated wave shrinks and decays. In the opposite case, it enlarges in all directions, as in Fig. 17.4c, e.

17.3 3 Dim Vortices

Here we show that, in 3 dimensions, a mechanism exists that can terminate a 3D vortex in the heart, using low-amplitude electric fields, which operates independently of vortex wave phase, corresponding to a VW of the full 2π radians in phase. This allows to hit the vulnerable window with every electric field pulse, making the scanning of vortex phase unnecessary.

The mechanism is based on the ability of an electric field pulse to transform the rectilinear filament of the scroll wave (i.e., the axis around which the scroll wave rotates) into a curved shape filament in an excitable layer (such as a heart wall), as illustrated in Fig. 17.5a. If a pulse of an electric field oriented parallel to the "I-shaped" filament, Fig. 17.5a(i), is applied, it depolarizes the rear surface. The filament becomes L-shaped, Fig. 17.5a(ii), since the filament is essentially the dividing line between the wavefront and waveback. In a thin layer, there is not enough room for rotation around the filament, and the scroll wave is terminated. In a thick layer, L-shaped filament relaxes within one rotation into a C-shaped filament (Fig. 17.5a(iii)). This C-shaped filament may be thought of as one-quarter of a scroll ring, assuming no-flow boundary conditions for the membrane potential on the system surfaces. Depending on system parameters, scroll rings, and therefore

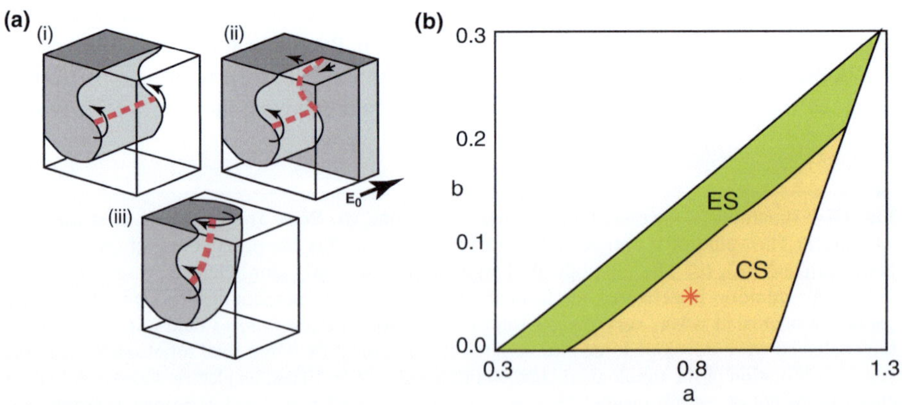

Fig. 17.5 **(a)** Illustration of (i) I-shaped, (ii) L-shaped and (iii) C-shaped filaments, and the scroll waves that rotate around them. **(b)** Location of our system parameters (*red star*) in Barkley parameter space. "CS" and "ES" label the contracting and expanding scroll regimes, respectively

C-shaped filaments, can shrink and disappear due to positive tension within the filament, resulting in the termination of the rotating wave.

The phase independence of this mechanism arises from the symmetry of the initial conditions with respect to the combination of a translation in time and a rotation around the scroll wave. Specifically, since the electric field is initially oriented along the scroll wave axis, application of the field at a time $t = t_0$ is identical to applying the field at a different time $t = t_1$ if the coordinate system is also rotated around the scroll wave axis through an angle of $2\pi(t_1 - t_0)/T_s$, where T_s is the scroll wave rotational period. Thus, the result of applying the electric field is independent of time, apart from any spatial features of the system that might break the symmetry. In particular, if the outcome of applying the electric field at any given time is termination of the scroll wave, we can also expect termination if the electric field is applied at any other time.

To study this mechanism, we conducted computer simulations in a 3D rectangular system meant to represent a portion of one of the walls of the heart. We used a simple forward Euler method on a rectangular grid to solve the monodomain Barkley model Eq. (17.1) with parameters $a = 0.8$, $b = 0.05$, $\epsilon = 0.02$, $D = 1.0$. The computational grid spacing and timestep were set to 0.167 and 1.6×10^{-3}, respectively. A scroll wave was initiated in the system with its filament oriented perpendicular to the front and rear surfaces of the system (Fig. 17.6a). These two surfaces may be thought of as portions of the endocardial and epicardial surfaces of the heart. The scroll wave was allowed to settle down over the course of at least three rotations prior to the delivery of a short-pulse, spatially uniform electric field.

Figure 17.6 shows the mechanism responsible for termination of a three-dimensional scroll wave in a thin-wall system. The scroll wave just prior to the application of the electric field pulse is displayed in Fig. 17.6a. When the pulse is applied at $t = 0$, with the electric field direction parallel to the scroll wave filament, the pulse immediately depolarizes the rear surface, as seen in Fig. 17.6b. The front surface is simultaneously hyperpolarized. The electric field pulse strength and duration used corresponded to 1.3 V/cm, 5 ms electric field pulse in a realistic heart. (This equivalence is based on an assumed threshold for wave initiation in a realistic cardiac setting of 0.2 V/cm for 5 ms, and the use in our system of a pulse field strength that is 6.5 times the threshold field.) The depolarization wavefront (red) subsequently propagates as a plane wave along the filament direction (black arrows) as shown in Fig. 17.6b, c. In this thin-layer system, the plane wave quickly depolarizes the entire space not occupied by the rotating wave or the region of hyperpolarization, leaving no place for the scroll wave wavefront to propagate. While the scroll wave moves away, in the direction indicated by black arrows in Fig. 17.6c, it loses the wavefront, and only the wave tail left, Fig. 17.6d. After, the whole medium becomes quiescent.

We also find that applying an electric field pulse 1.1 V/cm for 5 ms at any of the five time moments equally spaced in time by $0.2T_s$ resulted in termination of the scroll wave, while applying a slightly weaker pulse, 0.9 V/cm, at any of these five moments, fails to terminate the rotating wave. With termination occurring independently of the electric field application time, the vulnerable window (VW) for this thin-wall system is 2π radians.

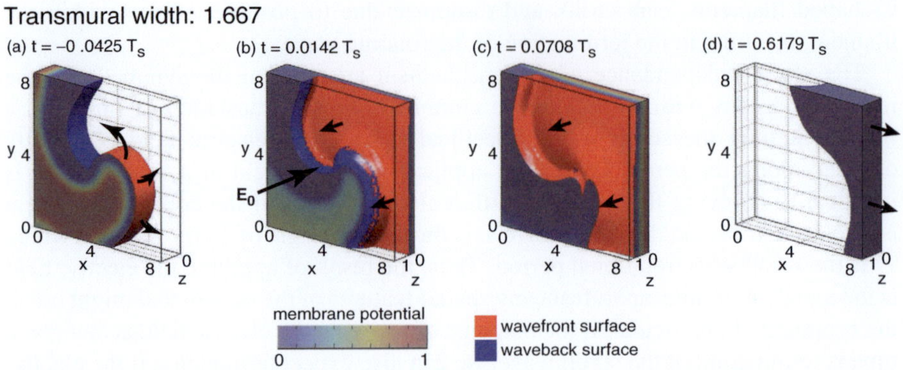

Fig. 17.6 A 3D mechanism for terminating a scroll wave in a thin-layer that results in a vulnerable window (VW) of 2π. Wall thickness: 1.667; electric field strength: 1.3 V/cm for 5 ms. **(a)** Scroll wave after four rotations. **(b)** Immediate effect of an electric field pulse applied at time $t = 0$. The electric field $\mathbf{E_0}$ (*long arrow*) is oriented parallel to the scroll wave filament. **(c)** The leading edge of the newly depolarized region (*red*) is a plane parallel to the *x-y* plane. It propagates as a plane wave in the *z*-direction (*black arrows*) filling the region not occupied by the scroll wave. **(d)** The scroll wave loses its wavefront, and therefore terminates shortly thereafter. We have observed that the 3D scroll wave terminates irrespective of its phase (0 to 2π) when the electric field is applied. *For all panels*: Regions occupied by action potential(s) ($u > b/a$) or its refractory period ($v > a/2 - b$) are shown as opaque. Muted colours represent the value of the membrane potential; see the colour bar

This important result, that the VW assumes the full range of available phases when the electric field pulse is oriented along the filament direction, is not just a property of thin layers. We illustrate this in Fig. 17.7 for a scroll wave in a system with a wall thickness four times that of Fig. 17.6. The electric field pulse again depolarizes the rear surface, which interferes with the portion of the rotating wave close to that surface. However, in this case, the remainder of the scroll wave is not immediately disturbed. Instead, the trailing (blue) surface of the rotating wave recedes, as the new, induced wave attempts to fill the excitable gap. The result of these two processes is the merging of the two waves into a single wave, which then rotates around a filament that is now L-shaped, shown as a white dashed line in panel (b) of Fig. 17.7, an effect previously demonstrated by Biktashev [28]. The filament quickly relaxes to curved, “C” shape (Fig. 17.7d, e) which, based on the theory of scroll ring dynamics [29, 30], is governed by the equation,

$$dR/dt = -\alpha/R \tag{17.6}$$

where R is the local radius of a curved filament, and the constant α, sometimes called filament tension, depends on the system parameters. In our case, with Barkley parameters $a = 0.8$ and $b = 0.05$, the system resides in the Contracting Scroll (CS) region in the parameter space, as shown in Fig. 17.5b [31]. In this regime, α is positive, and the filament shrinks due to positive filament tension. Shrinking

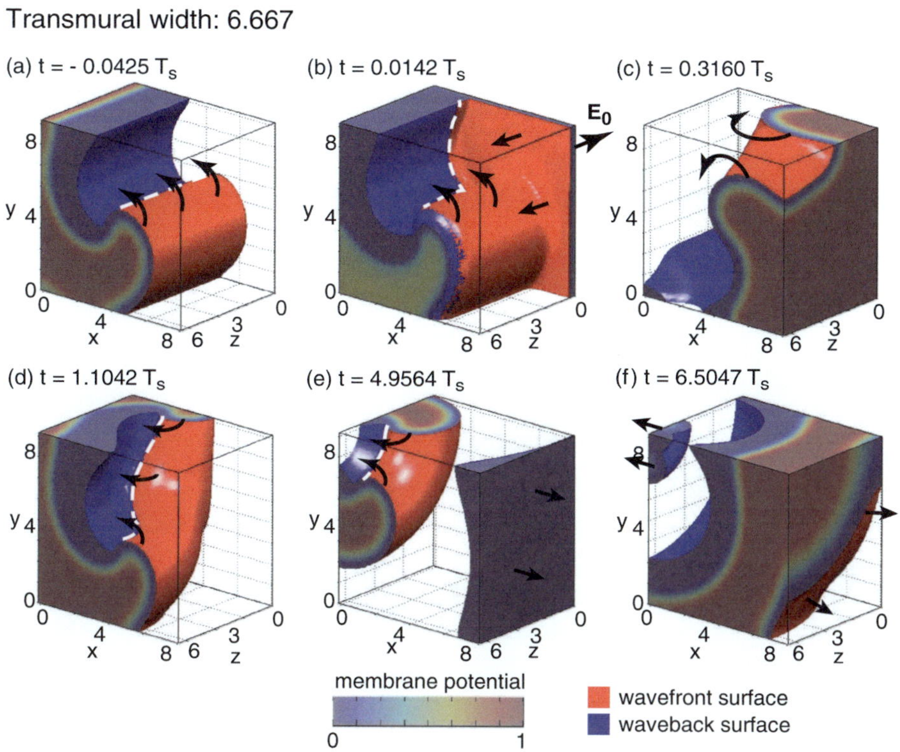

Fig. 17.7 Scroll wave termination in a thick layer also has VW $= \mathbf{2\pi}$. Wall thickness: 6.667; electric field strength: 1.1 V/cm for 5 ms. **(a)** Scroll wave prior to application of the electric field pulse. The *white dashed line* denotes the scroll wave filament. **(b)** The filament acquires an L-shape after the electric field pulse. **(c)** The filament becomes C-shaped (hidden, but visible in (**d**) and (**e**)) as the waveback (*in blue*) recedes. **(d)–(f)**: The curved filament shrinks, then disappears, due to the filament tension. Again, this mechanism operates, as illustrated, irrespective of when the electric field pulse is administered (i.e., a VW of 2π) for a pulse of sufficient strength

and self-annihilation of the filament is seen in Fig. 17.7b–f during the course of continued rotation. The disappearance of the filament then results in termination of the scroll wave.

We again find that the mechanism produces termination irrespective of the phase of the vortex when the electric field pulse is administered. In this case, we find that the termination of the scroll wave occurs when a field 2.2 V/cm is applied at any of five moments equally spaced in time by $0.2T_s$. The vulnerable window for scroll wave termination thus consists of all 100% of possible locations of the rotating wave in its rotation; thus, VW $= 2\pi$.

Finally, to contrast this method with conventional methods in 2D, we tried applying electric fields ranging in strength from 2.2 to 8.8 V/cm in the $\pm x$ and $\pm y$ directions. We also tried applying a field 2.2 V/cm in the $-y$ direction at ten different times, equally spaced within a spiral wave period. None of these trials produced scroll wave termination.

17.4 Experiment

Results of about 500 experiments with vortices termination in the isolated pig hearts are presented in Fig. 17.8. Fibrillation was induced and terminated as in [6, 7].

Figure 17.8a shows that the optimal pacing frequency $\hat{f} = 0.77$ is below the arrhythmia frequency ($\hat{f} < 1$) as it should be for terminating pinned vortices.

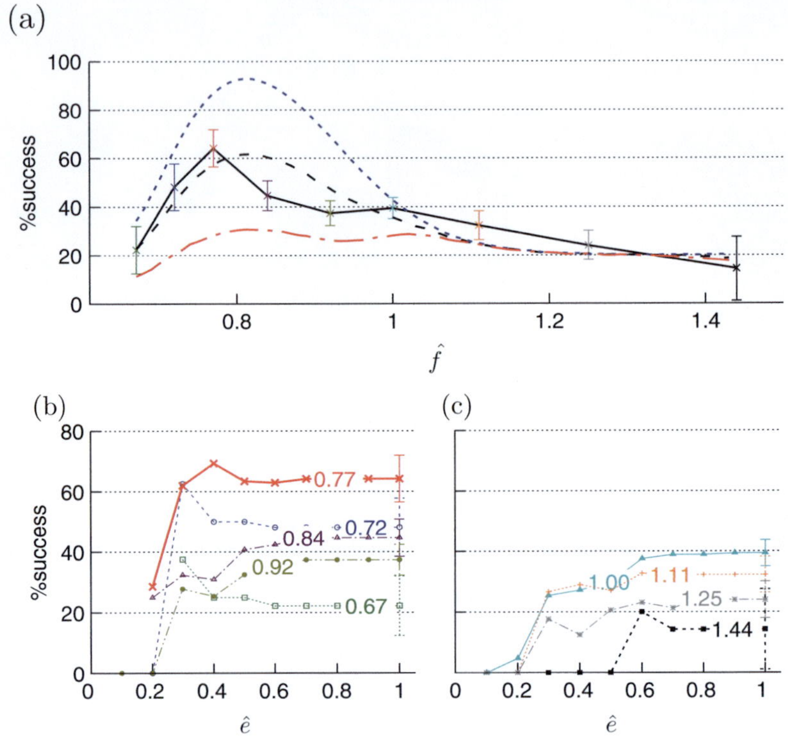

Fig. 17.8 Fibrillation termination in the isolated pig hearts. The success rate of defibrillation in 486 experiments by 5 biphasic E-pulses. **(a)** Success rate vs normalized frequency $\hat{f} = f/f_d$ where f_d is the dominant frequency of fibrillation. Error bars: the standard deviation. Dashed lines: theoretical prediction for termination of one pinned vortex (*blue*), two pinned vortices (*red*), when the slower vortex is enslaved by the faster one (*red*). In accordance with the mechanism explained near Eq. (17.3), the theoretical termination rate for two vortices is threefold lower than that for a single vortex. The experimental curve is between the theoretical curves for one and two pinned vortices. The *black dashed line* is the theoretical prediction for a mixture (50:50) of two identical vortices and two vortices with significantly different frequencies, as per Eq. (17.3). **(b, c)** Success rate for defibrillation energies not exceeding $\hat{e}$, for frequencies $\hat{f}$ shown near each curve. Normalized energy $\hat{e} = e/e_1$, where e_1 is the threshold E_{50} energy of defibrillation by 1 shock. Graphs (**b, c**) and the experimental curve in (**a**) are calculated from data in [32]. The optimal pacing frequency $\hat{f} = 0.77$ is below the arrhythmia frequency ($\hat{f} < 1$) as it should be for terminating pinned vortices. An interpretation of the experimental results is that about half of VF episodes was induced by pinned vortices of very close periods. These experiments evidence that pinned vortices, hidden from direct observation, are significant in fibrillation

Notice that elimination of a free, rather than pinned, vortex by inducing its drift via the mechanism described in [25] requires the pacing frequency to be above the arrhythmia frequency, $\hat{f} > 1$.

The theoretical estimation shown in dashed black line fits the experiment better than curves for one or two different vortices. It indicates that in about a half of fibrillation cases, the frequencies of the vortices were not much different, see Eq. 17.3.

This estimation was obtained in an axiomatic model which assumed that the strength of electric field is sufficient to ensure the normalized vulnerable window of about 0.24 (see below for more detail of the axiomatic model). Minimum energy for termination of a pinned vortex is achieved when the electric field strength is chosen so that the normalized vulnerable window $\overline{\mathrm{VW}}(E) = 1/N$, where N is the number of pacing pulses.

Maximum of the success rate is achieved when the pacing frequency $f = f_{\mathrm{best}}$, where f_{best} is the frequency for which the normalized scanning step $\mathbf{sn} = 1/N$.

When $f < f_{\mathrm{best}}$, i.e. $T > T_{\mathrm{best}}$, the scanning step $\mathbf{sn} > \overline{\mathrm{VW}}$, and the vulnerable window may be missed while scanning, thereby decreasing the success rate. When $f > f_{\mathrm{best}}$, so $T < T_{\mathrm{best}}$, the scanning step $\mathbf{sn} < \overline{\mathrm{VW}} = 1/N$, and not all phases are scanned. This also decreases the success rate.

As an estimate of the fastest vortex frequency we use the dominant frequency.

Below is a formal description of the axiomatic model used. Let $\phi_n^j \in [0, 1), j = 1, 2, n = 1, \ldots, N, N = 5$, describe the phase of jth vortex just after the delivery of the nth E-pulse, T_j be the own periods of the vortices, $T_2 > T_1$, and correspondingly $\mathbf{sn}_j = \hat{T} - \hat{T}_j = (T - T_j)/T_d$ are the scanning steps normalized by the measured dominant period, T_d. We postulate $\phi_{n+1}^j = (\phi_n^j + s) \mod 1$, subject to the following corrections: (1) if $\phi_{n+1}^j \in [1 - \overline{\mathrm{EG}}, 1)$, where $\overline{\mathrm{EG}}$ is the normalized duration of the excitable gap, then ϕ_{n+1}^j is replaced with 0: this describes resetting the jth phase by the E-pulse; (2) if $\phi_{n+1}^j \in [1 - \overline{\mathrm{EG}} - \overline{\mathrm{VW}}, 1 - \overline{\mathrm{EG}})$, where $\overline{\mathrm{VW}}$ is the normalized duration of the vulnerable window, then the jth vortex is considered terminated; (3) if neither vortex is terminated, then the slower vortice's phase is enslaved by the faster one's, $\phi_{n+1}^2 = (\phi_{n+1}^1 - D) \mod 1$, where D is a fixed phase delay; (4) if both vortices are terminated, iterations stop and E-pacing is deemed successful. Figure 17.8 shows results of Monte-Carlo simulations with random initial phases of vortices and normal distributions of parameters $\hat{T}_1 = 1 \pm 0.1$ and $\hat{T}_2 = 1.6 \pm 0.05$ (mean$\pm$standard deviation), with other parameters fixed at $\overline{\mathrm{EG}} = 0.4$, $\overline{\mathrm{VW}} = 0.2 = 1/N$, $D = 0.25$. Note that the normalized vulnerable window $\overline{\mathrm{VW}}$ depends on the strength of the electric field in the pacing pulses.

The three curves correspond to simulations with two-vortex initial conditions (red line), with initial conditions where only one vortex is present from the start (blue line), and a 50:50 mixture of the two (the black dashed line).

We see that the axiomatic model predicts that for two vortices, the success rate should be lower than for one vortex, for frequencies between the frequencies of the leading vortex and the led vortex. This is due to the cases then the leading vortex is terminated first, dominant frequency changes and the conditions for termination for the remaining, slower vortices are no longer optimal. As the figure shows, the actual experimental data best correspond to a mixture of one- and two-vortex cases.

However in the experiments the arrhythmia in all cases was VF rather than VT, so presumably involved more than one vortex. Note that in the axiomatic model, a case of one vortex is similar to that of two or more vortices of identical periods. Hence we conclude that the most likely interpretation of the available experimental data is that about half of VF episodes was induced by vortices of very close periods.

These experiments evidence that pinned vortices, hidden from direct observation, are significant in fibrillation.

17.5 Discussion

Here, we investigated two extreme cases: permanently pinned vortices and permanently free vortices. There is no sharp transition between them. In cardiac muscle, there are heterogeneities of all sizes, including those to which vortices pin weakly. A weakly pinned vortex is pinned for some time only, then leaves the pinning centre and moves as a free vortex, again for some time. When moving and meeting a pinning centre, it may pin to it, or may reach the boundary of the tissue and disappear. Such intermediate types of rotating waves can be more easily terminated while they are pinned.

Vortices termination can be induced also by other mechanisms different from vulnerability, e.g. pacing-induced drift of a free vortex [25], unpinning of weakly pinned vortices [33, 34] and by 3 dim mechanisms [35].

3 dim models are widely used in investigation of wave patterns induced by rotating waves, e.g. [14, 29]. Study of vortices termination in 2 dim models is a necessary step for developing understanding mechanisms of 3 dim vortices termination in the heart. A 3-dim mechanism of defibrillation was described in [28, 35, 36]. Termination vortices underlying fibrillation is only a small part of a problem preventing and curing the cardiac arrhythmias where combination of molecular and dynamics approaches is prominent [37].

In conclusion, we have shown mechanisms of terminating pinned and free vortices by electric field pulses when the geometric positions of their cores, and the phases of rotation are not known. We have demonstrated a new, low-energy, scroll wave termination mechanism that is an apparent solution to the problem of small vulnerable windows that are characteristic of other low-energy methods. When an electric field pulse is applied with field component oriented parallel to the scroll wave filament, we obtain wave termination with a vulnerable window of the full 2π radians using field strengths much smaller (1.1 V/cm for thin layers; 2.2 V/cm for wide layers) than those typically required by standard methods (5–6 V/cm for standard defibrillation). This reduces pulse energy requirements by an order of magnitude (since energy scales as E_0^2), thus minimizing pain, tissue damage and battery requirements. Since the VW is 2π, the new mechanism also eliminates the need to apply multiple electric field pulses to scan phases of all rotating waves. We therefore expect that the new mechanism will play an important role, alongside other mechanisms, in the future design of low-energy defibrillation stimulus protocols

These results form the physical basis for creation of new effective methods for termination vortices underlying fibrillation.

Acknowledgements The research leading to the results has received funding from Max Planck Gesellschaft, the European Community Seventh Framework Programme FP7/2007–2013 under Grant Agreement 17 No. HEALTH-F2-2009-241526, EUTrigTreat, and from EPSRC (UK) grant EP/I029664. We acknowledge support from the German Federal Ministry of Education and Research (BMBF) (project FKZ 031A147, GO-Bio), the German Research Foundation (DFG) (Collaborative Research Centres SFB 1002 Project C3 and SFB 937 Project A18), the German Center for Cardiovascular Research (DZHK e.V.), and EPSRC (UK) grant EP/N014391. US NIH grant no. R01HL089271.

Ethics The study was reviewed and approved by the ethics committee, permit no. 33.9-42052-04-11/0384, Lower Saxony State Office for Customer Protection and Food Safety.

Competing Interests We have no competing interests.

References

1. Wiener, N., Rosenblueth, A.: The mathematical formulation of the problem of conduction of impulses in a network of connected excitable elements, specifically in cardiac muscle. Arch. Inst. Cardiol. Mex. **16**, 205–265 (1946)
2. Belousov, B.P.: Periodically acting reaction and its mechanism. Sbornik Referatov po Radiotsionnoi Meditsine, [Collection of Abstracts on Radiation Medicine] p. 145 (1959)
3. Pumir, A., Krinsky, V.: Unpinning of a rotating wave in cardiac muscle by an electric field. J. Theor. Biol. **199**, 311–319 (1999)
4. Takagi, S., et al.: Unpinning and removal of a rotating wave in cardiac muscle. Phys. Rev. Lett. **93**, 058101 (2004)
5. Pumir, A., et al.: Wave emission from heterogeneities opens a way to controlling chaos in the heart. Phys. Rev. Lett. **99**, 208101 (2007)
6. Fenton, F.H., et al.: Termination of atrial fibrillation using pulsed low-energy far-field stimulation. Circulation **120**, 467–476 (2009)
7. Luther, S., et al.: Low-energy control of electrical turbulence in the heart. Nature **475**, 235–239 (2011)
8. Davidenko, J.M., Pertsov, A., Salomonsz, R., Baxter, W., Jalife, J.: Stationary and drifting spiral waves of excitation in isolated cardiac muscle. Nature **355**, 349–351 (1992)
9. Winfree, A.T.: Electrical turbulence in three-dimensional heart muscle. Science **266**, 1003–1006 (1994)
10. Gray, R.A., Pertsov, A.M. , Jalife, J.: Spatial and temporal organization during cardiac fibrillation. Nature **392**, 75–78 (1998)
11. Winfree, A., Strogatz, S.: Organizing centres for three-dimensional chemical waves. Nature **311**, 611–615 (1984)
12. Rappel, W.-J., Fenton, F., Karma, A.: Spatiotemporal control of wave instabilities in cardiac tissue. Phys. Rev. Lett **83**, 456–459 (1999)
13. Alonso, S. , Panfilov, A.: Negative filament tension at high excitability in a model of cardiac tissue. Phys. Rev. Lett. **100**, 218101 (2008)
14. Jiménez, Z. , Steinbock, O.: Stationary vortex loops induced by filament interaction and local pinning in a chemical reaction-diffusion system. Phys. Rev. Lett. **109**, 1–4 (2012)

15. Barkley, D.: Euclidean symmetry and the dynamics of rotating spiral waves. Phys. Rev. Lett. **72**, 164–167 (1994)
16. Mines, G.R.: On circulating excitations in heart muscles and their possible relation to tachycardia and fibrillation. Trans. R. Soc. Can. **8**, 43–52 (1914)
17. Pumir, A. et al.: Wave emission from heterogeneities opens a way to controlling chaos in the heart. Phys. Rev. Lett. **99**, 208101 (2007)
18. Sepulveda, N.G., Roth, B.J., Wikswo, J.P.: Current injection into a two-dimensional anisotropic bidomain. Biophys. J. **55**, 987–999 (1989)
19. Ripplinger, C.M., Krinsky, V.I., Nikolski, V.P., Efimov, I.R.: Mechanisms of unpinning and termination of ventricular tachycardia. Am. J. Physiol. **291**, H184–H192 (2006)
20. Bittihn, P., et al.: Far field pacing supersedes anti-tachycardia pacing in a generic model of excitable media. New J. Phys. **10**, 103012 (2008)
21. Bittihn, P., et al.: Phase-resolved analysis of the susceptibility of pinned spiral waves to far-field pacing in a two-dimensional model of excitable media. Philos. Trans. R. Soc. A **368**, 2221–2236 (2010)
22. Behrend, A., Bittihn, P. , Luther, S.: Predicting unpinning success rates for a pinned spiral in an excitable medium. Comput. Cardiol. **37** 345–348 (2010)
23. Pumir, A., Krinsky, V.: Unpinning of a rotating wave in cardiac muscle by an electric field. J. Theor. Biol. **199**, 311–319 (1999)
24. Shajahan, T., Berg, S., Luther, S., Krinsky, V., Bittihn, P.: Scanning and resetting the phase of a pinned spiral wave using periodic far field pulses. New J. Phys. **18**, NJP-104359 (2016)
25. Krinsky, V.I., Agladze, K.I.: Interaction of rotating waves in an active chemical medium. Physica D **8**, 50–56 (1983)
26. Mahajan, A., et al.: A rabbit ventricular action potential model replicating cardiac dynamics at rapid heart rates. Biophys. J. **94**, 392–410 (2008)
27. Tyson, J.J. , Keener, J.P.: Singular perturbation theory of traveling waves in excitable media (a review). Physica D **32**, 327–361 (1988)
28. Biktashev, V.N.: Evolution of vortices in active media. Ph.D. Thesis, Moscow Institute of Physics and Technology (1989)
29. Keener, J., Tyson, J.: The dynamics of scroll waves in excitable media. SIAM Rev. **34**, 1–39 (1992)
30. Panfilov, A.V., Rudenko, A.N. , Krinsky, V.I.: Vortical rings in three-dimensional active media with diffusion in two components. Biophys. **31**, 926–931 (1986)
31. Alonso, S., Sagués, F. , Mikhailov, A.S.: Taming winfree turbulence of scroll waves in excitable media. Science **299**, 1722 (2003)
32. Hornung, D.: Cardiac arrhythmia termination on the vascular and organ scale. Ph.D. Thesis, p.121, Georg-August Universität Göttingen (2013)
33. Isomura, A., Hoerning, M., Agladze, K., Yoshikawa, K.: Eliminating spiral waves pinned to an anatomical obstacle in cardiac myocytes by high-frequency stimuli. Phys. Rev. E **78**, 066216 (2008)
34. Pumir, A., et al.: Wave-train induced unpinning of weakly anchored vortices in excitable media. Phys. Rev. E **81**, 010901 (2010)
35. Otani, N., Krinski, V., Han, S., Carr, J. , Luther, S.: Modification of scroll wave filaments when electric fields are applied to the heart. In: SIAM-Life Sciences, July 11–16. SIAM, Boston (2016)
36. Zemlin, C., Mironov, S. , Pertsov, A.: Delayed success in termination of three-dimensional reentry: role of surface polarization. J. Cardiovasc. Electrophysiol. **14**, 257 (2003)
37. Weiss, J.N., et al.: Perspective: a dynamics-based classification of ventricular arrhythmias. J. Mol. Cell. Cardiol. **82**, 136 (2015)

Chapter 18
KDV Soliton Gas: Interactions and Turbulence

Efim Pelinovsky and Ekaterina Shurgalina

18.1 Introduction

Mikhail Rabinovich is one of the founders of the theory of nonlinear wave interaction. His main interest has been associated with the wave processes in active media, leading to competition between various waves and stochastization of dynamic processes, see e.g., [1–4]. Such processes are also possible in conservative systems, when a large number of interacting waves propagate in different directions and with different velocities. Their interference and interactions lead to a fast changing of wave pattern, and the description of wave field should be carried out within the framework of the statistical theory. Such theory called *weak wave turbulence* is described in the books [5, 6]. The main idea is that the wave process is described by the interaction of a large number of sinusoidal waves with independent phases in the linear approximation, and a weak nonlinearity induces weak phase correlation. Equations for the intensity are obtained by perturbation theory and statistical averaging. However, integrable systems have their own specifics which were formulated as the Fermi–Pasta–Ulam problem (see, for instance [1, 7]). Instead of the initial perturbation energy being distributed over the spectrum, after a certain time it is then concentrated in a small number of harmonics. This was discovered by the example of a vibrating string (Boussinesq equations that for unidirectional

E. Pelinovsky (✉)
Institute of Applied Physics, Nizhny Novgorod, Russia

Nizhny Novgorod State Technical University, Nizhny Novgorod, Russia

National Research University – Higher School of Economics, Nizhny Novgorod, Russia
e-mail: pelinovsky@hydro.appl.sci-nnov.ru

E. Shurgalina
Institute of Applied Physics, Nizhny Novgorod, Russia
e-mail: eshurgalina@mail.ru

I.S. Aranson et al. (eds.), *Advances in Dynamics, Patterns, Cognition*,
Nonlinear Systems and Complexity 20, DOI 10.1007/978-3-319-53673-6_18

waves lead to the Korteweg–de Vries equation). Thus it is clear that the wave turbulence can be very specific in integrable systems, discussed in [8, 9].

Solitons are an important part of the modern nonlinear physics providing the wave propagation over long distances. In integrable systems like Korteweg–de Vries (KdV) and nonlinear Schrodinger (NLS) equations there are a lot of approaches (inverse scattering method, Darboux and Backlund transforms, bilinear Hirota method) allowing to obtain rigorous solutions for description of soliton interaction. The soliton turbulence is a specific part of the wave turbulence theory, where kinetic equations describe the parameters of the associated scattering problem, not wave amplitudes or amplitudes of Fourier harmonics. Zakharov in 1971 first showed the fundamental role of pairwise soliton collisions within the Korteweg–de Vries equation framework. Later on, the kinetic theory for dense soliton gas was developed by El with his coauthors [10–13]; the theory was capable to describe various properties of soliton ensembles. Soliton turbulence in integrable systems is degenerated to some extent, because solitons are conserved in the interacting process which is why their characteristics (more precisely discrete eigenvalues of the associated spectral problem) do not change. That is why the nonlinear Fourier transform for KdV equation described sea waves on shallow water was developed. This allows exploring the "structure" of observed random waves [14–16]. These components of the wave field (analogues of cnoidal waves and solitons if they exist independently) do not change in time, but their superposition leads to random changes on the wave pattern. In practice it is not only important to know the parameters of solitons and cnoidal waves, but also the distribution and moments of the random wave field. Soliton ensembles have already been discussed in the literature numerically within this model [17–22]. In the framework of integrable Korteweg–de Vries equation soliton interaction is elastic, and therefore the parameters of the soliton field do not change [8, 10–13]. Nevertheless, their interaction with each other and with dispersion packages should lead to the change in the wave field characteristics, its extrema and statistical moments.

18.2 Mathematical Model

In this article we consider the interaction of random soliton ensembles. Korteweg de–Vries equation is selected as a mathematical model, which is the canonical equation for weakly nonlinear and weakly dispersive waves in nonlinear physics:

$$\frac{\partial u}{\partial t} + 6u\frac{\partial u}{\partial x} + \frac{\partial^3 u}{\partial x^3} = 0 \tag{18.1}$$

As is known, the KdV equation has an exact solution in the form of so-called solitary wave or soliton:

$$u(x,t) = A\mathrm{sech}^2\left[K\left(x - 4K^2 t - x_0\right)\right], \quad A = 2K^2. \tag{18.2}$$

Solution is determined by two independent parameters: amplitude A and phase (initial position of soliton) x_0. Multi-soliton solutions can also be written analytically [23].

18.3 Two-Soliton Interaction

As a first step, we investigated features of two-soliton interaction which can be considered as elementary act of soliton turbulence [24]. It was shown that the two-soliton interaction in the framework of the KdV equation leads to the decrease of the third and fourth moments of the nonlinear wave field (integrals from u^n over x) while the first and the second moments remain unchanged due to the conservation of the mass and momentum. The magnitudes of the relative variations of moments turn out to be no monotone functions of the soliton amplitude ratio A_2/A_1 each having a single maximum located at the point $A_2/A_1 \approx 0.32$, close to the boundary of the transition region between the exchange and overtaking scenarios of two-soliton interactions. The qualitative implication of this dynamical effect for the soliton turbulence theory will be a decrease of the skewness and kurtosis of the turbulent wave field and this will be shown below.

18.4 Numerical Modelling

The dynamics of irregular soliton trains is studied numerically using the pseudo-spectral method for a periodic spatial domain (details in [22]). For simplicity the explicit formula for initial condition is

$$u(x,t) = \sum_{i=1}^{N} u_i = \sum_{i=1}^{N} A_i \mathrm{sech}^2\left[K_i\left(x - 4K_i^2 t - x_i\right)\right], \quad A_i = 2K_i^2, \tag{18.3}$$

where N is a number of solitons in the computational domain. Phases x_i are chosen so that initially solitons are separated from each other ($\Delta x = 20$). In our experiments a random set of soliton amplitudes is selected and realizations differ from each other by the position of solitons in domain only. In the initial experiments with 100 solitons their amplitudes in the computational domain vary from $A_{\min} = 1$ to $A_{\max} = 3$. Two realizations of such fields are shown in Fig. 18.1.

Solitons have different amplitudes and different velocities; hence, they will interact over time. Figure 18.2 shows the evolution of the soliton field, represented in Fig. 18.1a, in x–t domain. The zoom of the figure is shown on the right figure. Different slopes and trajectories correspond to different soliton velocities. Soliton trajectories do not lie only on straight lines after interacting, which demonstrates the phase shift as a result of nonlinear soliton interaction. Similar conclusions about

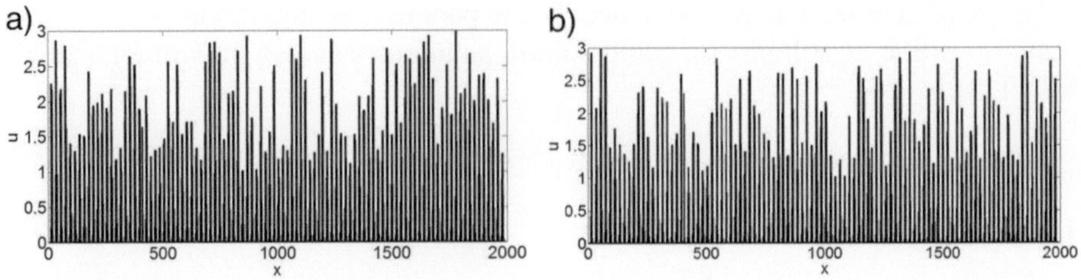

Fig. 18.1 Initial soliton ensembles, two realizations

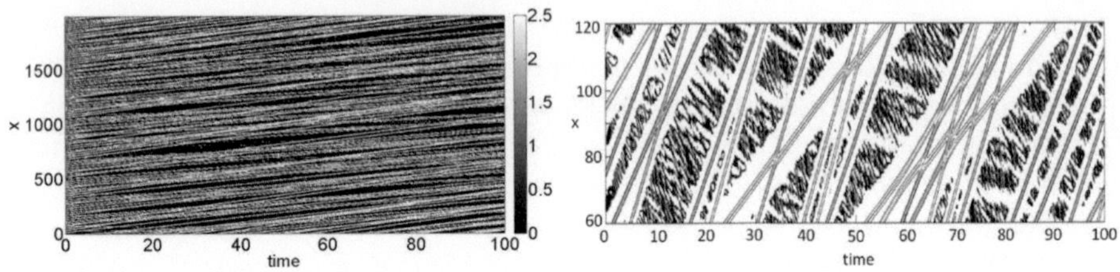

Fig. 18.2 *X–t* diagram of soliton field (zoom is presented at the *right panel*)

the trajectories soliton ensembles are made in [17–20]. On average, each soliton is involved in about 10–20 collisions during the simulation time. During this time, the average characteristics of the soliton gas reach a steady value.

18.5 The Effect of a Negative Soliton Velocity

A distinctive feature of KdV equation is that solitons move in one direction. However if there is an ensemble of solitons, interacting with each other, solitons can move in any direction including backwards (more precisely, with velocities less than the linear velocity of the long waves). An interest to this phenomenon has arisen because of the problem of soliton turbulence (soliton gas). The trajectories of solitons with random amplitudes from small to large within the KdV equation are shown in Fig. 18.3.

Small soliton moving backwards (light blue band) is clearly observed. A problem of defect propagation in the field of cnoidal waves within the KdV was considered in [25], where the expression for the velocity of the "defect" is obtained through the Weierstrass elliptic functions. It is mentioned that large-amplitude "defect" (compared to the average) moves to the right, and small-amplitude "defect" moves to the left. The same result is followed from kinetic equation [26]. Such results can be easily obtained from the kinematics by comparing the distance which a small soliton passes before the interaction with a large soliton with its phase shift during the interaction. Let us rewrite the soliton solution (18.2) in terms of velocity:

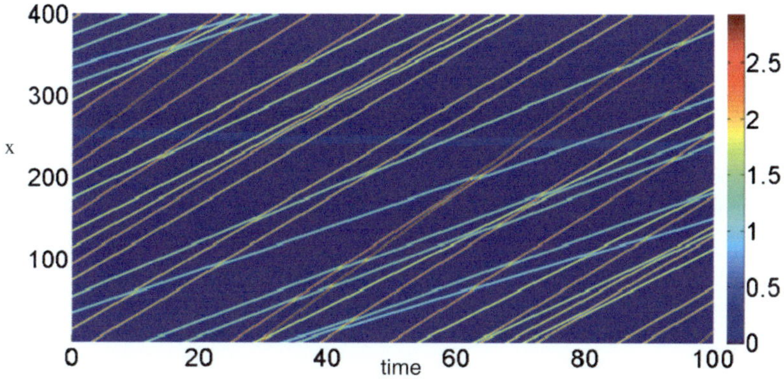

Fig. 18.3 Spatial-temporal diagram of soliton field which contains a soliton with negative velocity

$$u(x,t) = \frac{c}{2}\operatorname{sech}^2\left(\frac{\sqrt{c}}{2}(x - ct)\right). \tag{18.4}$$

In this case all further results will be also valid for the modified Korteweg–de Vries equation. The phase shift during the soliton interaction (with velocities $c_1,\ c_2$) for both equations is:

$$\Delta x_{1,2} = \pm\frac{2}{\sqrt{c_{1,2}}}\ln\left(\frac{\sqrt{c_2}+\sqrt{c_1}}{\sqrt{c_2}-\sqrt{c_1}}\right), \tag{18.5}$$

Δx is positive for larger soliton and negative for smaller one. In the case of the motion of a small soliton (defect) in the lattice consisting of large solitons moving with velocity c_2, and separated by a considerable distance (L) from each other the mean velocity of the trial soliton is

$$s_1 = c_1\frac{1 - \frac{c_2}{c_1}\frac{\Delta x_1}{L}}{1 - \frac{\Delta x_1}{L}}. \tag{18.6}$$

After substitution (18.5) to (18.6), assuming that the numerator is zero and defect is really "small" ($c_1 \ll c_2$), the condition of zero soliton velocity takes a simple form:

$$c_1 = \frac{4\sqrt{c_2}}{L}. \tag{18.7}$$

In the case of soliton gas the amplitudes and phases of the solitons are random, and there will be the interaction of many solitons between each other during the time. As it was shown in Fig. 18.3 the smallest soliton in a soliton gas can move backwards. For the explanation of this effect it is necessary to consider the large temporal interval, taking into account all possible soliton collisions, not only with defect but also between large ones. However if soliton gas density is small enough than the impact of "large" soliton interactions should be offset (some solitons accelerate and some solitons brake), while the negative phase shift after the collision of defect with large solitons accumulates. Therefore, if the phase shifts in large soliton interactions are neglected we can derive the criterion of the small soliton's negative velocity in a soliton gas:

$$c_1 = \frac{4\left\langle \frac{1}{\sqrt{c}} \right\rangle}{L\left\langle \frac{1}{c} \right\rangle}. \tag{18.8}$$

Taking the above-mentioned into account it can be concluded that a small soliton has a negative velocity in a lattice of regular solitons, when the criterion (18.7) is fulfilled, and in a random soliton gas—when the criterion (18.8) is fulfilled for both the Korteweg–de Vries and the modified Korteweg–de Vries equation.

18.6 Maxima of the Wave Field and Amplitude Distribution Function

Some conclusions about the dynamics of the wave fields can be found by analyzing the graph of field extreme (Fig. 18.4). As shown in [24], since pairs of soliton interactions lead to a decrease in the amplitude of the resulting impulse, the maximum value of the extremum does not exceed the amplitude of the largest soliton in the realization (in our case $A_{max} = 3$). The extremum of the field decreases during the interaction. This observation is fully in line with prediction, which follows from

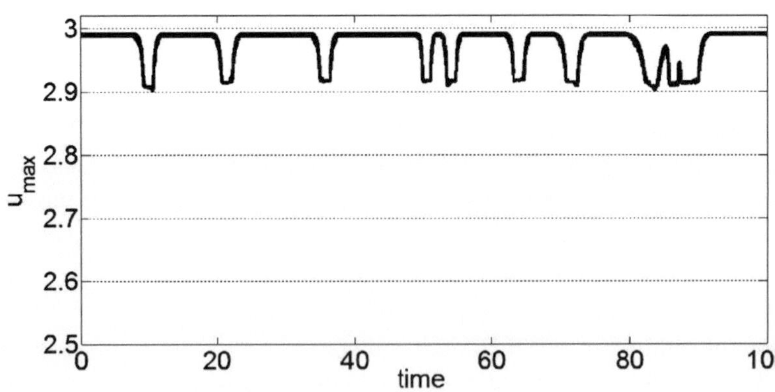

Fig. 18.4 Temporal variability of the maxima of the wave field

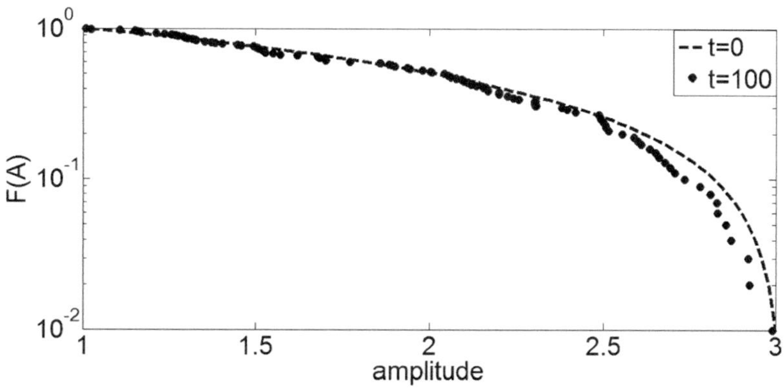

Fig. 18.5 Distribution function of soliton amplitudes at $t = 100$ (*dots*) and at the initial moment—uniform distribution (*line*)

the analysis of a two-soliton collision [24]. Larger waves do not occur in ensembles of unipolar solitons and it is also valid for unipolar mKdV soliton gas [22, 27].

These processes affect the distribution functions of the wave field and its statistical moments. At the initial moment the soliton amplitudes are chosen close to the uniform distribution (Fig. 18.5). The distribution function of the wave amplitude (local maxima of the wave field) varies in each realization over time, and the example of the distributions is shown in the same figure. Qualitative changes manifest in the same way: as the number of small amplitude impulses increases and the number of large waves decreases. As a result, the distribution function of the wave amplitudes becomes steeper in comparison to the initial distribution. However, in a field of purely solitons, this effect is weak, underlining the resilient nature of the soliton interaction and their ability to retain their parameters.

18.7 Statistical Characteristics of Wave Field

Let us discuss the statistical characteristics of the soliton gas. We have a random wave field, which depends on two variables: the coordinates x and time t, which are not very convenient for the analysis. For simplicity, we will consider the statistical characteristics averaged over the computational domain:

$$M(t) = \frac{1}{L}\int_0^L f(x,t)dx \tag{18.9}$$

(f—any characteristic of the wave field) which are functions of the current time. This procedure corresponds to the ergodic hypothesis, when the averaging over the ensemble of realizations is replaced by integration, in this case over the space.

At the same time $M(t)$ is a random function of time due to the random nature of the soliton interactions. Moments of a random function are the values averaged over the ensemble of realizations

$$\langle M\rangle = \frac{1}{n}\sum_{j=1}^{n} M_j(t) \tag{18.10}$$

This value for a large number of realizations, in the limit $n \to \infty$ becomes independent on time and determines the statistical moment of integral characteristics of the wave field. We can again use the ergodic hypothesis and change the averaging over ensemble to integration over time (which should be large enough). Below we will often call the integral characteristics (18.9) the moments of the wave field, as in (18.1), and we hope that in the text, the reader will not be confused between (18.9) and (18.1).

The statistical moments of rarefied solitonic gas may be estimated when the effect of soliton interaction is totally ignored (for instance, when solitons are well isolated in the initial condition [21, 22]. The integration can be carried out over an infinite limit and the last integral is trivial because of the narrowness of solitons in comparison with the size of the computational domain. Then, the mean field is

$$\langle u\rangle = \langle M_1\rangle = \frac{1}{L}\int_0^L u(x,0)dx \approx \frac{1}{L}\int_{-\infty}^{+\infty} u(x,0)dx = \frac{1}{L}\sum_{i=1}^{i=N}\int_{-\infty}^{+\infty} u_i(x,0)dx = \frac{4}{L}\sum_{i=1}^{i=N} K_i \tag{18.11}$$

The sum can easily be expressed in terms of the average value of K:

$$\langle u\rangle = \frac{4}{L}\sum_{i=1}^{i=N} K_i = 4\frac{N}{L}\langle K\rangle \tag{18.12}$$

where $\langle K\rangle$ is a statistical average over the ensemble of random amplitude solitons. $\langle u\rangle$ does not depend on the realization of the soliton gas, thus its value will not change in case of averaging over realizations. The coefficient N/L, included in (18.12), has a clear physical meaning of the density of the soliton gas

$$\rho = \frac{N}{L} \tag{18.13}$$

Finally, (18.12) can be transformed into

$$\langle u\rangle = 4\rho\langle K\rangle = 2\sqrt{2}\rho\langle A^{1/2}\rangle \tag{18.14}$$

As expected the mean grows with increasing of soliton gas density. Let us note also that $\langle A^{1/2}\rangle \neq \langle A\rangle^{1/2}$, thus the knowledge of the average soliton amplitude is not sufficient for the calculation of the average characteristics of the soliton gas.

Of course, the mean field, as well as any pedestal, in the framework of the Korteweg–de Vries can be eliminated by appropriate scaling, but in this case there is a number of difficulties in the transition to the water waves (definition of the unperturbed depth and, accordingly, the soliton width), and this will not be considered here.

The dispersion of the wave field is calculated similarly:

$$\sigma^2 = \langle [u - \langle u \rangle]^2 \rangle = \frac{16}{3}\rho\langle K^3\rangle - 16\rho^2\left[\langle K\rangle\right]^2 = \frac{8}{3\sqrt{2}}\rho\langle A^{3/2}\rangle - 8\rho^2\langle A\rangle. \quad (18.15)$$

From the positivity of the wave field dispersion there is a limit on the density of a soliton gas [13, 28]:

$$\rho < \rho_{\mathrm{cr}} = \frac{\langle A^{3/2}\rangle}{3\sqrt{2}\langle A\rangle} \quad (18.16)$$

The critical density is easily understood from the following considerations. Assuming that all the amplitudes are the same, the critical density is $\rho_{\mathrm{cr}} = K/3$. If we recall the definition of density as (18.13), the critical number of solitons is equal to $N_{\mathrm{cr}} = KL/3$. This condition corresponds to the situation when all solitons fill the domain "being very close to each other" and there is no place for a new soliton. In fact, the critical soliton density should be less because during the time solitons start to interact with each other, and the approximation of noninteracting solitons is no longer valid. It means that our formula works well only for a sufficiently rarefied gas. In this case the formula (18.15) can be simplified

$$\sigma^2 \approx \frac{16}{3}\rho\langle K^3\rangle = \frac{8}{3\sqrt{2}}\rho\langle A^{3/2}\rangle. \quad (18.17)$$

Asymptotic formulas of skewness and kurtosis valid for a rarified gas ($\rho \to 0$) take the following form

$$Sk \approx \frac{2\sqrt{3}}{5\sqrt{\rho}}\frac{\langle K^5\rangle}{(\langle K^3\rangle)^{3/2}} = \frac{\sqrt{3}\sqrt[4]{2}\langle A^{5/2}\rangle}{5\sqrt{\rho}(\langle A^{3/2}\rangle)^{3/2}} \quad (18.18)$$

$$Ku \approx \frac{18}{35\rho}\frac{\langle K^7\rangle}{(\langle K^3\rangle)^2} = \frac{9\sqrt{2}\langle A^{7/2}\rangle}{35\rho(\langle A^{3/2}\rangle)^2} \quad (18.19)$$

In (18.18) and (18.19) the soliton gas density is in denominator, thus skewness and kurtosis are abnormally large for a very rarefied gas and it is not a Gaussian process. Skewness is always positive because all solitons are positive. The large value of kurtosis indicates a greater probability of large wave appearance [29] that is clear in the framework of the linear theory, when the pulses can converge to a single point, overlapping each other. However, this conclusion is not obvious in

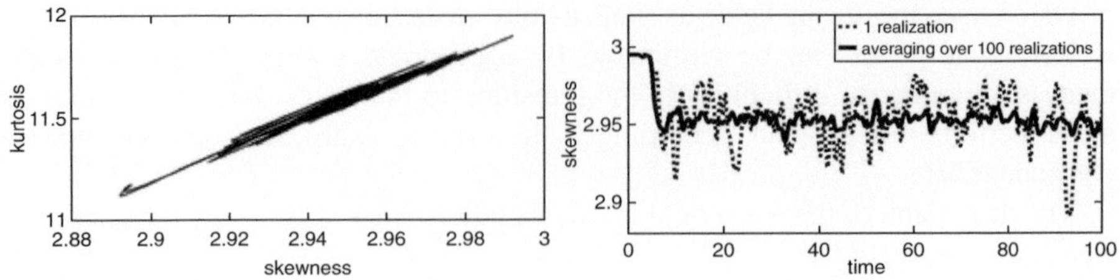

Fig. 18.6 The dependence of kurtosis versus skewness and temporal evolution of the skewness of soliton gas

the nonlinear theory, since it is known that even in the interaction of two unipolar solitons the resulting amplitude is always less than the amplitude of the maximum soliton (solitons "repel" from each other).

Left panel in Fig. 18.6 shows the dependence of kurtosis versus skewness for one realization of soliton gas consisted of 100 solitons during whole time. Almost linear law is clearly observed which is approximated by $u = 7.6x - 11$. This confirms the strong correlation between skewness and kurtosis. That is why we present here the time evolution of the third moment of the soliton gas only (Fig. 18.6, right panel). In one realization, the computed third and fourth moments have a higher variability. Averaged values over 100 realizations become stable enough after transitional period. The reason for this is the nature of the interaction of solitons, because such interactions lead to a decrease of the third and fourth moments. A finite sum of random variables is also a random variable, therefore by averaging over the realizations, we get only an estimation of skewness and kurtosis.

The third and fourth moments are positive. The averaged value of skewness is equal to 2.95 with standard deviation of 0.01. It is important to note that the average value of this ratio is less than the initial value of 3, demonstrating the contribution of the nonlinear interaction of solitons. The average value of kurtosis is equal to 11.6 with a standard deviation of 0.07, while the initial value is 11.9.

From the above it can be concluded that the interaction of solitons in the framework of the Korteweg–de Vries equation leads to a change of the statistical characteristics of the soliton gas (distribution function of the wave amplitudes, skewness and kurtosis). In each realization the soliton interactions do not lead to the formation of abnormally large waves, thus there are no freak waves.

However soliton gas can consist not only of solitons with the same polarity. For example, in the framework of the modified Korteweg–de Vries equation bipolar solitons can exist. In this case the dynamics of multi-soliton fields will be completely different. In [22, 27] was shown that such bipolar interactions lead to the increase of skewness and kurtosis of the whole soliton field; tails of amplitude distribution functions can grow significantly and finally in such bipolar soliton fields abnormally large waves can appear. It seems that the same results can be obtained in other models which allow the existence of solitons of different polarities.

18.8 Conclusions

Soliton turbulence (soliton gas) can play an important role in the whole wave dynamics. In this study we presented several analytical results and numerical experiments on solitonic gas turbulence in the framework of the integrable KdV equation. We have shown that in the framework of the KdV equation nonlinear soliton interactions lead to the decrease of the third and fourth moments (skewness and kurtosis) of the nonlinear wave field. The distribution function of the wave amplitudes becomes steeper in comparison to the initial distribution. The criterion of "zero" soliton velocity is presented.

Acknowledgements Different parts of the study were funded by RFBR (16-02-00167, 16-05-00049, 16-35-00175). The study of the effect of a negative soliton velocity was particularly funded by RFBR (16-32-60012). The authors also gratefully acknowledge the support from Volkswagen Foundation and President grant SC-6637.2016.5.

References

1. Rabinovich, M.I., Trubetskov, D.I.: Oscillations and Waves in Linear and Nonlinear Systems, p. 578. Kluwer Academic, Dorderecht (1989)
2. Gaponov-Grekhov, A.V., Rabinovich, M.I.: Nonlinearities in Action: Oscillations, Chaos, Order, Fractals, p. 191. Springer, New York (1992)
3. Pikovsky, A.S., and Rabinovich, M.I.: Stochastic behavior of dissipative systems, In: Novikov, S.P. (ed.) Soviet Scientific Reviews, Section C - Mathematical Physics Reviews, vol. 2, pp. 165–208. Harwood Academic, New York (1981)
4. Aranson, I.S., Gaponov-Grekhov, A.V., Gorshkov, K.A., Rabinovich, M.I.: Models of turbulence: chaotic dynamics of structures. In: Campbell, D.K. (ed.) Chaos/Soviet-American perspectives of Nonlinear Sciene, pp. 183–196. American Institute of Physics, New York (1990)
5. Zakharov, V.E., L'vov, V.S., Falkovich, G.: Kolmogorov Spectra of Turbulence, p. 264. Springer, Berlin (1992) 6
6. Nazarenko, S.: Wave Turbulence. Lecture Notes in Physics, p. 279. Springer, Berlin (2011) 26
7. Campbell, D.K., Rosenau, P., Zaslavsky, G. (eds.). A focus issue on "The "Fermi-Pasta-Ulam" problem—the first 50 years". Chaos **15**(1), 015101 (2005) 22
8. Zakharov, V. E.: Kinetic Equation for solitons. Sov. Phys. JETP, **33**(3), 538–540 (1971)
9. Zakharov, V.E.: Turbulence in integrable systems. Stud. Appl. Math. **122**, 219–234 (2009)
10. El, G.A., Kamchatnov, A.M.: Kinetic equation for a dense soliton gas. Phys. Rev. Lett. **95**, 204101 (2005)
11. El, G.A., Krylov, A.L, Molchanov, S.A., Venakides S.: Soliton turbulence as the thermodynamic limit of stochastic soliton lattices. Physica D **152–153**, 653–664 (2005)
12. El, G.A., Kamchatnov, A.M., Pavlov, M.V., Zykov, S.A.: Kinetic equation for a soliton gas and its hydrodynamic reductions. J. Nonlinear Sci. **21**, 151–191 (2011)
13. El, G.A.: Critical density of a soliton gas. Chaos **26**(2), 1–7 (2016)
14. Osborne, A.R.: Nonlinear Ocean Waves and the Inverse Scattering Transform, p. 976. Academic Press, New York (2010)
15. Osborne, A.R., Segre, E., Boffetta, G.: Soliton basis states in shallow-water ocean surface waves. Phys. Rev. Lett. **67**, 592–595 (1991)

16. Osborne, A.R., Serio, M., Bergamasco, L., Cavaleri, L.: Solitons, cnoidal waves and nonlinear interactions in shallow-water ocean surface waves. Physica D **123**, 64–81 (1998)
17. Salupere, A., Maugin, G. A., Engelbrecht, J., Kalda, J.: On the KdV soliton formation and discrete spectral analysis. Wave Motion **123**, 49–66 (1996)
18. Salupere, A., Peterson, P., Engelbrecht J.: Long-time behaviour of soliton ensembles. Part 1 – emergence of ensembles. Chaos Solitons Fractals **14**, 1413–1424 (2002)
19. Salupere, A., Peterson, P., Engelbrecht, J.: Long-time behaviour of soliton ensembles. Part 2 – periodical patterns of trajectories. Chaos Solitons Fractals **15**, 29–40 (2003)
20. Salupere, A., Peterson, P., Engelbrecht, J.: Long-time behavior of soliton ensembles. Math. Comput. Simul. **62**, 137–147 (2003)
21. Dutykh, D., Pelinovsky, E.: Numerical simulation of a solitonic gas in KdV and KdV–BBM equations. Phys. Lett. A **378**, 3102–3110 (2014)
22. Shurgalina, E.G., Pelinovsky, E.N.: Dynamics of Irregular Wave Ensembles in the Coastal Zone, Nizhny Novgorod State Technical University n.a. R.E. Alekseev. Nizhny Novgorod, p. 179 (2015)
23. Sawada, K., Kotera, T.: A Method for finding N-soliton solutions of the KdV equation and KdV-like equations. Prog. Theor. Phys. **51**(5), 1355–1367 (1974)
24. Pelinovsky, E.N., Shurgalina, E.G., Sergeeva, A.V., Talipova, T.G., El, G.A., Grimshaw, R.H.J.: Two-soliton interaction as an elementary act of soliton turbulence in integrable systems. Phys. Lett. A **377**(3–4) 272–275 (2013)
25. Kuznetsov, E.A., Mikhailov, A.V.: Stability of stationary waves in nonlinear weakly dispersive media. Sov. Phys. JETP **40**(5), 855–859 (1974)
26. Carbone, F., Dutykh, D., El, G.A.: Macroscopic dynamics of incoherent soliton ensembles: Soliton gas kinetics and direct numerical modelling. Eur. Phys. Lett. **113**(3), 30003–30008 (2016)
27. Shurgalina, E.G., Pelinovsky, E.N.: Nonlinear dynamics of a soliton gas: modified Korteweg-de Vries equation framework. Phys. Lett. A **380** (24), 2049–2053 (2016)
28. Shurgalina, E., Pelinovsky, E.: Dynamics of Random Ensembles of Free Surface Gravity Waves with Application to the Killer Waves in the Ocean, p. 116. Lambert Academic, Saarbrucken (2012)
29. Kharif, C., Pelinovsky, E., Slunyaev, A.: Rogue Waves in the Ocean, p. 216. Springer, Berlin (2009)

Chapter 19
Multi-Lump Structures in the Kadomtsev–Petviashvili Equation

Yury Stepanyants

В Непале есть столица Катманду.
Случайное, являясь неизбежным,
приносит пользу всякому труду.
Ведя ту жизнь, которую веду,
я благодарен бывшим белоснежным
листам бумаги, свернутым в дуду.
J. Brodsky,

"Twenty Sonnets to Mary Stewart".

19.1 Introduction

The spectrum of interests of Mikhail Rabinovich (MIR) is very broad, he is interested in many subjects and changed his fields of research many times. In his student years he published his own first scientific article "Process automation on the overhead thrust assembly line" in the prestigious Russian journal "Automation in Production". Then he contributed to the theory of relaxing oscillations, mode competition in nonlinear systems, development of the asymptotic theory of

Y. Stepanyants (✉)
University of Southern Queensland, West Street, Toowoomba, QLD 4350, Australia
e-mail: Yury.Stepanyants@usq.edu.au

I.S. Aranson et al. (eds.), *Advances in Dynamics, Patterns, Cognition*,
Nonlinear Systems and Complexity 20, DOI 10.1007/978-3-319-53673-6_19

oscillations, chaotic behaviour of simple systems,[1] strange attractors,[2] theory of turbulence,[3] etc. Nowadays his scientific interest is in the development of the theory of cognitive systems and consciousness. One of his former interests was in the pattern formation in complex systems. In particular, he has demonstrated jointly with his colleagues that nontrivial three-dimensional patterns can exist within the framework of Ginzburg–Landau, Swift–Hohenberg and other equations; several interesting examples of single patterns and their couplings were presented in journal papers and reviews [7, 8, 12, 36].

In 2008 MIR visited Australia where he was an invited speaker of the International conference on cognitive systems. One day we undertook a trip to one of the most picturesque places in Australia, Blue Mountains, near Sydney. MIR admired looking at the rocks called "Three Sisters" (see Fig. 19.1) and as a follow-up of our discussion on multi-dimensional patterns said: "Yurah, do you know any exact

Fig. 19.1 M.I. Rabinovich with the author against the background of the "Three Sisters" rock (Blue Mountains, Australia, 14 June 2008)

[1] At that time I was involved into one of the joint researches with him [11].

[2] MIR was actually the first to invent and realise jointly with S. V. Kiyashko and A. S. Pikovsky in 1980, the simplest example of a strange attractor [19], now known as the Chua attractor [10].

[3] MIR wrote a section about this in the classical book by L.D. Landau and E.M. Lifshitz, Hydrodynamics [22].

solution of nonlinear equations which resembles such beautiful patterns like this "Three Sisters" rock?". I answered: "Yes, such solutions and even more complex ones do exist within the framework of Kadomtsev–Petviashvili equation". I did not have a computer at that moment to show the pictures and details of relevant solutions and promised to demonstrate them some time later. But the days to follow were very busy for both of us and we did not have time to come back to that issue. Now it is a good time to fulfill my promise. This chapter is devoted to the exact solutions of the Kadomtsev–Petviashvili equation in the form of lumps and their bound states.

19.2 The Kadomtsev–Petviashvili Equation

In 1970 B.B. Kadomtsev and V.I. Petviashvili published a seminal paper [17] in which they derived a two-dimensional generalisation of the Korteweg–de Vries equation now known as the KP equation:

$$\frac{\partial}{\partial x}\left(\frac{\partial u}{\partial t} + c\frac{\partial u}{\partial x} + \alpha u\frac{\partial u}{\partial x} + \beta\frac{\partial^3 u}{\partial x^3}\right) = -\frac{c}{2}\frac{\partial^2 u}{\partial y^2}, \tag{19.1}$$

where c is the speed of long linear waves, α is the coefficient of nonlinearity, and β is the dispersion coefficient.

In that paper the authors investigated the stability of plane waves propagating along the x-axis and showed that for negative β plane waves with slightly modulated fronts are unstable with respect to self-focusing, whereas in the case of positive β the waves are stable against this phenomenon. That finding was later confirmed by Zakharov [42] who derived an exact formula for the growth rate of instability. Thus, in those and subsequent papers by many authors it was discovered that the properties of solutions of the KP equation essentially depend on the dispersive parameter β, therefore it is reasonable to distinguish two versions of this equation—KP1 equation for negative β and KP2 equation for positive β. Both versions of the KP equation are feasible in the physical context and can describe wave processes in different media.

Some time later Petviashvili discovered [33] numerically that the KP1 equation has solutions in the form of two-dimensional solitons called lumps and suggested an original numerical method for constructing stationary solutions of a certain class of nonlinear equations.[4] Then an analytical solution was found for the lumps [1, 26] and it was shown that after the interaction with each other their initial parameters completely restore even without a phase shift which usually appears when the plane solitons interact (see, e.g., [2, 16]).

[4] A rigorous substantiation of the convergence of the Petviashvili method was published in [31].

19.3 Lump Solutions of the Kadomtsev–Petviashvili Equation

Consider stationary solutions of the KP equation (19.1) which describe wave propagation at some angle with respect to the x-axis with velocity $\mathbf{V} = (V_x, V_y)$. Let us represent Eq. (19.1) in dimensionless form bearing in mind that $\beta < 0$:

$$\frac{\partial^2 \upsilon}{\partial X^2} + \frac{\partial^2 \upsilon}{\partial Y^2} - \nu \frac{\partial^2 \upsilon}{\partial X \partial Y} = \frac{\partial^2}{\partial X^2}\left(\upsilon^2 + \frac{\partial^2 \upsilon}{\partial X^2}\right), \tag{19.2}$$

$$\text{where} \quad \upsilon = \frac{\alpha u}{2(V_x - c)}, \quad \nu = V_y \sqrt{\frac{2}{c(c - V_x)}}, \tag{19.3}$$

$$X = \left[x - V_x t + \frac{V_y}{c}(y - V_y t)\right]\sqrt{\frac{V_x - c}{\beta}}, \quad Y = (y - V_y t)(c - V_x)\sqrt{\frac{2}{-\beta c}}. \tag{19.4}$$

In the simplest case of a lump propagating along the x-axis ($V_y = 0$, $\nu = 0$), it is described by the formula derived for the first time in [26]:

$$u(X, Y) = 12\frac{3 + Y^2 - X^2}{(3 + Y^2 + X^2)^2}. \tag{19.5}$$

Lump solutions play a fundamental role in the theory of KP1 equation. They describe nonlinear patterns in plasma [2, 16, 18, 34], on the surface of shallow water with dominating surface tension [4, 9, 18], in nonlinear optic media [32], in the Bose–Einstein condensate [27], in solids with inner microstructure [39], in thin elastic plates [35], etc. Due to the lump stability with respect to external perturbations [21], they can play a role of elementary wave excitations, and their ensembles with the nontrivial internal interaction between them (see below) may be regarded as a model of a strong wave turbulence.

The solution (19.5) in dimensionless variables is shown in Fig. 19.2.

For the lump propagating at an angle to the x-axis the generalised formula was actually derived in [26], but it was not analysed there in detail. Later the obliquely propagating lumps were rediscovered independently in [24, 25, 40] and were analysed in [37]. The solution is:

$$u(X, Y) = 12(4 - \nu^2)\frac{12 + (4 - \nu^2)\left[(4 - \nu^2)Y^2/4 - (X + \nu Y/2)^2\right]}{\left\{12 + (4 - \nu^2)\left[(4 - \nu^2)Y^2/4 + (X + \nu Y/2)^2\right]\right\}^2}, \tag{19.6}$$

where $-2 < \nu < 2$. An example of the lump solution for $\nu = 1.5$ is shown in Fig. 19.3.

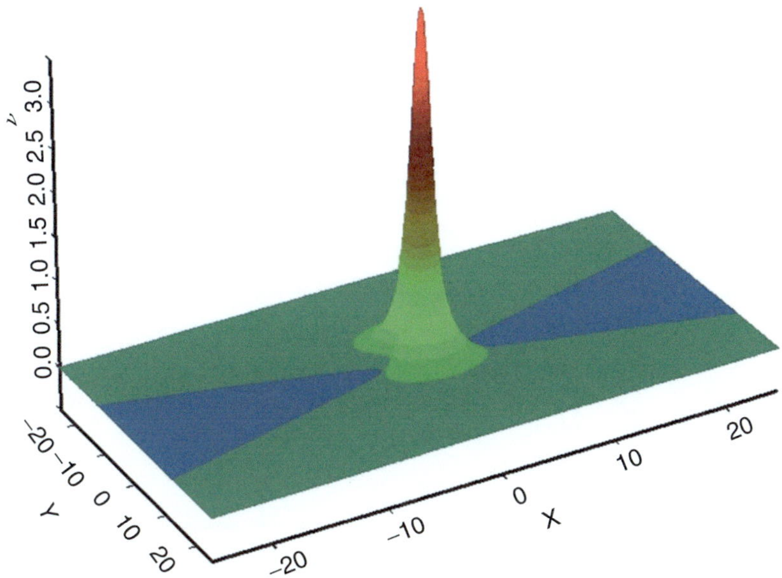

Fig. 19.2 3D profile of conventional KP1 lump as per Eq. (19.5)

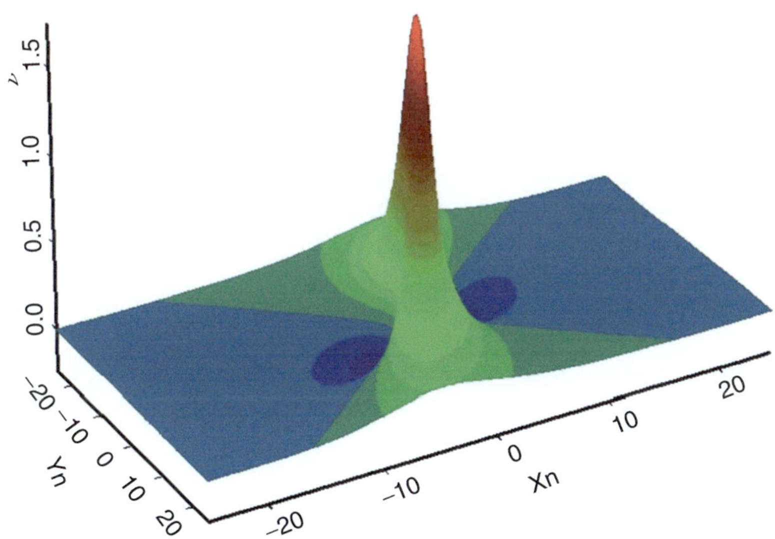

Fig. 19.3 3D profile of a skew lump as per Eq. (19.5) with $\nu = 1.5$

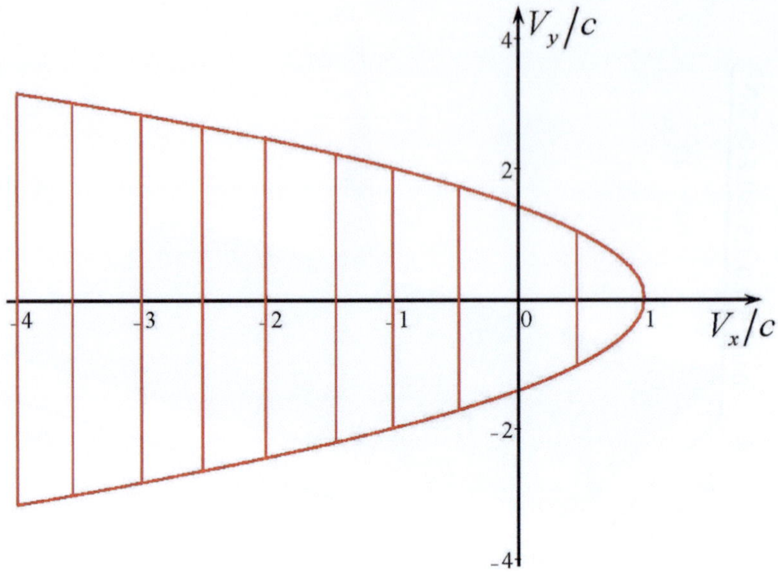

Fig. 19.4 The domain of possible velocity components (shaded) of a single lump as per Eq. (19.7)

The restriction for the parameter ν (see above) imposes the restriction on possible velocity components in dimensional variables:

$$V_x = c - \frac{V_y^2}{2c}. \tag{19.7}$$

The domain of possible velocity components of a single lump is shown in Fig. 19.4.

Obliquely propagating lumps have asymmetrical skew shapes; their structure depends on the parameter ν. Figure 19.5 shows the contour plots of skew lumps for a few values of ν.

19.4 Multi-Lump Solutions of the Kadomtsev–Petviashvili Equation

The wave field of a lump is non-monotonic in space, it contains two minima. Because of that two closely located lumps can form a bound state—a stationary moving bi-lump. Such a solution was obtained for the first time numerically [3], and then the analytical formulae were found for the bi-lumps propagating along the x-axis [29] and at the angle to the x-axis [37, 40]. The analytical formula for a bi-lump contains two free parameters, a and b, and can be written in the form:

$$\upsilon(X, Y) = 6\frac{\partial^2 f(X, Y)}{\partial X^2}, \quad \text{where} \tag{19.8}$$

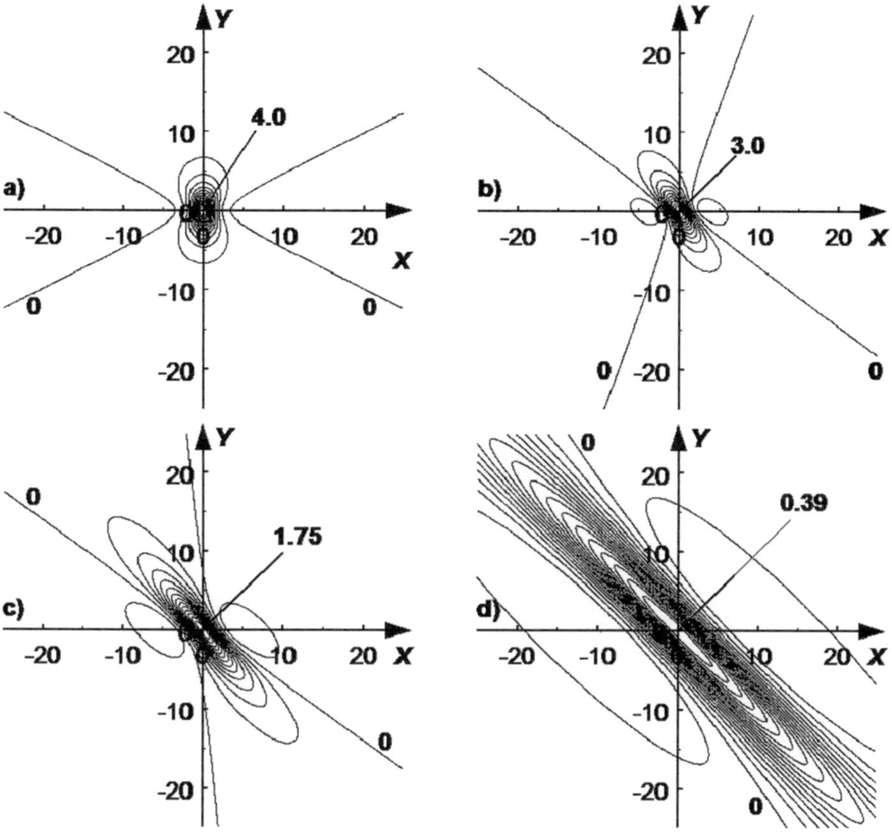

Fig. 19.5 Contour plots of lumps for different values of ν. (**a**) $\nu = 0.5$; (**b**) $\nu = 1.0$; (**c**) $\nu = 1.5$; and (**d**) $\nu = 1.9$. *Arrows* show maximum values, zero isolines are also indicated

$$\begin{aligned}
f(X,Y) = {} & \left(X^2+Y^2\right)^3 + \frac{100}{4-\nu^2}X^4 + 4\frac{17+14\nu^2-3\nu^4}{4-\nu^2}Y^4 + bX^3 + aY^3 + \nu^3X^3Y^3 \\
& - \frac{2000}{(4-\nu^2)^2}X^2 + 400\frac{19-6\nu^2}{(4-\nu^2)^2}Y^2 - \frac{4b}{4-\nu^2}X - 4\frac{5a+8b\nu-3b\nu^3}{(4-\nu^2)(1-\nu^2)}Y \\
& - \frac{2000\nu}{(4-\nu^2)^2}XY + 60\frac{6+\nu^2}{4-\nu^2}X^2Y^2 + 3\nu XY\Bigg[\left(X^2+Y^2\right)^2 + \nu XY\left(X^2+Y^2\right) \\
& + \frac{40}{3}\frac{5X^2+\left(9-\nu^2\right)Y^2}{4-\nu^2} - \frac{(a+b\nu)X+(b+a\nu)Y}{\nu(1-\nu^2)}\Bigg] \\
& + \frac{ab\nu(3-\nu^2)+a^2+b^2}{(1-\nu^2)^2(4-\nu^2)} + \frac{120000}{(4-\nu^2)^3}. \qquad (19.9)
\end{aligned}$$

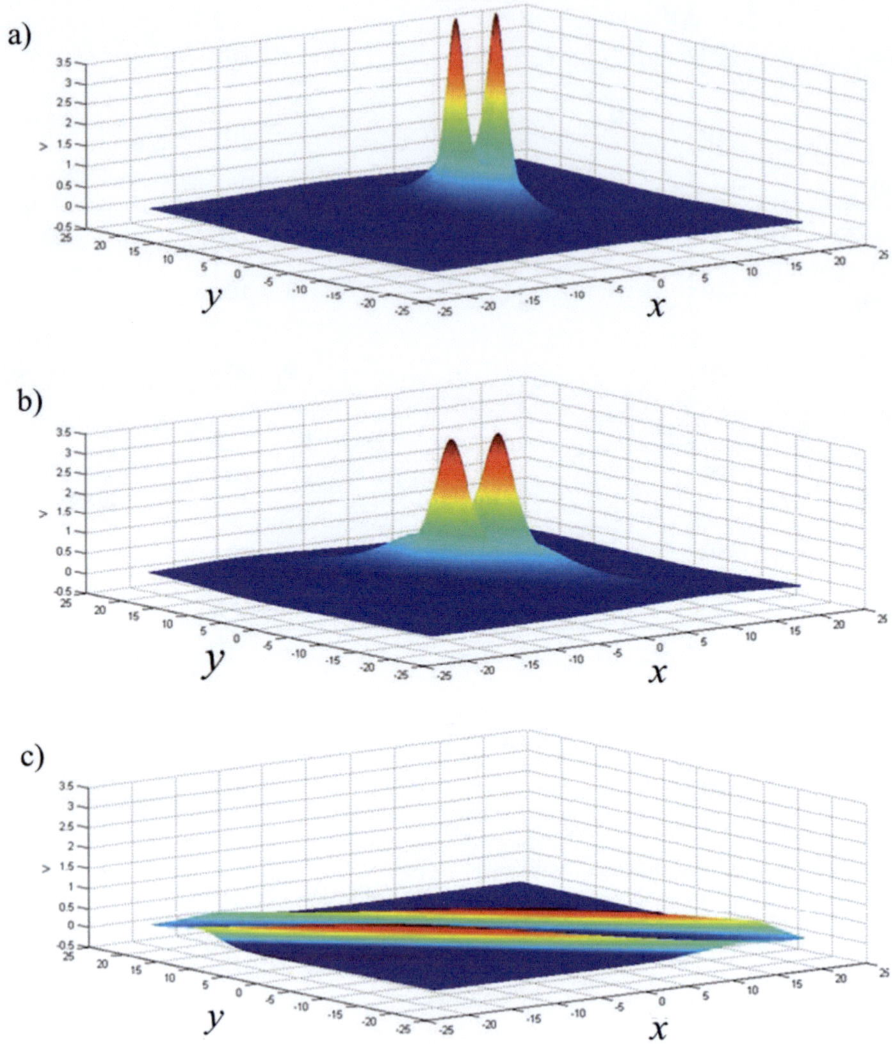

Fig. 19.6 Bi-lump solutions as per Eqs. (19.8) and (19.9) with $a = b = 0$. Panel (**a**) $\nu = 0$, panel (**b**) $\nu = 1$, panel (**c**) $\nu = 1.9$

Depending on the parameters a, b and ν, this solution can represent either a bi-lump or even a triple-lump as shown in Figs. 19.6 and 19.7.

In general, when ν approaches ± 2, the solution reduces to a very wide bi-lump of a small amplitude (see Fig. 19.6c), which completely vanishes in the limit $|\nu| = 2$.

As was shown in [29], the KP1 equation possesses a countable number of stationary multi-lump solutions which can be presented in terms of Eq. (19.8), where $f(X, Y)$ is a polynomial on X and Y of degree $p = m(m + 1)$, where m is an integer.

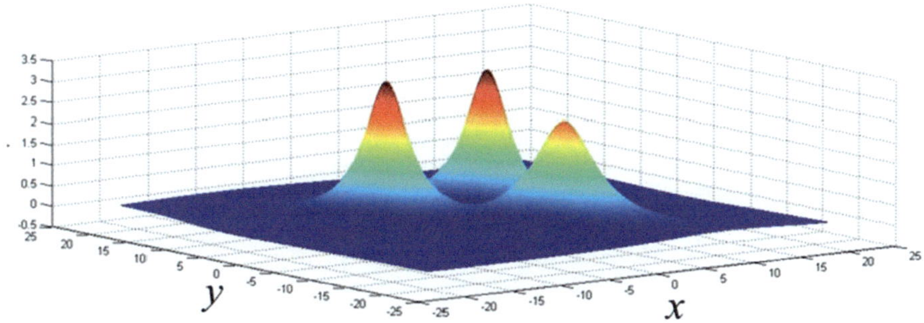

Fig. 19.7 Triple-lump solutions as per Eqs. (19.8) and (19.9) with $a = 10$, $b = 0$ and $v = 0.99$

For $m = 1$ ($p = 2$) we obtain a single lump solution; for $m = 2$ ($p = 6$)—bi-lump solutions (19.9); for $m = 3$ ($p = 12$)—triple-lump solutions; etc. The formulae become more and more complex when the degree of polynomial increases (the general problem of rational nonsingular solutions of the KP1 equation was considered in [20]). For simplicity we present only the next polynomial of degree 12 which describes a symmetrical triple-lump solution with $v = 0$:

$$
\begin{aligned}
f(X, Y) = {} & \left(X^2 + Y^2\right)^6 + 2\left(X^2 + Y^2\right)^3\left(49X^4 + 198X^2Y^2 + 29Y^4\right) + 5\left(147X^8\right. \\
& \left. + 3724X^6Y^2 + 7490X^4Y^4 + 7084X^2Y^6 + 867Y^8\right) + \frac{140}{3}\left(539X^6\right. \\
& \left. + 4725X^4Y^2 - 315X^2Y^4 + 5707Y^6\right) + \frac{1225}{9}\left(391314X^2\right. \\
& \left. - 12705X^4 + 4158X^2Y^2 + 40143Y^4 + 736890Y^2 + 717409\right). \quad (19.10)
\end{aligned}
$$

The 3D plot of this exact analytical solution shown in Fig. 19.8 resembles the "Three Sisters" rock in Fig. 19.1.

The stationary multi-lump structures can be interpreted within the theory of solitons as classical particles [13, 28]. According to that theory, solitons (and lumps, as the particular case) can be thought of as point particles with anisotropic masses $\hat{M}$ and described by the Newton equation:

$$
\hat{M}\frac{d^2\mathbf{r}}{dt^2} = -\nabla U(\mathbf{r}), \quad \text{where} \quad \hat{M} = \begin{pmatrix} 1 & 0 \\ 0 & -1 \end{pmatrix}, \qquad (19.11)
$$

where $\mathbf{r} = X\mathbf{i} + Y\mathbf{j}$.

The effective potential $U(\mathbf{r})$ is determined by external fields. In particular, when two lumps interact, the potential for one lump is determined by the field of the other lump:

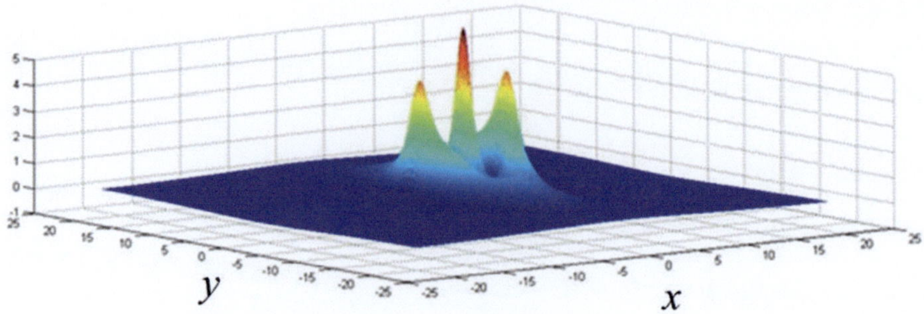

Fig. 19.8 Triple-lump solution as per Eqs. (19.8) and (19.10) —"Three Sisters"

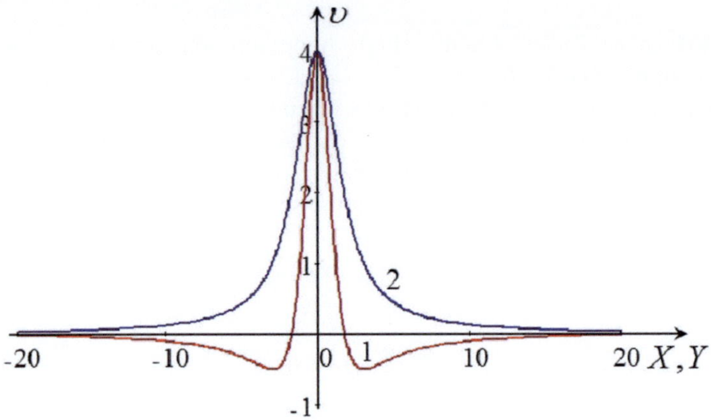

Fig. 19.9 The longitudinal (line 1) and transverse (line 2) cross-sections of a lump as per Eq. (19.5)

$$U(\mathbf{r}) = \frac{1}{24\pi} \iint \upsilon^2(\mathbf{r}')\upsilon(\mathbf{r}' + \mathbf{r})\, d\mathbf{r}', \tag{19.12}$$

where υ is determined by Eq. (19.5), and integrals are taken over the entire X, Y-plane. The typical cross-sections of a lump are shown in Fig. 19.9.

The local extrema of solitons give rise to bound states which can be stable or unstable with respect to small perturbations. The theory provides good qualitative and even quantitative agreement with the exact solutions, which allows us to interpret formal mathematical results in physical terms. Non-stationary dynamics of lumps is briefly outlined in the next section.

19.5 Normal and Anomalous Lump Scattering

Lump interactions were briefly studied in the papers [1, 26] (see also [2]) where lump solutions were constructed for the first time. Investigation of the multisoliton formulae describing the scattering of lumps obtained in those papers reveals a striking property of such formation. Lumps not only retain their shapes and initial parameters (amplitudes, velocities, sizes) on collisions, but even their phase shifts also turn out to be equal to zero. This does not mean, however, that the interaction of lumps is as trivial as the interaction of linear pulses in non-dispersive media. Detailed studies [15] have shown that the interaction even between two lumps by no means reduces to the superposition of their fields, and can lead to unexpected effects which are characterised by an infinite rather than zero phase shift.

In the particular case when two lumps of different amplitudes move one after another along the x-axis (the larger one initially moving behind the smaller one), their interaction looks as follows. The smaller lump splits into two which move at some angle to the x-axis absorbing energy from the larger lump. The latter one decreases and eventually vanishes, transferring its energy into the two newly created lumps. These new lumps separate first at a certain distance, but then draw together and line up again along the x-axis such that the larger lump becomes leading and the smaller one following behind it. Eventually the lumps completely restore their shapes and all other parameters. This is normal scattering. Figure 19.10 illustrates the variations of relative distances, ΔX and ΔY, between the lump maxima. The solid lines pertain to the exact analytical solution, and the dashed lines to the asymptotic theory of lumps as point classical particles described by Eq. (19.11).

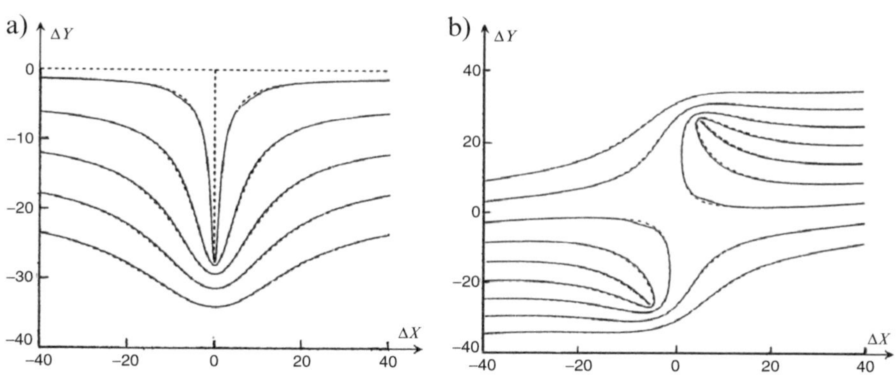

Fig. 19.10 Relative distances between lump maxima as per exact solution (*solid lines*) and within the framework of the asymptotic theory (*dashed lines*). Panel (**a**) shows two lumps with initial relative velocity $\Delta \mathbf{V} = (0.5, 0)$ (only the lower part of the figure is shown, whereas the figure is mirror symmetric with respect to the *dashed horizontal line*); panel (**b**) shows two lumps with initial relative velocity $\Delta \mathbf{V} = (0.5, 0.1)$

When two lumps moving along the x-axis have initially equal amplitudes and speeds, they start attracting each other, because their effective potentials are negative (see line 1 in Fig. 19.9). Then, as in the previous case, the first lump splits into two moving at an angle with respect to the x-axis and absorbing energy from the lump following behind. The latter decreases and eventually vanishes completely transferring its energy into the two newly created lumps. These new lumps of equal amplitudes continue separating and eventually move parallel to each other and the x-axis. Parameters of these new lumps are the same as the parameters of the initial lumps, whereas eventually they move side by side at the infinite distance between them. To a certain extent the phase shift in this case is infinite because the lumps never return to the initial x-axis; therefore, such interaction was called anomalous scattering [15]. On the other hand, the anomalous scattering can be treated as a particular case of normal scattering when the time of returning of the secondary lumps to the x-axis goes to infinity.

A detailed study of normal lump interaction was undertaken in [24, 40]. It was discovered that the process of interaction depends on the lump velocities and amplitudes. Figure 19.11 shows an example of oblique interaction when y-components of lump velocities have opposite signs.

According to the exact solution illustrated by Fig. 19.11, two lumps are moving from opposite y-directions to the horizontal x-axis along their original paths without noticing each other until they collide. This is true even when the two lumps have different amplitudes and speeds. However, when they interact they are far from being just a linear superposition of two independent lumps.

Another type of lump interaction occurs when two lumps initially located at the same side of the x-axis travel to the other side of this axis with different velocities so that their trajectories intersect at some point. The numerical results obtained in [24] are shown in Fig. 19.12. One can see that, when the lumps approach each other from the same side and have relatively small difference in their amplitudes, the larger one gradually reduces its amplitude and velocity, while the smaller one increases its amplitude and velocity. At some moment the lumps become of the same size at a certain nonzero minimum distance apart. After that, the lumps interchange their positions experiencing an abrupt phase change. Due to the nonlinear interaction, the front lump becomes taller until it recovers its initial amplitude, while the rear lump decreases and gradually recovers its original state. Before the collision the lumps attract each other and after the collision they repel.

Lump interaction is different when their relative amplitude difference is large enough. Two lumps can merge in this case and then separate and move along their own original paths. Figure 19.13 from [24] illustrates such interaction on the basis of the exact solution. Thus the interaction between two lumps strongly depends on their relative amplitude difference similarly to the KdV solitons. Strongly nonlinear overlapping interactions can occur when this difference becomes larger than a critical value which is not determined thus far. For the KdV equation it is known that the exchange-type interaction occurs when the soliton amplitude ratio $A_1/A_2 < 2.62$, and the overlapping-type interaction occurs when the soliton amplitude ratio $A_1/A_2 > 3.0$, whereas in the intermediate range, $2.62 < A_1/A_2 < 3$, a more complex interaction occurs [23].

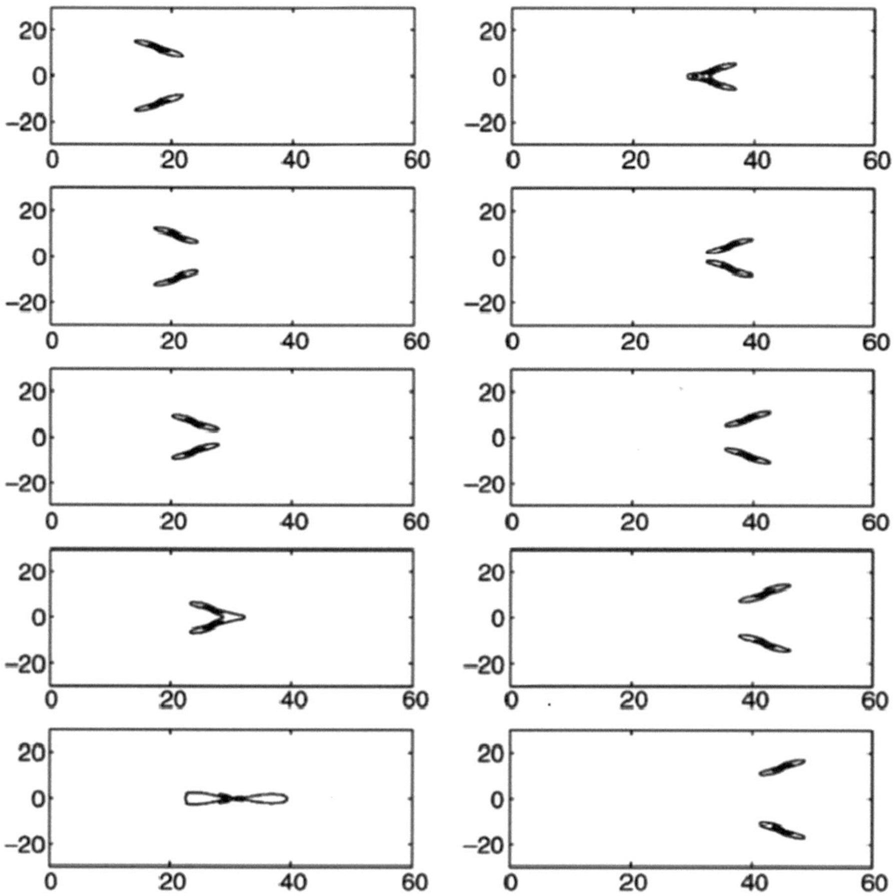

Fig. 19.11 Contour-plots of two lumps in the oblique interaction (case 1). *Left* (from *top to bottom*): $t = 1, 2, \ldots, 5$; *right* (from *top to bottom*): $t = 6, 7, \ldots, 10$

19.6 Conclusion

In this brief review we presented interesting features of two dimensional nonlinear formations described by the Kadomtsev–Petviashvili equation. This remarkable equation plays a specific role in the physics of nonlinear phenomena. It is completely integrable and possesses a reach class of solutions fully localised in space. Moreover, this equation is not just a mathematical toy, it is applicable to description of real physical phenomena in different media, as mentioned in the Introduction. It is worth noting that similar and even more complex multi-lump formations can exist not only within the framework of the integrable KP equation, but within many other model equations (see, e.g., [4–6, 9, 14, 38, 41]).

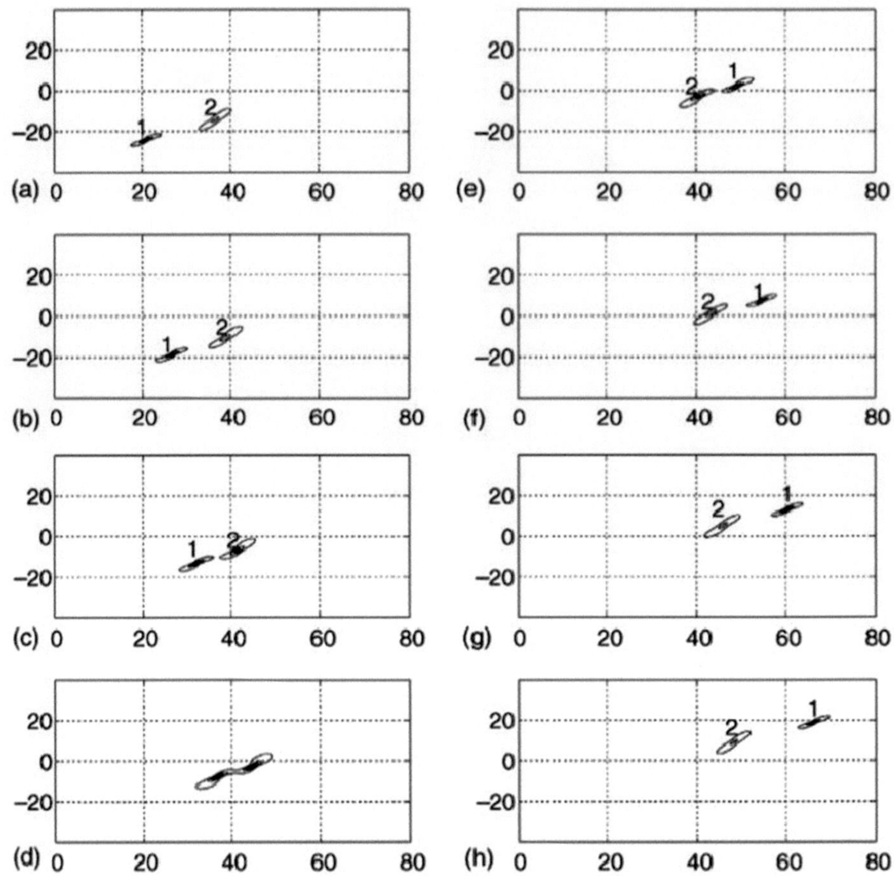

Fig. 19.12 Contour-plot of two lumps in the oblique interaction with relatively small amplitude difference (case 2). *Left* (from *top to bottom*): $t = 1, 2, 3$, and 4; *right* (from *top to bottom*): $t = 5, 6, 7$, and 8

The goal of this review was to attract the attention of young researchers to the physics of multidimensional patterns and to remind mature researchers about many unsolved problems in this area. Among such unsolved problems one can mention the stability of multidimensional formations, their possible role in the description of strong turbulence where lumps can represent elementary nonlinear excitations and form a gas of quasi-particles. As has been shown in [30] (see also [16] and the references therein), a chain of KP lumps can be formed in the course of instability of plane waves but the chain of lumps, in its turn, is unstable with respect to small modulations and can produce other chains of lumps, which are also unstable with respect to small modulations and so on. Therefore one can expect eventual stochastisation of lumps and formation of a random ensemble of nonlinear

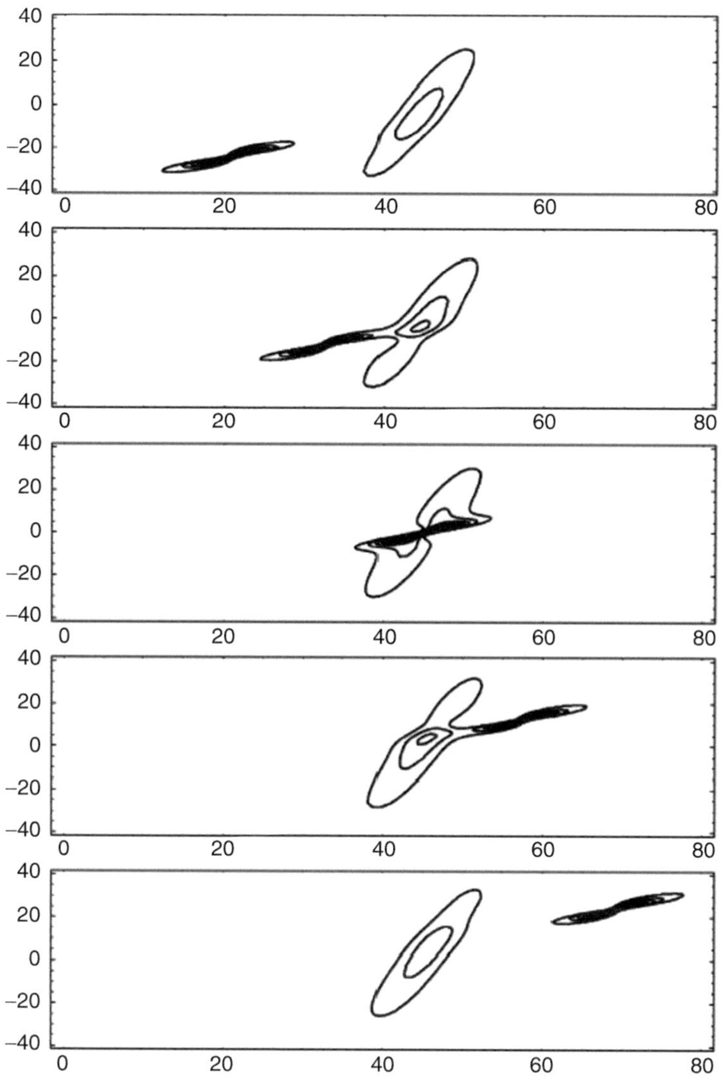

Fig. 19.13 Oblique interaction of two lumps with relatively big amplitude difference (case 2). From *top to bottom*: $t = 1, 3, 5, 7, 9$. In the middle frame, at $t = 5$, one can see overlapping of two lumps

interacting quasi-particles. Such phenomena were in the range of MIR's former interests (see, e.g., [7, 8, 12, 36]). I hope that someone of his numerous former or current students will advert to this intriguing problem.

In the conclusion I am wishing MIR many years of fruitful and productive work, as well as inexhaustible cheerfulness and health, which was always his distinctive feature, as one can see in Fig. 19.14.

Fig. 19.14 The "Three Brothers rock": M.I. Rabinovich (*centre*) with Anatoly Fabrikant (*left*) and the author (*right*). The photo from the Workshop on Plasma Turbulence, Sochi, Russia, May, 1986

О чем мечтать в моих годах?
Чтоб бегать под дождем,
Носить подругу на руках.
Чего еще мы ждем?
....
Ну дайте мне еще чуть-чуть,
Хотя бы вот полстолька.
Потом уж можно отдохнуть
На дружеской попойке.

M.I. Rabinovich

Acknowledgements The author is thankful to N.B. Krivatkina for her help with editing the paper.

References

1. Ablowitz, M.J., Satsuma, J.: Solitons and rational solutions of nonlinear evolution equations. J. Math. Phys. **19**, 2180–2186 (1978)
2. Ablowitz, M.J., Segur, H.: Solitons and the Inverse Scattering Transform. SIAM, Philadelphia (1981)

3. Abramyan, L.A., Stepanyants, Yu.A.: Two-dimensional multisolitons: stationary solutions of Kadomtsev–Petviashvili equation. Radiophys. Quantum Electron. **28**(1), 20–26 (1985)
4. Abramyan, L.A., Stepanyants, Yu.A.: The structure of two-dimensional solitons in media with anomalously small dispersion. Sov. Phys. JETP. **61**, 963–966 (1985)
5. Abramyan, L.A., Stepanyants, Yu.A.: Structure of two-dimensional solitons in the context of a generalised Kadomtsev–Petviashvili equation. Radiophys. Quantum Electron. **30**(10), 861–865 (1987)
6. Abramyan, L.A., Stepanyants, Yu.A., Shrira, V.I.: Multidimensional solitons in shear flows of boundary layer type. Sov. Phys. Dokl. **37**(12), 575–578 (1992)
7. Aranson, I.S. , Gorshkov, K.A., Lomov, A.S., Rabinovich, M.I.: Nonlinear dynamics of particle-like solutions of inhomogeneous fields. In: Gaponov-Grekhov, A.V., Rabinovich, M.I., Engelbrecht, J. (eds.) Nonlinear Waves. 3. Physics and Astrophysics, pp. 44–72. Springer, Berlin (1990)
8. Bazhenov, M., Bohr, T., Gorshkov, K., Rabinovich, M.I.: The diversity of steady state solutions of the complex Ginzburh–Landau equation. Phys. Lett. A **217**, 104–110 (1996)
9. Berger, K.M., Milewski, P.A.: The generation and evolution of lump solitary waves in surface-tension-dominated flows. SIAM J. Appl. Math. **61**(3), 731–750 (2000)
10. Bilotta, E., Pantano,P.: A gallery of Chua attractors. World Scientific Series on Nonlinear Science, Series A. vol. 61. World Scientific, Hackensack (2008)
11. Ezersky, A.B., Rabinovich M.I., Stepanyants, Yu.A., Shapiro, M.F.: Stochastic oscillations of a parametrically excited nonlinear chain. Sov. Phys. JETP **49**(3), 500–504 (1979)
12. Gaponov-Grekhov, A.V., Lomov, A.S., Osipov, G.V., Rabinovich, M.I.: Pattern formation and dynamics of two-dimensionla structures in nonequilibrium dissipative media. In: Gaponov-Grekhov, A.V., Rabinovich, M.I. (eds.) Nonlinear Waves. Dynamics and Evolution, pp. 61–83. Nauka, Moscow (in Russian) (1989)
13. Gorshkov, K.A., Ostrovsky, L.A.: Interaction of solitons in nonintegrable systems: direct perturbation method and applications. Physica D **3**, 428–438 (1981)
14. Gorshkov, K.A., Mironov, V.A., Sergeev, A.M.: Stationary bounded states of soliton formations. In: Gaponov-Grekhov, A.V., Rabinovich, M.I. (eds.) Nonlinear Waves. Selforganisation, pp. 112–128. Nauka, Moscow (in Russian) (1983)
15. Gorshkov, K.A., Pelinovsky, D.E., Stepanyants, Yu.A.: Normal and anomalous scattering, formation and decay of bound-states of two-dimensional solitons described by the Kadomtsev–Petviashvili equation. Sov. Phys. JETP **77**(2), 237–245 (1993)
16. Infeld, E., Rowlands, G.: Nonlinear Waves, Solitons and Chaos, 2nd edn. Cambridge University Press, Cambridge (2000)
17. Kadomtsev, B.B., Petviashvili, V.I.: On the stability of solitary waves in weakly dispersive media. DAN SSSR **192**(4), 753–756 (1970) (in Russian; Engl. transl.: Sov. Phys. Doklady, **15**, 539–541)
18. Karpman, V.I.: Nonlinear Waves in Dispersive Media. Nauka, Moscow (1973) (in Russian; Engl. transl.: Pergamon Press, Oxford, 1975)
19. Kiyashko, S.V., Pikovsky, A.S., Rabinovich, M.I.: Autogenerator of radio-frequency region with stocastic behaviour. Radiotekhnika i electronika (in Russian) **25**, 336–343 (1980)
20. Krichever, I.M.: On the rational solutions of Zakharov-Shabat equations and completely integrable systems of N particles on a line. Zap. Nauchn. Sem. LOMI **84**, 117–130 (in Russian) (1979)
21. Kuznetsov, E.A., Turitsyn, S.K.: Two- and three-dimensional solitons in weakly dispersive media. JETP **55**, 844–847 (1982)
22. Landau, L.D., Lifshitz, E.M.: Hydrodynamics. Course of Theoretical Physics, vol. 6, 5th edn. Fizmatlit, Moscow (2006) (in Russian) [Engl. transl.: *Fluid Mechanics*, 2nd ed., (Pergamon Press, Oxford, 1987)]
23. Lax, P.D.: Integrals of nonlinear equations of evolution and solitary waves. Commun. Pure Appl. Math. **21**, 467–490 (1968)
24. Lu, Z., Tian, E.M., Grimshaw, R.: Interaction of two lump solitons described by the Kadomtsev–Petviashvili I equation. Wave Motion **40**, 95–120 (2004)

25. Ma, W.-X.: Lump solutions to the Kadomtsev–Petviashvili equation. Phys. Lett. A **379**, 1975–1978 (2015)
26. Manakov, S.V. et al.: Two-dimensional solitons of the Kadomtsev–Petviashvili equation and their interaction. Phys. Lett. A **63**(3), 205–206 (1977)
27. Mironov, V.A., Smirnov, A.I., Smirnov, L.A.: Structure of vortex shedding past potential barriers moving in a Bose–Einstein condensate. JETP **110**, 877–889 (2010)
28. Ostrovsky, L.A., Gorshkov, K.A.: Perturbation theories for nonlinear waves. In: Christiansen, P., Soerensen M. (eds.), Nonlinear Science at the Down at the XXI Century, pp. 47–65. Elsevier, Amsterdam (2000)
29. Pelinovsky, D.E., Stepanyants, Y.A.: New multisoliton solutions of the Kadomtsev–Petviashvili equation. JETP Lett. **57**, 24–28 (1993)
30. Pelinovsky, D.E., Stepanyants, Y.A.: Self-focusing instability of plane solitons and chains of two-dimensional solitons in positive-dispersion media. JETP **77** (4), 602–608 (1993)
31. Pelinovsky, D.E., Stepanyants, Y.A.: Convergence of Petviashvili's iteration method for numerical approximation of stationary solutions of nonlinear wave equations. SIAM J. Numer. Anal. **42**(3),1110–1127 (2004)
32. Pelinovsky, D.E., Stepanyants, Y.A., Kivshar, Yu.S.: Self-focusing of plane dark solitons in nonlinear defocusing media. Phys. Rev. E **51**, 5016–5026 (1995)
33. Petviashvili, V.I.: Equation of an extraordinary soliton. Fizika Plazmy **2**(3), 469–472 (1976) (in Russian) (Engl. transl.: Soviet J. Plasma Phys. **2**, 257–258)
34. Petviashvili, V.I., Pokhotelov, O.V.: Solitary Waves in Plasmas and in the Atmosphere. Energoatomizdat, Moscow (1989) (in Russian) (Engl. transl.: Gordon and Breach, Philadelphia, 1992)
35. Potapov, A.I., Soldatov, I.N.: Quasi-plane beam of nonlinear longitudinal waves in a plate. Akust. Zh. **30**, 819–822 (1984)
36. Rabinovich, M.I., Ezersky, A.B., Weidman, P.D.: The Dynamics of Patterns. World Scientific, Singapore (2000)
37. Singh, N., Stepanyants, Y.: Obliquely propagating skew KP lumps. Wave Motion **64**, 92–102 (2016)
38. Stepanyants, Y.A., Ten, I.K., Tomita, H.: Lump solutions of 2D generalised Gardner equation. In: Luo, A.C.J., Dai, L., Hamidzadeh, H.R. (eds.), Nonlinear Science and Complexity, pp. 264–271. World Scientific, Singapore (2006)
39. Tauchert, T.R., Guzelsu, A.N.: An experimental study of dispersion of stress waves in a fiber-reinforced composite. Trans. ASME Ser. E J. Appl. Mech. **39**, 98–102 (1972)
40. Villarroel, J., Ablowitz, M.J.: On the discrete spectrum of the nonstationary Schrödinger equation and multipole lumps of the Kadomtsev-Petviashvili I equation. Commun. Math. Phys. **207**, 1–42 (1999)
41. Voronovich, V.V., Shrira, V.I., Stepanyants, Yu.A.: Two-dimensional models for nonlinear vorticity waves in shear flows. Stud. Appl. Math. **100**(1), 1–32 (1998)
42. Zakharov, V.E.: Instability and nonlinear oscillations of solitons. Pis'ma v ZhETF **22**(7), 364–367 (1975) (in Russian) (Engl. transl.: JETP Lett., 1975, **22**(7), 172–173)

Index

I.S. Aranson et al. (eds.), *Advances in Dynamics, Patterns, Cognition*,
Nonlinear Systems and Complexity 20, DOI 10.1007/978-3-319-53673-6

MIX
Papier aus verantwortungsvollen Quellen
Paper from responsible sources
FSC® C105338

If you have any concerns about our products,
you can contact us on
ProductSafety@springernature.com

In case Publisher is established outside the EU,
the EU authorized representative is:
Springer Nature Customer Service Center GmbH
Europaplatz 3, 69115 Heidelberg, Germany

Printed by Libri Plureos GmbH
in Hamburg, Germany